Plant Pathogen Interactions: 2nd Edition

Plant Pathogen Interactions: 2nd Edition

Guest Editors

Jessie Fernandez
Sixue Chen

MDPI

Basel • Beijing • Wuhan • Barcelona • Belgrade • Novi Sad • Cluj • Manchester

Guest Editors
Jessie Fernandez
Department of Microbiology
and Cell Science
University of Florida
Gainesville, FL
USA

Sixue Chen
Department of Biology
University of Mississippi
Oxford, MS
USA

Editorial Office
MDPI AG
Grosspeteranlage 5
4052 Basel, Switzerland

This is a reprint of the Special Issue, published open access by the journal *International Journal of Molecular Sciences* (ISSN 1422-0067), freely accessible at: https://www.mdpi.com/journal/ijms/special_issues/ON2G9U32N7.

For citation purposes, cite each article independently as indicated on the article page online and as indicated below:

Lastname, A.A.; Lastname, B.B. Article Title. *Journal Name* **Year**, *Volume Number*, Page Range.

ISBN 978-3-7258-7907-6 (Hbk)
ISBN 978-3-7258-7908-3 (PDF)
https://doi.org/10.3390/books978-3-7258-7908-3

Cover image courtesy of Sixue Chen

Contents

About the Editors

Jessie Fernandez

Jessie Fernandez, Department of Microbiology and Cell Science at the University of Florida, where she is interested in understanding how plant pathogens grow inside host cells, suppress or evade host defenses, and ultimately cause disease. Her research focuses on the molecular and cellular strategies used by fungal pathogens, particularly the rice blast fungus *Magnaporthe oryzae*, to colonize host tissue, with a strong emphasis on effector biology and how these molecules reprogram host cells. In parallel, her lab investigates biocontrol and beneficial microbiomes as sustainable strategies to protect crops from disease. Jessie Fernández has published more than 25 peer-reviewed papers on plant-pathogen interactions, nutrient acquisition and metabolism during infection, effector function, and microbial ecology. She also serves as an editor for *Molecular Plant-Microbe Interactions (MPMI)* and *Fungal Genetics and Biology (FGB)*, contributing to the advancement and dissemination of research in plant pathology and fungal biology.

Sixue Chen

Sixue Chen, Department of Biology at the University of Mississippi. He is interested in understanding how plants respond to biotic and abiotic stresses. His lab is known for application of single-cell type proteomics and metabolomics technologies in elucidating plant molecular networks in multiple stress responses, including those to bacteria pathogens (e.g., *Pseudomonas syringe* DC 3000 *pv* tomato). Sixue Chen's research has led to more than 350 publications. He serves as Associate Editors/Board Members of *BBA-Proteins and Proteomics, Frontiers in Plant Proteomics, International Journal of Molecular Sciences, Journal of Proteomics* etc. He is an elected Honorary Fellow of the American Association for the Advancement of Science, USA.

Preface

Plant pathogens represent a significant threat to agricultural systems, causing major damage to the food industry worldwide. Pathogens are constantly adapting to evade or suppress plant defense responses to cause disease. Despite this, plants have evolved sophisticated approaches to recognize and restrict the pathogen to the infection site. A unique and intimate association between plant and pathogens is created as they are in a constant arms race to coexist or compete for survival in nature. Deciphering how plant–pathogen interactions are established is not only an essential aspect in plant pathology but also extremely important for crop improvement, sustainability, and global food security.

Leading by Dr. Jessie Fernandez and Dr. Sixue Chen, this Reprint covers original research and review articles that present recent advances in the field of Plant-Pathogen Interactions, with a focus on but not limited to the molecular mechanisms underlining disease progression, effector biology, plant immunity, and virulence factors.

Jessie Fernandez and Sixue Chen
Guest Editors

International Journal of
Molecular Sciences

MDPI

Editorial

Special Issue "Plant–Pathogen Interactions 2.0"

Jessie Fernandez [1,*] and Sixue Chen [2,*]

[1] Department of Microbiology and Cell Science, University of Florida, Gainesville, FL 32610, USA
[2] Department of Biology, University of Mississippi, Oxford, MS 38677, USA
* Correspondence: jessie.fernandez@ufl.edu (J.F.); schen8@olemiss.edu (S.C.)

Plant–pathogen interactions encompass a highly dynamic network of molecular events that define whether a plant successfully resists microbial attack or succumbs to disease. These interactions are driven by a continuous exchange of signals between hosts and microbes, beginning with the plant's perception of potential invaders through surface receptors and intracellular sensors, and extending through to hormone-regulated signaling, transcriptional reprogramming, and metabolic adjustment [1–3]. Pathogens counter these defenses using diverse strategies, including the following: they deploy effectors that suppress immune responses, secrete enzymes that degrade structural barriers like the plant cell wall, rewire host molecular/cellular processes, such as protein trafficking and signaling, and exploit the host's resources to support their growth and propagation [2,4]. At the same time, non-pathogenic and beneficial microbes colonizing the rhizosphere, phyllosphere, and endosphere are capable of activating overlapping defense pathways, modulating stress responses, and enhancing overall plant resilience [3,5,6]. The studies collated in this Special Issue illustrate many facets of this molecular 'tug of war', spanning the manipulation of salicylic acid (SA) metabolism and peptide-triggered immune regulation, the coordinated control of bacterial virulence systems, fungal reliance on protein-targeting and cell-wall-degrading enzymes, the viral hijacking of host cytoskeletal and nuclear components, and microbiome-mediated enhancement of plant defense and stress tolerance. Together, these nine contributions (eight research articles and one review) provide a timely and integrated view of plant–pathogen interactions, advancing our understanding of susceptibility genes, pathogen regulatory networks, effector biology, cell-wall degradation, viral movement, beneficial microbe-mediated defense activation, and rhizosphere-driven resilience. The contributions are expanded upon below.

Recent advances in identifying plant susceptibility factors illustrate that pathogens often exploit host metabolic pathways to suppress immunity. A compelling example is the characterization of peanut SA 5-hydroxylases *Ah*S5H1 and *Ah*S5H2, which catalyze the conversion of SA into 2,5-dihydroxybenzoic acid, thereby lowering SA levels and weakening SA-dependent defense signaling (Contribution 1). When expressed in *Arabidopsis thaliana*, both enzymes suppress SA-responsive *AtPR1* and *AtPR2* induction and increase susceptibility to *Pseudomonas syringae* pv. tomato DC3000. Differential tissue expression patterns in peanut further support their coordinated roles in modulating defense. These findings emphasize that susceptibility arises not only from pathogen aggression but also from host genes, the normal biochemical functions of which can be co-opted to weaken immunity.

The ability of plants to initiate multilayered immune responses is likewise exemplified by research on endogenous elicitor peptides. Immune activation mediated by the peach peptide *Pp*Pep2 involves a broad-pattern-triggered immunity-like response that is shaped in part by microRNAs, as revealed by miRNA-Seq analysis identifying 33 differentially expressed miRNAs. Most of the microRNAs change at 24 h after treatment, and they belong to families such as miR156, miR167, miR168, miR169, miR390, miR395, miR396, and miR482 (Contribution 2). Several of these miRNAs are associated with defense- and stress-related processes. For example, miR482 is linked to resistance (R) gene regulation, and miR395 is associated with sulfate-mediated protection against oxidative stress (Contribution 2). The authors further show that many predicted miRNA target transcripts exhibit opposite expression patterns following *Pp*Pep2 treatment, and that numerous targets are also regulated at the mRNA level in previously published *Pp*Pep2 transcriptome data [7]. Together, these findings indicate coordinated relationships between peptide-triggered gene expression and miRNA-mediated post-transcriptional regulation, demonstrating additional layers of complexity within plant immune signaling.

Insights into pathogen strategies are also expanded through work on bacterial regulatory networks. The identification of TfmR as a TetR-family transcriptional regulator in *Xanthomonas oryzae* pv. *oryzicola* illustrates how pathogens coordinate virulence through global control points (Contribution 3). This regulator positively regulates both the diffusible signal factor quorum-sensing system and the Type III secretion pathway by directly binding to a conserved promoter motif of *rpfG* and *hrpX*. Loss of TfmR impairs motility, extracellular polysaccharide production, hypersensitive response induction, and virulence, while constitutive expression of *rpfG* or *hrpX* partially restores pathogenicity (Contribution 3). These findings reveal that TfmR acts as an upstream integrator linking quorum-sensing cues with secretion system activation to optimize infection processes.

Multiple fungal studies provide complementary mechanistic perspectives on virulence determinants. In *Fusarium graminearum*, the role of the guided entry of tail-anchored (TA) proteins (GET) pathway in pathogenicity is demonstrated through the characterization of *Fg*GET3, an ortholog of the yeast (*Saccharomyces cerevisiae*) GET3 ATPase (Contribution 4). Functional conservation with yeast GET3 is demonstrated by the ability of *Fg*GET3 to restore normal function in a *S. cerevisiae* Δ*get3* mutant. Deletion of *FgGET3* results in pleiotropic developmental defects, including reduced hyphal growth, fragmented vacuoles, altered abiotic stress responses, markedly reduced conidiation, delayed conidial germination, and decreased deoxynivalenol mycotoxin production (Contribution 4). The Δ*FgGET3* mutant also exhibits significantly weakened virulence on wheat spikes. Together, these results underscore the significance of the GET pathway in targeting TA proteins and maintaining intracellular homeostasis, thereby supporting fungal development, secondary metabolism, and successful infection.

Polygalacturonase activity during barley infection by *Pyrenophora graminea* also emerges as a key virulence determinant. Four PG-encoding genes (*PgPG1-PgPG4*) were identified, with *PgPG1* showing the strongest inducible expression during infection (Contribution 5). Functional analysis via RNAi-silenced and overexpression strains revealed that silencing *PgPG1* reduced disease incidence by ~40–63%, whereas overexpression increased incidence by ~10–13%. *Pg*PG1 was confirmed to be secreted and, when transiently expressed in *Nicotiana benthamiana*, induced host cell death, supporting roles in pectin degradation and host response modulation (Contribution 5). Together, these findings reinforce the importance of cell-wall-degrading enzymes in fungal pathogenicity.

Effector biology continues to be a central theme in plant pathology, and this Special Issue includes a significant contribution in this area. A secreted CFEM-domain effector, *FvCFEM12*, is shown to be highly expressed during maize infection, to exhibit secretory activity, and to suppress Bax- and INF1-induced cell death in *N. benthamiana*, indicating strong immunosuppressive capabilities (Contribution 6). Heterologous expression of *FvCFEM12* in maize leaves using *P. syringae* strain D36E further compromises immune responses. Interaction with the maize wall-associated receptor kinase *Zm*WAK17ET is demonstrated, and silencing of this kinase increases susceptibility, highlighting its function as a host target of the effector (Contribution 6). Deletion of *FvCFEM12* impairs fungal virulence, alters colony and hyphal morphology, and reduces tolerance to cell-wall stressors, revealing a dual role that links fungal cell-wall integrity and physiology with immune evasion.

Plant–virus interactions are also represented. Using an in vivo co-immunoprecipitation strategy with an epitope-tagged alfalfa mosaic virus (AMV) movement protein (MP) expressed during genuine infection, Villar-Álvarez et al. (2024) developed a method to identify host factors associated with 30K family MPs in planta. This approach revealed 121 host protein candidates interacting with the AMV MP, including histone H2B, actin, 14-3-3A, the translation initiation factor eIF4A, and a peroxidase (POX) (Contribution 7). In vivo protein–protein interaction assays in *N. tabacum* show that H2B, actin, 14-3-3A, and eIF4A interact directly with multiple 30K family MPs, while POX is proposed to be part of a protein complex associated with these interactors (Contribution 7). Together, these findings point to a conserved reliance on core host components such as cytoskeletal elements, chromatin-associated proteins, and translation machinery to support intracellular trafficking, plasmodesmata-mediated movement, and viral spread.

This Special Issue further highlights how beneficial microbes can influence immunity and plant resilience. Genomic and transcriptomic analyses of *Burkholderia contaminans* strain BJ3 reveal a broad suite of plant growth-promoting traits, including pathways for indole 3-acetic acid (IAA) production, non-ribosomal peptides, volatile organic compounds, siderophores, hydrolytic enzymes, spermidine, nitrogen fixation, phosphate solubilization, and 1-aminocyclopropane-1-carboxylate deaminase (Contribution 8). In *A. thaliana*, BJ3 strengthens root architecture, increases lignin and antioxidant accumulation, accelerates flowering, and markedly improves tolerance to both salt and drought stress. Transcriptomic data shows activation of SA, ethylene, and IAA signaling networks, together with the induction of defense metabolites such as glucosinolates and camalexin, illustrating the capacity of beneficial microbes to simultaneously enhance stress adaptation and immune strength (Contribution 8).

A broader ecological perspective is provided through a conceptual review synthesizing rhizosphere-mediated strategies that enhance agroecosystem resilience (Contribution 9). Beneficial microbes, intercropping, green manures, and biochar amendments are described as tools that influence microbial community structure, soil carbon dynamics, nutrient cycling, and plant health. By integrating microbial ecology with agronomic management, the review highlights promising approaches for sustainable disease suppression, reduced reliance on chemical inputs, and enhanced plant resilience in the face of environmental change.

In summary, the articles collected in this Special Issue on "Plant-Pathogen Interactions 2.0" reflect the current landscape of research at the interface of molecular plant biology, pathology, and microbiome science. Together, these nine papers present a richly detailed and multidimensional view of plant–pathogen interactions. The studies demonstrate that pathogen infection relies on interconnected networks of regulatory proteins, effectors, metabolic pathways, and structural components, while plant defense emerges from hormonal balance, miRNA-mediated modulation, receptor signaling, and contributions from beneficial microbes. By combining molecular genetics, cell biology, systems-level approaches, and ecological perspectives, this Special Issue captures the diversity of contemporary research directions in the field and highlights promising avenues for developing durable resistance, microbiome-based protection strategies, and innovative solutions to crop health challenges.

Funding: Research on plant–pathogen interactions in the Fernandez laboratory has been supported by a University of Florida Research Opportunity Seed Fund; that conducted in the Chen laboratory has been supported by NSF Plant Genome Research Program 2318746, the Division of Molecular and

Cellular Bioscience 2423128, and the United States Department of Agriculture National Institute of Food and Agriculture, grant no. 2020-67013-40658/project accession no. 1024092 to S.C.

Acknowledgments: The Guest Editors would like to express sincere gratitude to the authors, peer reviewers, and editorial team, whose contributions have made this Special Issue possible. Additionally, they also want to thank the Managing Editor, for her great support. They would highly appreciate the continuous support for the next Special Issue, "Plant–Pathogen Interactions 3.0".

Conflicts of Interest: The authors declare no conflicts of interest.

List of Contributions

1. Liang, B.; Bai, Y.; Zang, C.; Pei, X.; Xie, J.; Lin, Y.; Liu, X.; Ahsan, T.; Liang, C. Overexpression of the First Peanut-Susceptible Gene, *Ahs5h1* or *Ahs5h2*, Enhanced Susceptibility to *Pst* Dc3000 in Arabidopsis. *Int. J. Mol. Sci.* **2023**, *24*, 14210.
2. Foix, L.; Pla, M.; Martín-Mur, B.; Esteve-Codina, A.; Nadal, A. The Pppep2-Triggered Pti-Like Response in Peach Trees Is Mediated by Mirnas. *Int. J. Mol. Sci.* **2024**, *25*, 13099.
3. Chang, Z.; Ma, Z.; Su, Q.; Xia, X.; Ye, W.; Li, R.; Lu, G. The Transcriptional Regulator Tfmr Directly Regulates Two Pathogenic Pathways in *Xanthomonas oryzae* Pv. Oryzicola. *Int. J. Mol. Sci.* **2024**, *25*, 5887.
4. Liu, C.; Lei, L.; Zhu, J.; Chen, L.; Peng, S.; Zhang, M.; Zhang, Z.; Tang, J.; Chen, Q.; Kong, L.; et al. Fgget3, an Atpase of the Get Pathway, Is Important for the Development and Virulence of *Fusarium graminearum*. *Int. J. Mol. Sci.* **2024**, *25*, 12172.
5. Si, E.; Guo, M.; Liu, H.; Li, C.; Wang, J.; Yao, L.; Meng, Y.; Ma, X.; Li, B.; Yang, K.; et al. The Essentials of *Pgpg1*, a Polygalacturonase-Encoding Gene for the Invasion of *Pyrenophora graminea* to *Hordeum vulgare*. *Int. J. Mol. Sci.* **2025**, *26*, 2401.
6. Li, H.; Ishfaq, S.; Liang, X.; Wang, R.; Wei, H.; Guo, W. A Novel Cfem Effector in *Fusarium verticillioides* Required for Virulence Involved in Plant Immunity Suppression and Fungal Cell Wall Integrity. *Int. J. Mol. Sci.* **2025**, *26*, 4369.
7. Villar-Álvarez, D.; MLeastro, O.; Pallas, V.; Sánchez-Navarro, J.Á. Identification of Host Factors Interacting with Movement Proteins of the 30k Family in *Nicotiana tabacum*. *Int. J. Mol. Sci.* **2024**, *25*, 12251.
8. Chang, Y.-L.; Chang, Y.-C.; Kurniawan, A.; Chang, P.-C.; Liou, T.-Y.; Wang, W.-D.; Chuang, H.-W. Employing Genomic Tools to Explore the Molecular Mechanisms Behind the Enhancement of Plant Growth and Stress Resilience Facilitated by a *Burkholderia rhizobacterial* Strain. *Int. J. Mol. Sci.* **2024**, *25*, 6091.
9. Asghar, W.; Craven, K.D.; Swenson, J.R.; Kataoka, R.; Mahmood, A.; Farias, J.G. Enhancing the Resilience of Agroecosystems through Improved Rhizosphere Processes: A Strategic Review. *Int. J. Mol. Sci.* **2025**, *26*, 109.

References

1. Dodds, P.N.; Rathjen, J.P. Plant Immunity: Towards an Integrated View of Plant–Pathogen Interactions. *Nat. Rev. Genet.* **2010**, *11*, 539–548. [CrossRef] [PubMed]
2. Jones, J.D.G.; Staskawicz, B.J.; Dangl, J.L. The Plant Immune System: From Discovery to Deployment. *Cell* **2024**, *187*, 2095–2116. [CrossRef] [PubMed]
3. Akter, T.; Maqsood, H.; Castilla, N.; Song, W.; Chen, S. Systems Biology Applications in Revealing Plant Defense Mechanisms in Disease Triangle. *Int. J. Mol. Sci.* **2025**, *26*, 7318. [CrossRef] [PubMed]
4. Kalicharan, R.E.; Fernandez, J. Triple Threat: How Global Fungal Rice and Wheat Pathogens Utilize Comparable Pathogenicity Mechanisms to Drive Host Colonization. *Mol. Plant-Microbe Interact.* **2025**, *38*, 173–186. [CrossRef] [PubMed]
5. Khawula, S.; Daniel, A.I.; Nyawo, N.; Ndlazi, K.; Sibiya, S.; Ntshalintshali, S.; Nzuza, G.; Gokul, A.; Keyster, M.; Klein, A.; et al. Optimizing Plant Resilience with Growth-Promoting *Rhizobacteria* under Abiotic and Biotic Stress Conditions. *Plant Stress* **2025**, *17*, 100949. [CrossRef]

6. Johnson, T.; Kemmerer, L.; Garcia, N.; Fernandez, J. *Bacillus subtilis* Strain Ud1022 as a Biocontrol Agent against *Magnaporthe oryzae*, the Rice Blast Pathogen. *Microbiol. Spectr.* **2025**, *13*, e00797-25. [CrossRef] [PubMed]
7. Foix, L.; Nadal, A.; Zagorščak, M.; Ramšak, Ž.; Esteve-Codina, A.; Gruden, K.; Pla, M. Prunus persica Plant Endogenous Peptides Pppep1 and Pppep2 Cause Pti-Like Transcriptome Reprogramming in Peach and Enhance Resistance to Xanthomonas arboricola pv. pruni. *BMC Genom.* **2021**, *22*, 360. [CrossRef] [PubMed]

International Journal of
Molecular Sciences

MDPI

Article

Overexpression of the First Peanut-Susceptible Gene, *AhS5H1* or *AhS5H2*, Enhanced Susceptibility to *Pst* DC3000 in Arabidopsis

Bingbing Liang [1], Yuanjun Bai [1,2], Chaoqun Zang [1], Xue Pei [1], Jinhui Xie [1], Ying Lin [1], Xiaozhou Liu [1], Taswar Ahsan [1] and Chunhao Liang [1,*]

1 Institute of Plant Protection, Liaoning Academy of Agricultural Sciences, Shenyang 110161, China; liangbb2121@163.com (B.L.)
2 Institute of Rice Research, Liaoning Academy of Agricultural Sciences, Shenyang 110101, China
* Correspondence: liangchunhao99@126.com

Abstract: Salicylic acid (SA) serves as a pivotal plant hormone involved in regulating plant defense mechanisms against biotic stresses, but the extent of its biological significance in relation to peanut resistance is currently lacking. This study elucidated the involvement of salicylic acid (SA) in conferring broad-spectrum disease resistance in peanuts through the experimental approach of inoculating SA-treated leaves. In several other plants, the salicylate hydroxylase genes are the typical susceptible genes (*S* genes). Here, we characterized two SA hydroxylase genes (*AhS5H1* and *AhS5H2*) as the first *S* genes in peanut. Recombinant AhS5H proteins catalyzed SA in vitro, and showed SA 5-ydroxylase (S5H) activity. Overexpression of *AhS5H1* or *AhS5H2* decreased SA content and increased 2,5-DHBA levels in Arabidopsis, suggesting that both enzymes had a similar role in planta. Moreover, overexpression of each *AhS5H* gene increased susceptibility to *Pst* DC3000. Analysis of the transcript levels of defense-related genes indicated that the expression of *AhS5H* genes, *AhNPR1* and *AhPR10* was simultaneously induced by chitin. Overexpression of each *AhS5H* in Arabidopsis abolished the induction of *AtPR1* or *AtPR2* upon chitin treatment. Eventually, *AhS5H2* expression levels were highly correlated with SA content in different tissues of peanut. Hence, the expression of *AhS5H1* and *AhS5H2* was tissue-specific.

Keywords: salicylic acid; susceptible genes; salicylate hydroxylase; *Arachis hypogaea*; Arabidopsis

1. Introduction

The cultivated peanut (*Arachis hypogaea* L.) is a highly popular and widely consumed oil crop across the globe. It is cultivated in over 100 countries [1]. However, there are numerous types of pathogens that pose a significant threat to peanut cultivation and overall crop yield. The control of pathogens in peanut cultivation often involves the use of large quantities of fungicides, which can pose significant risks to human health and the environment.

Salicylic acid (SA), also known as 2-hydroxybenzoic acid, is a plant hormone. It serves as a mediator for plant defense responses against both biotic and abiotic stresses. In addition, SA plays a crucial role in regulating various physiological and biochemical processes in plants [2,3]. SA can be generated in plants via two enzymatic pathways, that is, isochorismate (ICS), and phenylalanine ammonia lyase (PAL)-mediated pathways, which require the same primary metabolite chorismate [2]. Studies on mutants deficient in SA biosynthesis or signaling revealed that SA is essential to PTI and ETI in local tissue and systemic acquired resistance (SAR) over long distances [4–6].

An increase in SA levels activates the transcription of pathogenesis-related (*PR*) genes in Arabidopsis [6,7]. Plants accumulating low SA levels show decreased oxidative stress and increased shoot and seed number, suggesting that SA has a negative role in plant growth and development [3,8–10].

In peanut, the root endophyte *Phomopsis liquidambaris* B3 effectively protected peanut against *Fusarium oxysporum* by activating defense, which positively correlated with SA [11]. The biological importance of SA in peanut resistance remains deficient.

Although the dominant resistance (*R*) genes have been used to develop resistant varieties through the ages, resistance mediated by a single *R* gene frequently lacks durability due to the loss or mutation of the *R* gene's cognate molecule (effector) of pathogens [12]. The strategy of combining multiple *R* genes in one genotype, also known as "gene stacking", is a more effective approach for maintaining crop resistance [13,14]. Another strategy for acquiring resistant plants is to inactivate the so-called susceptibility genes (*S*), which promote infection and mediate compatible interactions with pathogens [15]. One well-known *S* gene is *Mildew Locus O* (*MLO*), whose mutations are linked to resistance to powdery mildew in many crops [16–22].

No *S* genes were known to confer resistance to pathogens in peanuts, but *S* genes are usually conserved among plant species [23]. In Arabidopsis, the *downy mildew resistant* (*dmr6*) mutant has increased resistance against the oomycete pathogen *Hyaloperonospora parasitica*, and DMR6 was cloned as a member of 2-oxoglutarate-dependent dioxygenases (2OGDs) [24]. DMR6 and its close homolog DLO1 (DMR6-like oxygenase1) act as partially redundant, but distinct, suppressors of immunity [25]. Later, DMR6 and DLO1/S3H1 were identified as SA 5-hydroxylase (S5H) and 3-hydroxylase (S3H), respectively [26,27]. In the *dmr6 dlo1/s3h1* double mutant, the SA level is further elevated compared with that of the *dmr6* single mutant, leading to immunity to pathogens and strong growth retardation [25,27]. In the last few years, orthologs of the *DMR6* and *DLO* genes have been identified and proved to be *S* genes in several crops such as tomato, banana, grapevine and rice [28–31].

This paper demonstrates the significant role of salicylic acid in enhancing peanuts' broad-spectrum disease resistance through the inoculation of leaves treated with SA. We subsequently characterized the primary SA hydroxylases, AhS5H1 and AhS5H2, from peanut. The function of salicylate hydroxylases were confirmed both in vitro and in transgenic Arabidopsis plants. Furthermore, the significance of salicylate hydroxylase genes in the defense response was confirmed through the detection of defense-related gene expression by applying chitin treatment. The biochemical functions of AhS5H1 and AhS5H2 were highly similar. However, the expression of *AhS5H1* and *AhS5H2* in peanuts was specific to certain tissues.

2. Results

2.1. Peanut Resistance to Multiple Pathogens Was Induced by Exogenous SA Treatment

Although SA is a plant immune signal essential for many species of plants, the role played in peanut resistance to pathogens remained unclear. Therefore, we inoculated peanut (Baisha1016) with multiple common peanut pathogens after treatment with 1 mM salicylic acid at 24 hpt (hours post treatment). Plants inoculated without treatment were used as controls.

In comparison to the control samples, it was observed that plants subjected to SA treatment exhibited significantly reduced disease lesions upon exposure to *Phoma arachidicola* LN1, a highly virulent strain of the peanut web blotch fungus (Figure 1a). Furthermore, it is noteworthy to mention that the fungal biomass observed within the leaves exhibited a decrease when compared to the control group, as depicted in Figure 1b. Early leaf spot resistance was assessed through the application of spores via foliar spraying on plants that were three weeks old. The control peanuts exhibited a higher susceptibility to the *Cercospora arachidicola* SY7 isolate, while leaves treated with SA displayed a reduced number of diseased spots (Figure 1c,d). In a similar vein, the plants underwent inoculation with *Elsinoë arachidis* C01, a pathogenic agent known for inducing peanut scab. The observed reduction in the number of disease spots on the leaves of plants treated with SA was found to be statistically significant compared to the control plants. This decrease in disease spots was also accompanied by a decrease in fungal biomass, as depicted in Figure 1e,f.

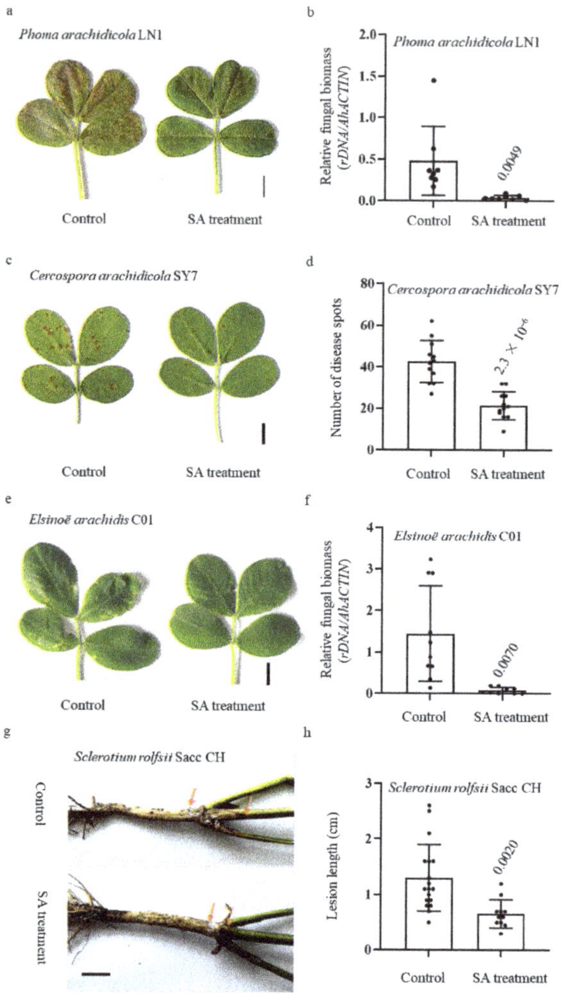

Figure 1. Peanut resistance to multiple pathogens was induced by exogenous SA treatment. Disease symptoms and severity statistics are shown. Disease phenotypes (**a**) and severity of *P. arachidicola* (**b**). Three-week-old plants were inoculated with *P. arachidicola* LN1 isolate (2×10^6 conidia/mL) by foliar spraying; evaluation of disease severity and photography taken were conducted at 7 d post inoculation. The relative fungal biomass was evaluated by qRT-PCR using *P. arachidicola* 26S rDNA and peanut *AhACTIN* gene (n = 8). Disease phenotypes (**c**) and severity of *C. arachidicola* SY7 (**d**). Three-week-old plants were inoculated with *C. arachidicola* SY7 isolate (5×10^5 conidia/mL) by foliar spraying; evaluation of disease severity and photography taken were conducted at 8 d post inoculation. Disease numbers were determined (n = 12). Disease phenotypes (**e**) and severity of *E. arachidis* (**f**). One-month-old plants were inoculated with *E. arachidis*. Evaluation of disease severity and photography taken were conducted at 10 d post inoculation. The relative fungal biomass was evaluated by qRT-PCR using *E. arachidis*. 26S rDNA and peanut *AhACTIN* gene (n = 10). Disease phenotypes (**g**) and lesion lengths of *Sclerotium rolfsii* Sacc CH (**h**). Four-week-old plants were inoculated with *S. rolfsii* Sacc CH isolate. A slice of filter paper containing the *S. rolfsii* Sacc CH was pinned around the bottom of the stem. Lesion lengths were measured 6 d after the inoculation (n = 15). p value evaluated using Student's t-test is above the boxplot. Bar = 1 cm.

It is often reported that SA promotes resistance against both local defense response and SAR [32]. Hence, the basal stem of SA-treated and control plants was inoculated with *S. rolfsii* Sacc CH, a virulent peanut stem rot strain, to investigate its important role in peanut SAR. We observed that the lesion lengths of SA-treated plants were significantly shorter than the control ones (Figure 1g,h). These results collectively indicate that peanut resistance to multiple pathogens was induced by exogenous SA treatment.

2.2. Phylogenetic Analysis and Characterization of SA Hydroxylases in Peanut

In several other species of plant, orthologs of *DMR6* or *DLO1* regulate SA homeostasis and broad-spectrum resistance to pathogens [27–31]. Therefore, we took into account the necessity of screening Arabidopsis S3H or S5H homologs in peanut.

Based on conserved domains of the conserved 2-oxoglutarate-dependent dioxygenases, we used the protein sequences of Arabidopsis S3H and S5H to blast the NCBI database (https://www.ncbi.nlm.nih.gov/ accessed on 22 January 2022). As shown in Figure 2a, two homologs were identified in peanut and named AhS5H1 (XP_025631840), AhS5H2 (XP_025683565). AhS5H1 displayed high similarity to Arabidopsis S3H and its close homolog DLO2 [25,26] due to the phylogenetic analysis, and AhS5H2 was similar to S5H [27].

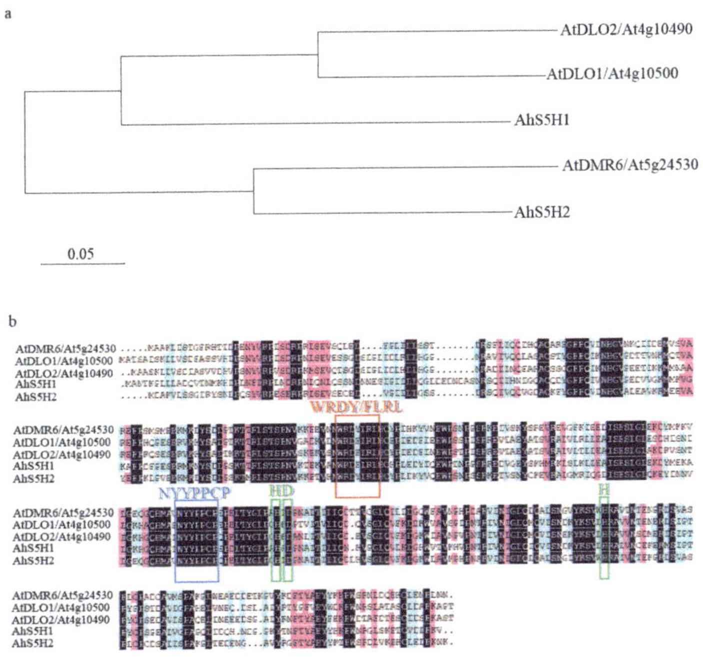

Figure 2. Phylogenetic tree and sequence alignment. (**a**) The alignment was colored with the default DNAMAN color scheme according to the amino acid chemical properties. The ordering of the sequences was based on pairwise similarity. Motifs that were important for the catalytic function were marked by colored boxes: the HDH motif (green), the NYYPPCP motif (blue), and the WRDY/FLRL motif (red)—specific to the DMR6 and DLO proteins. (**b**) The evolutionary history was inferred using the Neighbor-Joining method. The optimal tree with the sum of branch length = 1.15622734 was shown. The tree was drawn to scale, with branch lengths in the same units as those of the evolutionary distances used to infer the phylogenetic tree. The evolutionary distances were computed using the Poisson correction method and were in the units of the number of amino acid substitutions per site. Evolutionary analyses were conducted in MEGA7.

The open reading frame (ORF) of AhS5H1 and AhS5H2 was validated to be 1059 bp and 1014 bp. Respectively, the amino acid sequences of AhS5H1 and AhS5H2 were 353 and 338 in length.

Figure 2b illustrates the identification of conserved patterns observed in both (2OG)-Fe(II) oxygenases and the DMR6-DLO clade. These patterns are represented by colored boxes and signify common motifs and specific residues. The HDH motif is responsible for binding the catalytic iron (Fe II), whereas the NYYPPCP motif plays a crucial role in interacting with the 2-oxoglutarate substrate. The WRDY/FLRL motif has been postulated to play a role in the binding of SA [33]. As a result, AhS5H1 and AhS5H2 were chosen as candidate SA hydroxylases for further study.

2.3. *Identification of the Salicylic Acid 5-Hydroxylases In Vitro*

To test our hypothesis, we treated peanut seedlings (Baisha1016) with 1 mM SA. As was shown in Figure 3a,b, the expression of *AhS5H1* and *AhS5H2* was strongly induced at the time period tested in contrast to the control seedlings.

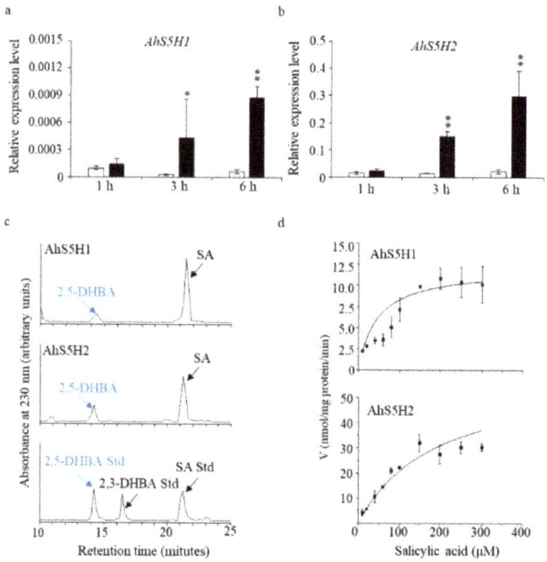

e

Protein	K_m(μM)	V_{max}*	K_{cat}(min^{-1})	R^2
AhS5H1	44.72 ±6.95	12.02 ± 1.51	40.07±5.03	0.89
AhS5H2	190.52 ± 26.33	60.40 ± 6.99	201.33±23.27	0.98

Figure 3. Identification of the salicylic acid 5-hydroxylases in vitro. Expression levels of *AhS5H1* and AhS5H2 in peanuts after SA treatment (**a**,**b**). Three-week-old seedlings were treated with 1 mM SA in 5 mM MES buffer for 1 h, 3 h and 6 h. For the mock treatment, the seedlings received the same volume of DMSO solvent. Gene expression was determined by qRT-PCR using *AhACTIN* as the reference gene. Values are means ± SD ($n = 3$). Asterisks indicate statistically significant differences compared with the corresponding mock using Student's *t*-test (*, $p < 0.05$; **, $p < 0.01$). (**c**) HPLC profiles of the 30 min reaction of the recombinant AhS5H1 and AhS5H2 proteins on the SA substrate. Authentic 2,5-dihydroxyl benzoic acid (2,5-DHBA) was used as a standard. Kinetics curves (**d**) and kinetic parameters (**e**). * nmol/mg protein/min. K_m for Michaelis constant, V_{max} for maximum reaction rates, K_{cat} for catalytic rate constant. Kinetic parameters and curves were obtained from the reactions at pH 6.8 and 40 °C for 30 min. The data are presented as means ± SD ($n = 3$).

In order to elucidate the biochemical role of the two anticipated SA hydroxylases in an in vitro setting, the recombinant proteins of AhS5H1 and AhS5H2 were isolated from *Escherichia coli* and subjected to enzymatic activity assays. The enzymatic assays revealed that both of the candidate SA hydroxylases demonstrated exclusive SA 5-hydroxylase activities (Figure 3c,d). Figure 3e presents the biochemical parameters pertaining to the AhS5H proteins. The observed maximum reaction rates (V_{max}) for AhS5H1 and AhS5H2 enzymes on substrate SA were determined to be 12.02 ± 1.51 and 60.40 ± 6.99 nmol/mg protein/min, respectively. The K_m values of AhS5H1 and AhS5H2 were determined to be 44.72 ± 6.95 and 190.52 ± 26.33 μM, respectively. AhS5H2 exhibited a higher level of enzymatic activity, indicating a greater degree of robustness in its enzyme function. Based on our collective analysis, it can be deduced that AhS5H1 and AhS5H2 serve as the primary functional SA 5-hydroxylase enzymes in the peanut species.

2.4. Overexpression of the Candidate Peanut Salicylate Hydroxylase Genes Reduced Salicylic Acid Level in Arabidopsis

The full-length *AhS5H1* and *AhS5H2* coding sequences were cloned and placed under the Cauliflower mosaic virus 35S promoter. The obtained constructs (35S::AhS5H1-GFP and 35S::AhS5H2-GFP) and control (35S::GFP) were transformed into *A. tumefaciens* strain GV3101 to generate transgenic Arabidopsis. Two lines of Arabidopsis for each of the two genes were obtained.

The expression of *AhS5H1* and *AhS5H2* was observed to be significantly enhanced in their respective lines that were subjected to overexpression, as depicted in Figure S1. The levels of SA, serving as the substrate for AhS5H1 and AhS5H2, exhibited a significant decrease in comparison to control and wild-type (Col-0) plants at 40 DAG (days after germination) (Figure 4a). Simultaneously, there was a significant enhancement in the production of 2,5-dihydroxybenzoic acid (2,5-DHBA) (Figure 4b). The results indicated that both AhS5H1 and AhS5H2 exhibit hydroxylase activity in plants.

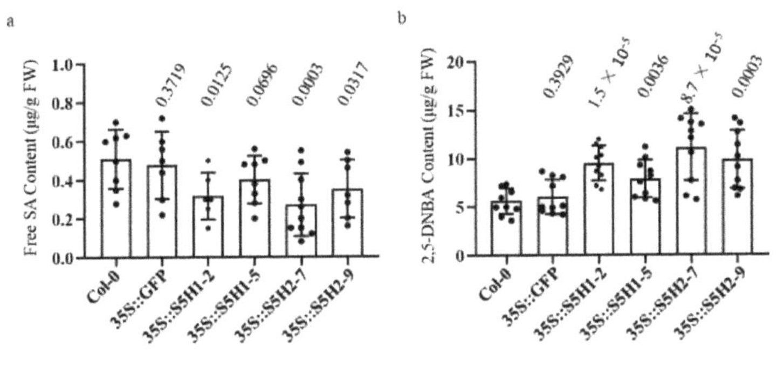

Figure 4. Phenolic accumulation of transgenic Arabidopsis plants. Accumulation of SA (**a**) and 2,5-DHBA (**b**). The amounts of compounds were determined by HPLC. Values are given as means ± SD ($n \geq 10$). p value evaluated using Student's t-test is above the boxplot. Prefix 35S for overexpressing gene.

2.5. Overexpression of the Candidate Peanut Salicylate Hydroxylase Genes Enhanced Susceptibility to Pst DC3000 in Arabidopsis

We also assessed the pathogen resistance of WT, control (35S::GFP), *AhS5H1* overexpression lines (35S::S5H1-2 and 35S::S5H1-5) and AhS5H2 overexpression lines (35S::S5H2-7 and 35S::S5H2-9). The transgenic and wild-type plants were inoculated with *Pst* DC3000, a common bacterial pathogen, by foliar spraying of a bacterial suspension on three-week-old plants. All the *AhS5H1* and *AhS5H2* overexpression lines displayed enhanced susceptibility to *Pst* DC3000 compared to the WT or control lines. (Figure 5a,b), which is consistent with the phenotypes of the *DMR6* overexpression lines [25,27].

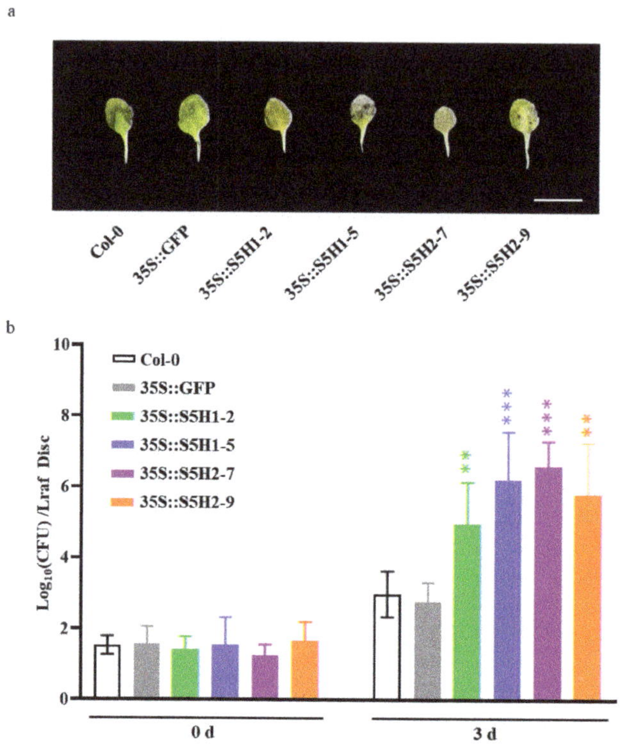

Figure 5. Disease phenotypes and severity of *Pst* DC3000. (**a**) Disease symptoms of three-week-old Arabidopsis plants from the WT, 35S::GFP and *AhS5H* overexpression lines 3 days after *Pst* DC3000 suspension (OD_{600} = 0.05) infiltration. Bar = 1 cm (**b**) The *AhS5H* overexpression lines in the wild-type background were more susceptible to *Pst* DC3000 than the wild-type or 35S::GFP line. CFU, Colony forming units. The data are presented as means ± SD (n = 8). **, $p < 0.01$ and ***, $p < 0.001$ (Student's t-test).

2.6. The Expression of AhS5H1 and AhS5H2 Was Defense Associated

To elucidate the potential alterations in the transcription of SA 5-hydroxylase genes during the defense response, we employed quantitative real-time polymerase chain reaction (qRT-PCR) to determine the relative transcript levels in both peanut and transgenic Arabidopsis. *AhNPR1* and *AhPR10* are genes that play a role in the defense mechanisms of peanut plants, specifically in the salicylic acid (SA) signaling pathway [11]. The expression levels of *AhNPR1* and *AhPR10* were observed to exhibit a statistically significant increase subsequent to chitin treatment, as depicted in Figure 6a,b. In the interim, the transcriptional activity of the *AhS5H1* and *AhS5H2* genes exhibited an up-regulated pattern during the corresponding time period, as shown in (Figure 6c,d).

To gain more information, a set of defense-related genes, such as *AtPR1* and *AtPR2* [34], were subjected to qRT-PCR analysis using six biological replicates for validation. Expression levels were quantified in the wild-type (WT) samples, control (35S::GFP) samples, as well as the overexpression lines of *AhS5H1* (35S::S5H1-2 and 35S::S5H1-5) and *AhS5H2* (35S::S5H2-7 and 35S::S5H2-9). Both tested genes exhibited clear induction in response to chitin stimulation in both wild-type (WT) and control plant samples, when compared to the overexpression lines (Figure 6e,f). The integrated dataset revealed that the expression of *AhS5H1* and *AhS5H2* genes exhibited alterations in the response to defense mechanisms.

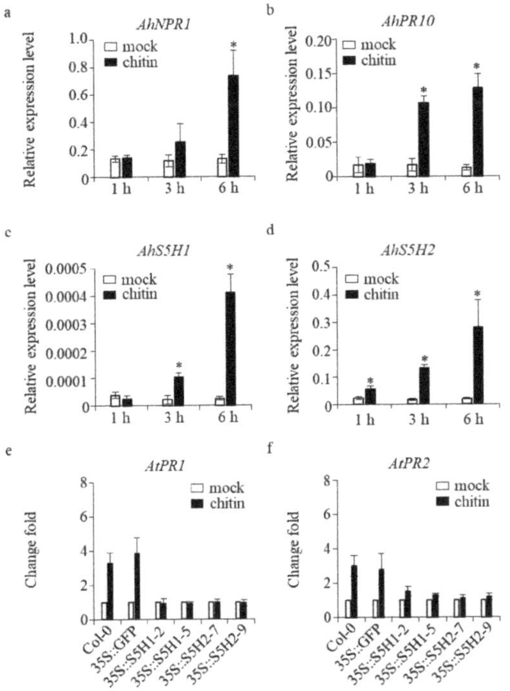

Figure 6. The expression of *AhS5H1* and *AhS5H2* was defense-associated. Induction of defense-related genes (**a**,**b**) and *AhS5Hs* (**c**,**d**) by chitin. Three-week-old peanut plants grown in soil were treated with 100 μg/mL chitin and the same volume of DMSO solvent for the mock treatment. Transcript level was determined by qRT-PCR using *AhACTIN* as the reference gene. Values are given as means ± SD (n = 3). Asterisks indicate statistically significant differences compared with the mock (Student's t-test, *, $p < 0.05$). Average fold change in *AtPR1* (**e**) or *AtPR2* (**f**) expression in WT, 35S::GFP and *AhS5H* overexpression lines of Arabidopsis. Three-week-old seedlings was treated with 100 μg/mL chitin. Gene expression in mock samples was set as 1 and the samples treated with chitin were compared with their own mock samples, unless otherwise indicated. Values are given as means ± SD (n = 6).

2.7. The Expression of AhS5H1 and AhS5H2 in Peanut Was Tissue-Specific

As indicated by the results above. The AhS5H1 and AhS5H2 were highly similar in biochemical function and response to chitin treatment. Therefore, in order to find the differences between the two genes, we collected the roots, stems, leaves and flowers of peanut at the flowering stage for the detection of transcript levels.

The expression levels of *AhS5H2* were highest in leaves, followed by roots, stems, and flowers (Figure 7b). Interestingly and surprisingly, the order of *AhS5H1* expression levels was completely reversed(Figure 7a), suggesting that the expression of *AhS5H1* and *AhS5H2* in peanut was tissue-specific.

The determination of phenolics showed that free SA content was highest in leaves among the tissues tested(Figure 7c), followed by roots, stems, and flowers. So was the level of 2,5-DHBA (Figure 7d). The relationships between the expression level of salicylate hydroxylase genes and SA content are shown in Table S2. *AhS5H2* expression levels were highly correlated with SA content (r = 0.9600) in roots, stems, leaves, and flowers. In contrast, *AhS5H1* was highly negatively correlated (r = −0.8038). It is possible that *AhS5H2* prefers to be expressed in tissues with high SA content in order to exert a stable function due to its higher K_m value.

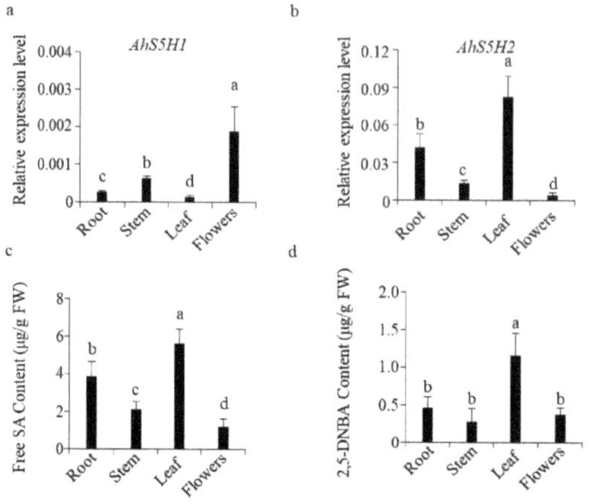

Figure 7. Expression of *AhS5Hs* and accumulation of SA and 2,5-DHBA. *AhS5H1* (**a**) and *AhS5H2* (**b**) expression in different tissues of peanut. Roots, leaves, stems and flowers were collected from plants grown in the paddy field at stage of flowering. Transcript level was determined by qRT-PCR using *AhACTIN* as the reference gene. The accumulation of SA (**c**) and 2,5-DHBA (**d**). Values are means ± SD (n = 3). Columns marked with different letters (a–d) indicate significant differences, as analyzed by the SPSS software version 22.0 (Duncan's multiple range test, α = 0.05).

3. Discussion

The roles of SA during immune responses and senescence have been widely investigated in different plant species [35–39]. However, the biological importance of SA in peanut resistance remains deficient. Our results showed that the resistance of peanuts to many common pathogens could be induced by the SA. Therefore, the role of the SA-signaling pathway in peanut disease resistance deserves further investigation.

Phytochemical protection can easily cause environmental safety problems. In contrast, the acquisition of resistant materials seems more worthy of adaptation. The development of resistant varieties typically relies on the deployment of dominant resistance (*R*) genes, whose products mediate the recognition and protection against specific pathogen strains. However, resistance mediated by a single *R* gene frequently lacks durability, because pathogens can easily lose or mutate their *R* gene's cognate molecule (effector) [12]. Resistance is not easily overcome by combining multiple *R* genes in a genotype. It implies that extra effort is needed to find more *R* genes [13,14]. On the contrary, it is more efficient to look for the so-called *S* genes that are conserved in various species of plants [40]. No *S* genes are yet known to confer resistance to pathogens in peanuts. In this work, we identified two *S* genes (*AhS5H1* and *AhS5H2*) with some similar characteristics to those previously described in rice and Arabidopsis [25–27,30] (Figure 2). Both of the proteins (AhS5H1 and AhS5H2) converted SA to 2,5-DHBA in vitro (Figure 3), but also showing differences in the kinetic properties. For example, both OsS5H-1 and OsS5H-2 in rice are inhibited by the substrate (SA), showing very low K_{si} values (9.10 and 1.60 μM) [41]. In contrast, AhS5H1 and AhS5H2 were not inhibited by SA, showing a hyperbolic response when SA levels were increased (Figure 3d). This result suggests that the role of rice enzymes in hydroxylating SA is probably more significant at low SA concentrations, while the peanut enzymes may have a major role at higher SA concentrations.

To further analyze the roles of *AhS5H1* and *AhS5H2*, we generated overexpressing lines in Columbia-0 (Col-0) background Arabidopsis. In this study, overexpression of *AhS5H1* or *AhS5H2* decreased SA content, suggesting that both enzymes could have a similar role in planta. In the meantime, the content of 2,5-DHBA was significantly increased in the corresponding overexpression lines. The results of the phenolic content assay were consistent with those of the *AtDMR6* overexpression lines in previous reports [24,25]. Therefore, the functions of AHS5H1 and AHS5H2 were highly similar to those of AtDMR6.

The inactivation of orthologs of *AtDMR6* caused an increase in the SA content in plants. It always confers resistance to more than one pathogen species or to most races or strains of the same pathogen [28,31,42]. In this study, SA treatment was used to simulate the increase in endogenous SA in peanut, and our results indicated that peanut resistance to multiple pathogens was induced by exogenous SA. The phenotype of increased resistance was similar to that of those mutants. Moreover, *Ah5H1* and *AhS5H2* overexpression lines displayed low SA levels and were more susceptible to *Pst* DC3000. This also proves that the candidate S genes, *AhS5H1* and *AhS5H2*, are important for their effect on disease resistance.

Systemic acquired resistance (SAR) is a defense reaction that can be aroused when plants are infected by pathogens. It is effective against bacterial, fungal and virus pathogens [43–45] through the concerted activation of pathogenesis-related (*PR*) genes [46,47]. Generally, pathogen-induced SAR relies on an activated SA-dependent pathway [43], while *PR10* encodes SA-inducible *PR* [48]. Additionally, none expressor of pathogenesis-related genes 1 (*NPR1*) has been identified as an important component for the SA-regulated resistance [49]. Both *AhNPR1* and *AhPR10* are associated with *Phomopsis liquidambaris* B3-induced SA-dependent signaling in suppressing root rot [11]. Our results confirmed that the levels of *AhNPR1* and *AhPR10* expression were significantly induced after being treated with chitin. Simultaneously, the *Ah5H1* and *AhS5H2* were remarkably up regulated. In Arabidopsis, the SAR against pathogens has been associated with the accumulation of salicylic acid (SA) and the expression of the pathogenesis-related proteins AtPR1, and AtPR2 [38]. Overexpression of each *AhS5H* in Arabidopsis abolished the induction of *AtPR1* or *AtPR2* upon chitin treatment (Figure 6e,f). Our study demonstrated that expression of *AhS5H1* and *AhS5H2* was affected during defense responses.

AhS5H1 and AhS5H2 were highly similar in biochemical function and response to chitin treatment. However, the transcriptional levels of *AhS5H1* and *AhS5H2* were tissue-specific. Compared with AhS5H1, AhS5H2 had a higher K_m value. This means that a higher content of SA is required to achieve an optimal enzymatic reaction rate for AhS5H2. The detection of compounds also attested that the expression of *AhS5H2* was higher in tissues containing more adequate SA.

In conclusion, the results presented here demonstrate that peanut plants have two salycilic hydroxylase enzymes. *AhS5H1* and *AhS5H2* were identified as candidate susceptibility genes in peanuts for the first time. Both hydroxylases have similar biochemical functions. Meanwhile, the expression of *AhS5H1* and *AhS5H2* was defense-associated and tissue-specific in peanut.

4. Materials and Methods

4.1. Plant Materials and Treatments

The peanut variety Baisha1016 (*Arachis hypogaea* cv. Baisha1016, as a hyper-susceptible variant to all the pathogens involved in this study) was used for all inoculations. At 28 °C in darkness, pre-germinated seeds were sown in a controlled climate room that was maintained at 28 °C, with a 12 h photo-phase. The Arabidopsis plants were grown on potting soil at 21 °C with 16 h of light and 75% relative humidity. The Arabidopsis and peanut seedings were treated with 5 mM MES (4-morpholine ethanesulfonic acid, pH 5.8) buffer containing 1 mM SA or 100 µg/mL chitin, and the same volume of DMSO solvent was used as the control.

4.2. Phylogenetic Analysis

The molecular weights were deduced using Lasergene ediseq ver.7.1 software. Homologs of DMR6 and DLO proteins were retrieved from GenBank databases using protein blast (available online: https://blast.ncbi.nlm.nih.gov/Blast.cgi, accessed on 18 January 2022). The multiple sequence alignment was carried out via DNAMAN6.0, and a phylogenetic dendrogram was generated with MEGA 7.0 using the neighbor-joining method.

4.3. Pathogen Inoculation

After 24 h of SA treatment, peanuts were inoculated with the pathogens, and untreated plants served as controls. Three-week-old peanut plants were inoculated with a virulent *P. arachidicola* LN1 strain by spraying the spore suspension (2×10^6 conidia/mL containing 0.1% Tween 20) as described by [50]. Disease severity was evaluated by PCR amplification of the relative biomass. Quantification was performed with primers from *P. arachidicola* 26S rDNA and the *AhACTIN* gene. Template DNAs from the inoculated leaves were collected 7 days after the inoculation.

To evaluate resistance against Early Leaf Spot, three-week-old peanut plants were inoculated with *C. arachidicola* SY7 by foliar spraying (5×10^5 conidia/mL spores containing 0.005% Silwet L-77), in which the *C. arachidicola* SY7 strain was isolated from the naturally infected peanut leaves. The number of disease spots was counted as described previously by Zanão Júnior et al. [51].

Baisha1016 susceptible to *E. arachidis* served as a host. Mycelium suspension was sprayed onto the one-month-old peanut leaves, and then incubated in a chamber under constant light conditions at 25 °C for lesion formation [52]. Disease severity was also evaluated by PCR amplification of the relative biomass. Quantification was performed with primers from *E. arachidis* 26S rDNA and the *AhACTIN* gene. The inoculated leaves were collected 10 days after the inoculation.

Subsequently, four-week-old peanut plants were inoculated with *Sclerotium rolfsii* Sacc CH isolate grown on PDA with slices of filter paper (about 10 mm × 4 mm) on the agar for five days. The filter paper, grown over by CH, was pinned around the bottoms of peanut. The lengths of the disease lesion were measured at 6 dpi (days post inoculation).

The Arabidopsis inoculation protocol was performed as described by Zeilmaker et al. [25]. To measure the growth of *P. syringae pv tomato* DC3000, Arabidopsis plants at the age of three weeks were subjected to a bacterial suspension (optical density 0.05) supplemented with 0.02% silwet L-77. Leaf samples (four plants per line; three leaves per plant) were collected for enumeration of colonies at 0 and 3 days after inoculation.

4.4. Protein Expression and Enzyme Assays

The sequences of *AhS5H1* and *AhS5H2* were amplified from cDNAs of *Arachis hypogaea* L. leaves. Both were cloned into a modified pGEX-tag vector with 3 × Myc at the C-terminus of the recombinant protein [53]. After transformation of each plasmid into *Escherichia coli* BL21 (DE3), protein expression was induced by the addition of 0.2 mM IPTG grown at 28 °C for 6 h. Recombinant proteins were purified using Glutathione Sepharose 4B (GE Healthcare, Chicago, IL, USA).

The enzymatic activity assay was performed as described previously [30]. Briefly, a total volume of 100 μL reaction mixture contains 5 μg recombinant protein in the reaction buffer (1 mM 2-oxoglutaric acid, 1 mM sodium ascorbate, 0.4 mM $FeSO_4$, 0.1 mg/mL catalase, 5 mM DTT, 50 mM phosphate buffer at pH 6.8) with different concentrations of SA. The reaction was incubated at 40 °C for 30 min and stopped by adding two volumes of acetonitrile and boiling for 1 min. The supernatant was analyzed by HPLC.

4.5. Determination of Metabolites

The leaves and other tissues of the peanut from wild-type and transgenic Arabidopsis plants were collected and stored at −80 °C. The samples were extracted with 90% aqueous methanol containing 0.1% formic acid as described previously [54].

Chemicals were separated by a DIKMA (Beijing, China)C18 column (250 × 4.6 mm, 5 μm) on a Waters 2695 separation module (Waters, Shanghai, China). The elution conditions were at a flow rate of 1 mL/min with a gradient program of 12% acetonitrile for 15 min up to 75% in 8 min, then to 95% in 1 min, then the column was washed and equilibrated to the initial conditions. The absorption at 230 nm was detected and concentrations were calculated by the peak area of samples according to a standard curve.

4.6. *Reverse Transcribed Quantitative (qRT-PCR) Analysis*

After removing possible DNA contamination, two micrograms of total RNA were reverse transcribed with random hexamers and oligo(dT)18 primers using M-MLV reverse transcriptase (Takara, Kusatsu-shi, Japan). The relative transcript levels were quantified using SYBR Green PCR Master Mix (Takara) and normalized to *AhACTIN* or *AtACTIN2*. The relative expression level of each gene was analyzed using the delta-delta C_t method. Gene-specific primers used in qRT-PCR are listed in Supplementary Table S1.

5. Conclusions

Two salicylate hydroxylase genes were successfully predicted and identified within the peanut genome, marking a significant milestone as the inaugural discovery of susceptible genes in this particular crop. The salicylate hydroxylase exhibited analogous biochemical functions, thereby resulting in a reduction of the SA content within transgenic Arabidopsis. However, their expression exhibited tissue specificity.

Supplementary Materials: The supporting information can be downloaded at https://www.mdpi.com/article/10.3390/ijms241814210/s1.

Author Contributions: B.L. and C.L. conceptualized and supervised the project. B.L. and C.Z. performed most of the experiments. Y.B., X.P., J.X. and Y.L. evaluated peanut resistance to pathogens. X.L. provided technical support for data analysis. B.L. and T.A. drafted the manuscript. All authors contributed to the review and editing of the manuscript. All authors have read and agreed to the published version of the manuscript.

Funding: This work was supported by The Collaborative Innovation Project of Liaoning Academy of Agricultural Sciences (2022XTCX0502003), The Peanut Germplasm Innovation Project and Basic Research Expenses of Liaoning Academy of Agricultural Sciences (2023HQ1903).

Institutional Review Board Statement: Not applicable.

Informed Consent Statement: Not applicable.

Data Availability Statement: The data presented in this study are available upon request from B.L.

Acknowledgments: We thank Han Wang (China Agricultural University) for useful discussions, Rujun Zhou (Shenyang Agricultural University) for providing the *Elsinoë arachidis* C01 strain, and He Chen for field management.

Conflicts of Interest: The authors declare no competing interest.

References

1. FAO. 2020. Available online: https://www.fao.org/faostat/en/#data/QCL (accessed on 16 August 2023).
2. Vlot, A.C.; Dempsey, D.A.; Klessig, D.F. Salicylic acid, a multifaceted hormone to combat disease. *Annu. Rev. Phytopathol.* **2009**, *47*, 177–206. [CrossRef]
3. Rivas-San Vicente, M.; Plasencia, J. Salicylic acid beyond defence: Its role in plant growth and development. *J. Exp. Bot.* **2011**, *62*, 3321–3338. [CrossRef]
4. Gaffney, T.; Friedrich, L.; Vernooij, B.; Negrotto, D.; Nye, G.; Uknes, S.; Ward, E.; Kessmann, H.; Ryals, J. Requirement of salicylic acid for the induction of systemic acquired resistance. *Science* **1993**, *261*, 754–756. [CrossRef]
5. Gao, X.; Zhang, J.Q.; Zhang, X.; Zhou, J.; Jiang, Z.; Huang, P.; Tang, Z.; Bao, Y.; Cheng, J.; Tang, H. Rice qGL3/OsPPKL1 functions with the GSK3/SHAGGY-like kinase OsGSK3 to modulate brassinosteroid signaling. *Plant Cell* **2019**, *31*, 1077–1093. [CrossRef]
6. Peng, Y.; Yang, J.; Li, X.; Zhang, Y. Salicylic acid: Biosynthesis and signaling. *Annu. Rev. Plant Biol.* **2021**, *72*, 761–791. [CrossRef]
7. Mou, Z.; Fan, W.; Dong, X. Inducers of plant systemic acquired resistance regulate NPR1 function through redox changes. *Cell* **2003**, *113*, 935–944. [CrossRef]
8. Herrera-Vasquez, A.; Salinas, P.; Holuigue, L. Salicylic acid and reactive oxygen species interplay in the transcriptional control of defense genes expression. *Front. Plant Sci.* **2015**, *6*, 171. [CrossRef]
9. Lamb, C.; Dixon, R.A. The oxidative burst in plant disease resistance. *Annu. Rev. Plant Physiol. Plant Mol. Biol.* **1997**, *48*, 251–275. [CrossRef]
10. Abreu, M.E.; Munne-Bosch, S. Salicylic acid deficiency in *NahG* transgenic lines and *sid2* mutants increases seed yield in the annual plant *Arabidopsis thaliana*. *J. Exp. Bot.* **2009**, *60*, 1261–1271. [CrossRef]

11. Sun, K.; Xie, X.G.; Lu, F.; Zhang, F.M.; Zhang, W.; He, W.; Dai, C.C. Peanut preinoculation with a root endophyte induces plant resistance to soil-borne pathogen *Fusarium oxysporum* via activation of salicylic acid-dependent signaling. *Plant Soil.* **2021**, *460*, 297–312. [CrossRef]
12. Dangl, J.L.; Horvath, D.M.; Staskawicz, B.J. Pivoting the plant immune system from dissection to deployment. *Science* **2013**, *341*, 746–751. [CrossRef] [PubMed]
13. Fuchs, M. Pyramiding resistance-conferring gene sequences in crops. *Curr. Opin. Virol.* **2017**, *26*, 36–42. [CrossRef] [PubMed]
14. Mundt, C.C. Pyramiding for resistance durability: Theory and practice. *Phytopathology* **2018**, *108*, 792–802. [CrossRef] [PubMed]
15. Van Schie, C.C.; Takken, F.L. Susceptibility genes 101: How to be a good host. *Annu. Rev. Phytopathol.* **2014**, *52*, 551–581. [CrossRef]
16. Jørgensen, I.H. Discovery, characterization and exploitation of Mlo powdery mildew resistance in barley. *Euphytica* **1992**, *63*, 141–152. [CrossRef]
17. Büschges, R.; Hollricher, K.; Panstruga, R.; Simons, G.; Wolter, M.; Frijters, A.; Van Daelen, R.; Van der Lee, T.; Diergaarde, P.; Groenendijk, J.; et al. The barley *Mlo* gene: A novel control element of plant pathogen resistance. *Cell* **1997**, *88*, 695–705. [CrossRef]
18. Bai, Y.; Pavan, S.; Zheng, Z.; Zappel, N.F.; Reinstädler, A.; Lotti, C.; De Giovanni, C.; Ricciardi, L.; Lindhout, P.; Visser, R.; et al. Naturally occurring broad-spectrum powdery mildew resistance in a Central American tomato accession is caused by loss of *Mlo* function. *Mol. Plant-Microbe Interact.* **2008**, *21*, 30–39. [CrossRef]
19. Humphry, M.; Reinstädler, A.; Ivanov, S.; Bisseling, T.; Panstruga, R. Durable broad-spectrum powdery mildew resistance in pea *er1* plants is conferred by natural loss-of-function mutations in *PsMLO1*. *Mol. Plant Pathol.* **2011**, *12*, 866–878. [CrossRef]
20. Zheng, Z.; Nonomura, T.; Appiano, M.; Pavan, S.; Matsuda, Y.; Toyoda, H.; Wolters, A.M.A.; Visser, R.G.; Bai, Y. Loss of Function in *Mlo* orthologs reduces susceptibility of pepper and tomato to powdery mildew disease caused by *Leveillula taurica*. *PLoS ONE* **2013**, *8*, e70723. [CrossRef]
21. Wang, Y.; Cheng, X.; Shan, Q.; Zhang, Y.; Liu, J.; Gao, C.; Qiu, J.L. Simultaneous editing of three homoeoalleles in hexaploid bread wheat confers heritable resistance to powdery mildew. *Nat. Biotechnol.* **2014**, *32*, 947–951. [CrossRef]
22. Pessina, S.; Angeli, D.; Martens, S.; Visser, R.G.; Bai, Y.; Salamini, F.; Velasco, R.; Schouten, H.J.; Malnoy, M. The knock-down of the expression of *MdMLO19* reduces susceptibility to powdery mildew (*Podosphaera leucotricha*) in apple (*Malus domestica*). *Plant Biotechnol. J.* **2016**, *14*, 2033–2044. [CrossRef]
23. Dong, O.X.; Ronald, P.C. Genetic engineering for disease resistance in plants: Recent progress and future perspectives. *Plant Physiol.* **2019**, *180*, 26–38. [CrossRef]
24. Van Damme, M.; Huibers, R.P.; Elberse, J.; Van den Ackerveken, G. Arabidopsis *DMR6* encodes a putative 2OG-Fe(II) oxygenase that is defense-associated but required for susceptibility to downy mildew. *Plant J.* **2008**, *54*, 785–793. [CrossRef] [PubMed]
25. Zeilmaker, T.; Ludwig, N.R.; Elberse, J.; Seidl, M.F.; Berke, L.; Van Doorn, A.; Schuurink, R.C.; Snel, B.; Van den Ackerveken, G. DOWNY MILDEW RESISTANT 6 and DMR6-LIKE OXYGENASE 1 are partially redundant but distinct suppressors of immunity in Arabidopsis. *Plant J.* **2014**, *81*, 210–222. [CrossRef] [PubMed]
26. Zhang, K.W.; Halitschke, R.; Yin, C.X.; Liu, C.J.; Gan, S.S. Salicylic acid 3-hydroxylase regulates Arabidopsis leaf longevity by mediating salicylic acid catabolism. *Proc. Natl. Acad. Sci. USA* **2013**, *110*, 14807–14812. [CrossRef] [PubMed]
27. Zhang, Y.J.; Zhao, L.; Zhao, J.; Li, Y.; Wang, J.; Guo, R.; Gan, S.S.; Liu, C.J.; Zhang, K. *S5H/DMR6* Encodes a salicylic acid 5-hydroxylase that fine-tunes salicylic acid homeostasis. *Plant Physiol.* **2017**, *175*, 1082–1093. [CrossRef]
28. Thomazella, D.P.T.; Seong, K.; Mackelprang, R.; Dahlbeck, D.; Geng, Y.; Gill, U.S.; Qi, T.; Pham, J.; Giuseppe, P.; Lee, C.Y.; et al. Loss of function of a *DMR6* ortholog in tomato confers broad-spectrum disease resistance. *Proc. Natl. Acad. Sci. USA* **2021**, *118*, e2026152118. [CrossRef]
29. Tripathi, J.N.; Ntui, V.O.; Shah, T.; Tripathi, L. CRISPR/Cas9-mediated editing of *DMR6* orthologue in banana (*Musa* spp.) confers enhanced resistance to bacterial disease. *Plant Biotechnol. J.* **2021**, *19*, 1291–1293. [CrossRef]
30. Liang, B.B.; Wang, H.; Yang, C.; Wang, L.Y.; Qi, L.L.; Guo, Z.J.; Chen, X.J. Salicylic acid is required for broad-spectrum disease resistance in rice. *Int. J. Mol. Sci.* **2022**, *23*, 1354. [CrossRef]
31. Pirrello, C.; Malacarne, G.; Moretto, M.; Lenzi, L.; Perazzolli, M.; Zeilmaker, T.; Van den Ackerveken, G.; Pilati, S.; Moser, C.; Giacomelli, L. Grapevine *DMR6-1* is a candidate gene for susceptibility to downy mildew. *Biomolecules* **2022**, *12*, 182. [CrossRef]
32. Han, S.Y.; Zhou, X.M.; Shi, L.; Zhang, H.Y.; Geng, Y.; Fang, Y.J.; Xia, H.; Liu, H.; Li, P.C.; Zhao, S.Z.; et al. AhNPR3 regulates the expression of *WRKY* and *PR* genes, and mediates the immune response of the peanut (*Arachis hypogaea* L.). *Plant J.* **2022**, *110*, 735–747. [CrossRef] [PubMed]
33. Zeilmaker, T. Functional and Applied Aspects of the DOWNY MILDEW RESISTANT 1 and 6 Genes in Arabidopsis. Ph.D. Thesis, Utrecht University, Utrecht, The Netherlands, 2012.
34. Liu, W.X.; Zhang, F.C.; Zhang, W.Z.; Song, L.F.; Wu, W.H.; Chen, Y.F. Arabidopsis Di19 functions as a transcription factor and modulates *PR1*, *PR2* and *PR5* expression in response to drought stress. *Mol. Plant* **2013**, *6*, 1487–1502. [CrossRef] [PubMed]
35. Jones, J.D.G.; Dangl, J.L. The plant immune system. *Nature* **2006**, *444*, 323–329. [CrossRef] [PubMed]
36. Zhang, Y.; Li, X. Salicylic acid: Biosynthesis, perception, and contributions to plant immunity. *Curr. Opin. Plant Biol.* **2019**, *50*, 29–36. [CrossRef] [PubMed]
37. Shimono, M.; Koga, H.; Akagi, A.; Hayashi, N.; Goto, S.; Sawada, M.; Kurihara, T.; Matsushita, A.; Sugano, S.; Jiang, C.J.; et al. Rice WRKY45 plays important roles in fungal and bacterial disease resistance. *Mol. Plant Pathol.* **2012**, *13*, 83–94. [CrossRef] [PubMed]

38. Nawrath, C.; Métraux, J.P. Salicylic acid induction-deficient mutants of Arabidopsis express PR-2 and PR-5 and accumulate high levels of camalexin after pathogen inoculation. *Plant Cell* **1999**, *11*, 1393–1404.
39. Delaney, T.P.; Uknes, S.; Vernooij, B.; Friedrich, L.; Weymann, K.; Negrotto, D.; Gaffney, T.; Gut-Rella, M.; Kessmann, H.; Ward, E.; et al. A central role of salicylic acid in plant disease resistance. *Science* **1994**, *266*, 1247–1250. [CrossRef]
40. Zaidi, S.S.; Mukhtar, M.S.; Mansoor, S. Genome editing: Targeting susceptibility genes for plant disease resistance. *Trends Biotechnol.* **2018**, *36*, 898–906. [CrossRef]
41. Zhang, Y.J.; Yu, Q.L.; Gao, S.L.; Yu, N.N.; Zhao, L.; Wang, J.B.; Zhao, J.Z.; Huang, P.; Yao, L.B.; Wang, M.; et al. Disruption of the primary salicylic acid hydroxylases in rice enhances broad-spectrum resistance against pathogens. *Plant Cell Environ.* **2022**, *45*, 2211–2225. [CrossRef]
42. Kieu, N.P.; Lenman, M.; Wang, E.S.; Petersen, B.L.; Andreasson, E. Mutations introduced in susceptibility genes through CRISPR/Cas9 genome editing confer increased late blight resistance in potatoes. *Sci. Rep.* **2021**, *11*, 4487. [CrossRef]
43. Van Loon, L.C. Plant responses to plant growth-promoting rhizobacteria. *Eur. J. Plant Pathol.* **2007**, *119*, 243–254. [CrossRef]
44. Vlot, A.C.; Klessig, D.F.; Park, S.W. Systemic acquired resistance: The elusive signal(s). *Curr. Opin. Plant Biol.* **2008**, *11*, 436–442. [CrossRef]
45. Mukhtar, F.B.; Mohammed, M.; Ajeigbe, A.H. Effect of benzyl amino purine (BAP), coconut milk (CM) and manure applications on leaf senescence and yield in photoperiod sensitive cowpea variety (Kanannado). *Afr. J. Plant Sci.* **2009**, *3*, 142–146.
46. Loake, G.; Grant, M. Salicylic acid in plant defence-the players and protagonists. *Curr. Opin. Plant Biol.* **2007**, *10*, 466–472. [CrossRef] [PubMed]
47. Sels, J.; Mathys, J.; De Coninck, B.M.; Cammue, B.P.; De Bolle, M.F. Plant pathogenesisrelated (PR) proteins: A focus on PR peptides. *Plant Physiol. Biochem.* **2008**, *46*, 941–950. [CrossRef]
48. Park, C.J.; Kim, K.J.; Shin, R.; Park, J.M.; Shin, Y.C.; Paek, K.H. Pathogenesis-related protein 10 isolated from hot pepper functions as a ribonuclease in an antiviral pathway. *Plant J.* **2004**, *37*, 186–198. [CrossRef] [PubMed]
49. Wu, Q.; Wang, Z.X.; Tang, Y.Y.; Yu, H.T.; Ding, Y.F.; Yang, C.D.; Cui, F.G.; Zhang, J.C.; Wang, C.T. Molecular cloning and characterization of *NPR1* gene from *Arachis hypogaea*. *Mol. Biol. Rep.* **2014**, *41*, 5247–5256. [CrossRef]
50. Zhang, X.; Xu, M.L.; Wu, J.X.; Dong, W.B.; Chen, D.X.; Wang, L.; Chi, Y.C. Draft genome sequence of *Phoma arachidicola* Wb2 causing peanut web blotch in China. *Curr. Microbiol.* **2019**, *76*, 200–206. [CrossRef] [PubMed]
51. Júnior, L.A.Z.; Rodrigues, F.Á.; Fontes, R.L.F.; Korndörfer, G.H.; Neves, J.C. Rice Resistance to Brown Spot Mediated by Silicon and its Interaction with Manganese. *J. Phytopathol.* **2009**, *157*, 73–78. [CrossRef]
52. Jiao, W.L.; Liu, L.; Zhou, R.J.; Xu, M.X.; Xiao, D.; Xue, C.Y. Elsinochrome phytotoxin production and pathogenicity of *Elsinoë arachidis* isolates in China. *PLoS ONE* **2019**, *14*, e0218391. [CrossRef]
53. Liu, J.Q.; Chen, X.J.; Liang, X.X.; Zhou, X.G.; Yang, F.; Liu, J.; He, S.Y.; Guo, Z.J. Alternative splicing of rice WRKY62 and WRKY76 transcription factor genes in pathogen defense. *Plant Physiol.* **2016**, *171*, 1427–1442. [CrossRef] [PubMed]
54. Liang, X.; Chen, X.; Li, C.; Fan, J.; Guo, Z. Metabolic and transcriptional alternations for defense by interfering OsWRKY62 and OsWRKY76 transcriptions in rice. *Sci. Rep.* **2017**, *7*, 2474. [CrossRef] [PubMed]

International Journal of *Molecular Sciences*

MDPI

Article

The PpPep2-Triggered PTI-like Response in Peach Trees Is Mediated by miRNAs

Laura Foix [1,2], Maria Pla [2], Beatriz Martín-Mur [3,4], Anna Esteve-Codina [3,4] and Anna Nadal [2,*]

1 BETA Technological Center (TECNIO Network), University of Vic-Central University of Catalonia (UVic-UCC), Carretera de Roda 70, 08500 Vic, Spain; laura.foix@udg.edu
2 Institute for Agricultural and Food Technology, Universitat de Girona, C/ Maria Aurèlia Capmany, 61, 17003 Girona, Spain; maria.pla@udg.edu
3 Centre Nacional d'Anàlisi Genòmica (CNAG), C/ Baldiri Reixac 4, 08028 Barcelona, Spain; beatriz.martin@cnag.eu (B.M.-M.); anna.esteve@cnag.eu (A.E.-C.)
4 Parc Científic de Barcelona, Universitat de Barcelona, C/ Baldiri Reixac, 4, 08028 Barcelona, Spain
* Correspondence: anna.nadal@udg.edu

Abstract: Plant diseases diminish crop yields and put the world's food supply at risk. Plant elicitor peptides (Peps) are innate danger signals inducing defense responses both naturally and after external application onto plants. Pep-triggered defense networks are compatible with pattern-triggered immunity (PTI). Nevertheless, in complex regulatory pathways, there is crosstalk among different signaling pathways, involving noncoding RNAs in the natural response to pathogen attack. Here, we used *Prunus persica*, PpPep2 and a miRNA-Seq approach to show for the first time that Peps regulate, in parallel with a set of protein-coding genes, a set of plant miRNAs (~15%). Some PpPep2-regulated miRNAs have been described to participate in the response to pathogens in various plant–pathogen systems. In addition, numerous predicted target mRNAs of PpPep2-regulated miRNAs are themselves regulated by PpPep2 in peach trees. As an example, peach miRNA156 and miRNA390 probably have a role in plant development regulation under stress conditions, while others, such as miRNA482 and miRNA395, would be involved in the regulation of resistance (R) genes and sulfate-mediated protection against oxygen free radicals, respectively. This adds to the established role of Peps in triggering plant defense systems by incorporating the miRNA regulatory network and to the possible use of Peps as sustainable phytosanitary products.

Keywords: plant elicitor peptide (Pep); plant defense; *Prunus*; miRNA sequencing; differential network analysis; gene set enrichment analysis

1. Introduction

Plants have a sophisticated defense system that is triggered by the perception of molecular patterns associated with pathogens (PAMPs) and endogenous danger signals (DAMPs) through transmembrane pattern recognition receptors (PRRs) [1,2].

Plant elicitor peptides (Peps) are approximately 23 to 40 amino acids long secondary DAMPs cleaved out of precursor proteins (PROPEPs) that are released to the apoplast following cell disruption due to pathogen attack or wounding. They are recognized by specific Leucine-Rich Repeat Receptor-Like Kinase (LRR-LRK) transmembrane Pep receptors (PEPRs), which result in the induction and amplification of pattern-triggered immunity (PTI, i.e., the first line of inducible plant defense). PTI involves common signaling molecules, e.g., Ca^{2+}, ROS and hormones (primarily salicylic acid, SA, ethylene, ET and jasmonic acid, JA), and extensive transcriptional reprogramming [2–6]. The overexpression and exogenous application of Peps at nanomolar concentrations increases resistance to various bacterial or fungal pathogens and herbivore attack [7–11] in ex vivo experiments and in planta [7,9,10,12–14]. As an example, the preventive treatment of peach plants with 1 μM doses of the *Prunus persica* peptide PpPep2 resulted in about 50% reduction in the symptoms following massive infection [3] with the bacterial pathogen *Xanthomonas arboricola* pv. *pruni* (*Xap*) [7].

PROPEP and PEPR orthologue genes are widely present within the angiosperms [15,16]. Sequences from the same plant family cluster together and family-specific motifs have been identified. Thus, there is inter-family incompatibility in their activity. The signal transduction pathways, however, are analogous.

Transcriptome deep sequencing analyses in *Arabidopsis* and *P. persica* (using AtPep1, and PpPep1 and PpPep2, respectively) showed that the topical application of Peps mimics the PTI natural response [7,17]. This includes the quick induction of PEPRs, PRRs and intracellular receptors, regulatory genes such as those related to hormone signaling (ET, JA, ABA) and transcription factors, calcium signaling proteins, plant disease resistance R proteins, etc., and the slower induction of pathogenesis-related proteins and cell wall-related genes.

There is a clear parallelism in the transcriptome dynamics caused by exogenous application of Peps and pathogen attacks, mostly affecting the same types of genes, although the latter showed an overall slower response, possibly due to the infection progress [7]. Thus, PTI induction is postulated as the Pep mode of action to achieve plant protection.

Complementary to protein coding genes, noncoding RNAs (ncRNAs) play essential regulatory roles in plants, particularly in development and in the response to abiotic and biotic stresses [18–21]. Micro RNAs (miRNAs) are 20–24 nucleotides small endogenous ncRNAs. They are transcribed from MIR genes into a stem-loop structured primary miRNA (pri-miRNA), which can be processed to shorter precursor pre-miRNAs and subsequently into miRNA/miRNA* duplexes. MIR transcription depends on a complex regulatory mechanism that involves numerous factors and microprocessor elements. miRNA/miRNA* duplexes are loaded on ARGONAUTE1 (AGO1) to constitute the RNA-induced silencing complex (miRISC), which is translocated to the cytosol. After that, the miRNA* is removed from miRISC, leaving the functional miRNA to target complementary mRNAs for degradation and/or translational silencing, thus contributing to gene regulation [22–30]. miRNAs can be functional within the same cell and can also be exported to distant cells.

miRNAs have been involved in plant immunity to bacteria and other pathogens [20,29,31–35] by regulating PTI and/or effector-triggered immunity (ETI). This includes pathogen perception through plant receptors, signal transduction and downstream immune responses, e.g., reactive oxygen species (ROS) accumulation, transcription factors and pathogenesis-related (PR) gene expression, callose deposition and plant hormone regulation. Some miRNAs are known to silence the auxin signaling pathway, which results in the activation of PTI [36,37] while some others contribute to fine-tuning the defense response [19].

The exogenous application of the PAMP flg22 onto *Arabidopsis* plants has an effect on the levels of various miRNAs, in coherence with miRNAs having a role in PAMP signaling [36,37]. Given the similarity of downstream events in response to PAMPs and DAMPs, we hypothesized that exogenous treatments with DAMPs might as well have an effect on the levels of certain miRNAs. We thus envisaged further characterization of the miRNA response to treatment with the DAMP PpPep2 onto *P. persica*, taken as an example. Peps are good candidates to establish natural, environmentally friendly and targeted culture management strategies. Improved knowledge on its mechanism of action is an important step towards transference of this technology. Here, we analyzed miRNA profiles of peach trees upon PpPep2 treatment using the application conditions that provide optimal protection against massive *Xap* infection.

2. Results and Discussion

2.1. miRNA-Seq Characterization of Peach Response to PpPep2

To assess the possible role of miRNAs in the response of peach to preventive treatment with PpPep2, we treated the leaves of juvenile plants with 1 μM of this peach peptide and carried out miRNA-Seq after 1 and 24 h using untreated plants as control. We previously showed that this is the optimal dose to protect peach trees against *Xap* [10] and that peach leaf transcriptome is strongly modulated 1 and 24 h after PpPep2 application [7]. One hour after PpPep2 application, there was a strong transcriptional regulation affecting up to 5% peach genes (i.e., differentially expressed genes, DEGs), with more than 90% upregulated. One day after PpPep2 treatment, most of these genes were downregulated to recover their levels before treatment, and there was ca. 1% novel DEGs. Transcriptional differences between the 24 h and the 48 h time points were less significant.

Nine miRNA libraries were constructed that corresponded to three experimental replicates of each, the control and 1 and 24 h time points, each consisting of nine plants and five leaves per plant. Data supporting this analysis are available in the Gene Expression Omnibus (GEO) repository, record GSE214135 (https://www.ncbi.nlm.nih.gov/geo/query/acc.cgi?acc=GSE214135; accessed on 1 December 2024).

Quality control and mapping statistics of the miRNA analysis are summarized in Table S1. We obtained 117, 267 and 647 raw reads that corresponded to an average of 12, 11 and 16 million reads 1 and 24 h after PpPep2 application and control, respectively. Moreover, 92% of pre_miRNA genes annotated in the *P. persica* NCBIv2.38 genome produced any read pair in any analyzed sample, while 16 of 210 annotated genes (i.e., 8%) did not produce any read pair. A total of 157 of the annotated pre_miRNAs (i.e., 75%) were consistently detected in all three replicates of at least one condition and reached values up to ~10^5 normalized counts (Table S2).

The processed data's principal component analysis (PCA) (Figure 1) reveals that the main miRNA transcriptomic changes can be attributed to the Pep2 treatment and time-course condition. Untreated and the two Pep2-treated samples are widely separated across the main component (PC1), which explains up to 48.7% of data variability. It might be speculated that the variability within 1 h of biological replicates observed in PC2 (Figure 1) is linked to quick changes in miRNA levels occurring around this time.

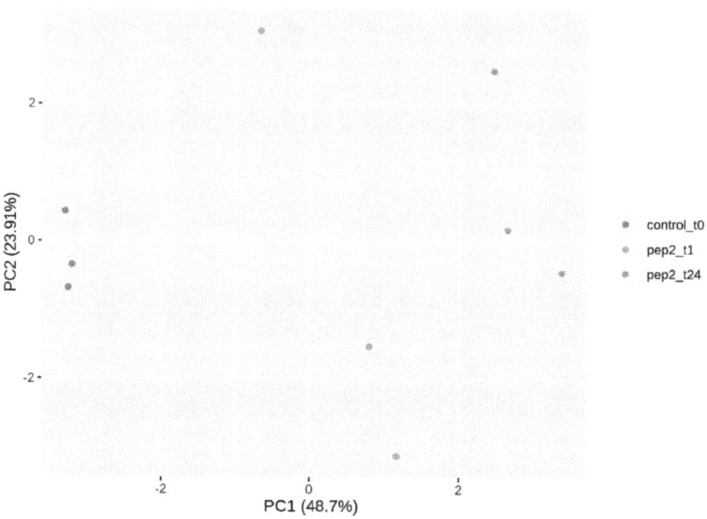

Figure 1. Principal component analysis (PCA) of log2 normalized expression data from the top 50 most variable miRNAs (DESeq2 1.20.0 package, [38]). Two principal components, PC1 and PC2, are represented. They have Eigenvalues above 1 and explain 48.7% and 23.91% of the variability, respectively. Colors represent the different peptide treatments: orange, time zero control (t0); green, 1 h PpPep2 (t1); and blue, 24 h PpPep2 (t24). Three biological replicates per sample are shown.

Differential expression analysis was performed to compare samples obtained 1 h or 24 h after peptide treatment with control samples. There were 33 differentially expressed miRNAs overall (DEMs, adj. $p < 0.05$) including the ones found in the comparisons of both time points against the control group (Figure 2, Table S2), which represents up to

15% of annotated *P. persica* miRNAs in the Ensembl Plants database. Thus, miRNAs seem to be involved in the peach response to PpPep2, which is in coherence with their role in plant immunity.

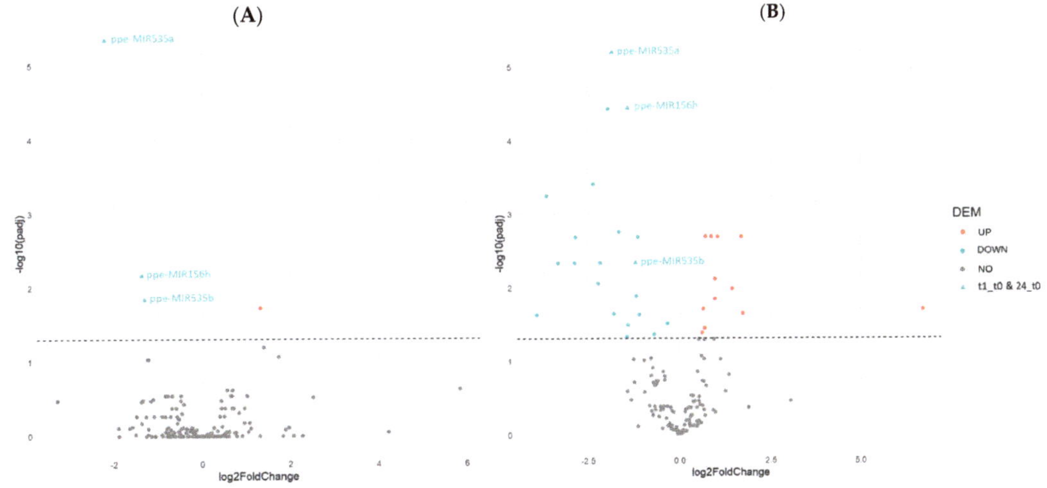

Figure 2. Volcano plots showing miRNA expression (expressed as log2FoldChange) 1 h after PpPep2 application in comparison to untreated control (**A**) and 24 h after PpPep2 application in comparison to untreated control (**B**); and the adj. *p*-value of the same comparisons (expressed as log10[adj. *p*]). The dashed line delimits differentially expressed miRNAs (DEMs), shown in red (upregulated DEMs, log2FC > 1) and blue (downregulated DEMs, log2FC < 1), from unregulated miRNAs (gray) using adj. $p < 0.05$. DEMs regulated at both comparisons are named and represented in triangles.

Except for one DEM, all were regulated 24 h after peptide application, with only three being regulated as early as one hour after the treatment. The remaining DEM was transiently upregulated one hour after PpPep2 application (Figure 2, Table 1). In response to PpPep2, 40% of DEMs were upregulated and 60% were downregulated. This is in contrast with the quick and massive mRNA changes observed in peach leaves in response to the same PpPep2 treatment. One hour after PpPep2 application, there was a peak in regulation affecting 1255 peach genes (ca. 5% of the total *P. persica* genes), with the vast majority upregulated and regaining normal levels within the next 24 (in some cases 48) h. Regulation was first detected 24 and 48 h after peptide application for a much smaller number of DEGs (i.e., 1.2% and 0.8% peach genes, respectively) [7]. The same pattern was described in *Arabidopsis* after treatment with the corresponding AtPep2, regulating up to 17% of genes two hours upon peptide application and 2.5% of genes ten hours later [17].

The 33 DEMs belong to 18 families: ppe-MIR156, ppe-miR159, ppe-miR162, ppe-MIR166, ppe-MIR167, ppe-miR168, ppeMIR169, ppe-miR171, ppe-MIR172, ppe-MIR390, ppe-MIR395, ppe-miR396, ppe-miR403, ppe-MIR482, ppe-miR530, ppe-MIR535, ppe-miR7122 and ppe-miR8127 (Table 1 and Table S2). It is known that miRNA molecules derived from many of these DEMs take part in the response to diseases caused by bacteria (e.g., miR156, miR159, miR167, miR169, miR390, miR482), fungi (miR156, miR168, miR396, miR482), viruses (e.g., miR159, miR171, miR395, miR482) and parasitic nematodes (e.g., miR167a) (see a review in [39] and references therein). This is consistent with their involvement in the PpPep2-driven defense response. Despite the known role of PAMPs and DAMPs in eliciting PTI, as well as the role of miRNAs in fine-tuning the defense response, only Li and colleagues (2010) [19] identified miRNAs involved in PAMP-triggered plant innate immunity using flg22 and *Arabidopsis thaliana*. When flg22 was applied, a number of miRNAs were found to be regulated and to bind AGO1. Here, we show that some of these flg22-regulated miRNAs (e.g., miR156, miR167, miR168, miR169, miR396) were also regulated by the DAMP PpPep2 in peach. Thus, there is some correlation between the miRNA response to PAMPs and DAMPs. Nevertheless, we identified a number of DEMs specific to PpPep2, notably those belonging to the ppe-MIR395 family but also those belonging to the ppe-miR171, miR7122, ppe-miR530, ppe-MIR535 and ppe-MIR166 families.

Table 1. *P. persica* leaf DEMs 1 and 24 h after PpPep2 topic application. For every DEM, the Ensembl (ID) and miRBase (miRNA) IDs are shown, together with their predicted mature sequences, average normalized counts in control (t0) and PpPep2-treated samples for one (t1) and 24 (t24) h and expression changes at every time point compared to t0 (log2 fold-values, t1_t0 and t24_t0).

ID	miRNA	Predicted Mature Sequence		Normalized Counts			t1_t0	t24_t0
		Strand	Sequence 5′-3′	t0	t1	t24		
ENSRNA049996926	ppe-miR156g	5p	UUGACAGAAGAUAGAGAGCAC	1	93	170	6.0	6.8
ENSRNA049996633	ppe-miR171c	5p	UGAUUGAGCCGUGCCAAUAUC	4	16	16	2.1	2.1
ENSRNA049995981	ppe-miR7122b	3p 5p	CCGUGUUUCCUUGUAUAAAG UUAUACAAUGAAAUCACGGUCG	115	301	368	1.4	1.7
ENSRNA049995989	ppe-miR7122a	3p 5p	GCCGUGUUUCUUUGUAUAAAG UUAUACAAUGAAAUCACGGCCG	2506	5006	6695	1.0	1.4
ENSRNA049996187	ppe-miR482a	3p 5p	UUUCCGAAACCUCCCAUUCCAA GGGUGAGAGGUUGCCGGAAAGA	324	501	620	0.6	0.9
ENSRNA050003821	ppe-miR396b	5p	UUCCACAGCUUUCUUGAACUU	8203	12,003	14,798	0.5	0.9
ENSRNA049996209	ppe-miR482f	3p	UCUUUCCUACUCCACCCAUUCC	3215	4544	6556	0.5	1.0
ENSRNA049996042	ppe-miR403	3p	UUAGAUUCACGCACAAACUCG	1307	1823	2510	0.5	0.9
ENSRNA049996194	ppe-miR482e	3p	UUGCCUAUUCCUCCCAUGCCAA	534	608	825	0.2	0.6
ENSRNA049996910	ppe-miR162	3p	UCGAUAAACCUCUGCAUCCAG	2901	3101	4694	0.1	0.7
ENSRNA049996592	ppe-miR171a	3p	UGAUUGAGCCGUGCCAAUAUC	74	99	111	0.4	0.6
ENSRNA050003799	ppe-miR159	3p	UUUGGAUUGAAGGGAGCUCUA	82	106	128	0.4	0.6
ENSRNA049997214	ppe-miR167a	5p	UGAAGCUGCCAGCAUGAUCUA	200	495	284	1.3	0.5
ENSRNA049996319	ppe-miR8127	3p 5p	UUCAAAGGGUACAUCCACAGU CAACUGUGGACAUACCCUUUG	1745	1803	1372	0.0	−0.3
ENSRNA049996833	ppe-miR168	5p	UCGCUUGGUGCAGGUCGGGAA	788	648	485	−0.3	−0.7
ENSRNA049997189	ppe-miR535b	5p	UGACGACGAGAGAGAGCACGC	466	178	199	−1.4	−1.2
ENSRNA049996546	ppe-miR482c	3p 5p	UUCCCAAGCCCGCCCAUUCCAA UUCCCAAGCCCGCCCAUUCCAA	4406	2720	1915	−0.7	−1.2
ENSRNA049996130	ppe-miR395c	3p	CUGAAGUGUUUGGGGGAACUC	108	62	49	−0.8	−1.1
ENSRNA049996345	ppe-miR172b	3p	AGAAUCUUGAUGAUGCUGCAU	113	116	43	0.0	−1.4
ENSRNA050003783	ppe-miR390	5p	AAGCUCAGGAGGGAUAGCGCC	416	299	152	−0.5	−1.4
ENSRNA049996145	ppe-MIR395n	3p	CUGAAGUGUUUGGGGGAACUC	233	185	84	−0.3	−1.5
ENSRNA049996894	ppe-miR166e	3p	UCGGACCAGGCUUCAUUCCCC	2574	1091	801	−1.2	−1.7
ENSRNA049996107	ppe-miR395a	3p 5p	CUGAAGUGUUUGGGGGGACCC GUUCCCUCAAACACUUCAUU	6318	3680	1792	−0.8	−1.8
ENSRNA049997194	ppe-miR535a	5p	UGACAACGAGAGAGAGCACGC	167	37	43	−2.2	−1.9
ENSRNA049996563	ppe-miR156h	5p	UUGACAGAAGAUAGAGAGCAC	693	289	188	−1.3	−1.9
ENSRNA049996028	ppe-miR169d	5p	UGAGCCAAGGAUGACUUGCCA	48	27	10	−0.8	−2.2
ENSRNA049996100	ppe-miR395l	3p	CUGAAGUGUUUGGGGGAACUC	1193	558	252	−1.1	−2.2
ENSRNA050003167	ppe-miR530	5p	UCUGCAUUUGCACCUGCACCU	391	178	76	−1.1	−2.4
ENSRNA049996115	ppe-miR395j	3p	CUGAAGUGUUUGGGGGAACUC	40	21	5	−0.9	−3.0
ENSRNA049996137	ppe-miR395i	3p	CUGAAGUGUUUGGGGGAACUC	50	19	6	−1.4	−3.1
ENSRNA049996172	ppe-miR395k	3p	CUGAAGUGUUUGGGGGAACUC	376	233	36	−0.7	−3.4
ENSRNA049996180	ppe-miR395f	3p	CUGAAGUGUUUGGGGGAACUC	53	20	4	−1.4	−3.8
ENSRNA049996069	ppe-miR395m	3p	CUGAAGUGUUUGGGGGAACUC	17	7	1	−1.3	−3.9

Color codes correspond to statistically up—(red) and down—(blue) regulated miRNAs in response to PpPep2, DESeq2 v1.26.0 adj. $p < 0.05$. Color intensity indicates fold change. Numbers in gray font indicates not statistically significant. Note that the last letter in every miRNA code differentiates members within the same miRNA family. Table S2 shows normalized counts and statistics for all samples and replicates.

2.2. Analysis of Peach miRNA Response to PpPep2 on the Level of Processes

Mature DEM sequences are what constitute functional miRNA molecules. The mature DEM sequences were deduced using a combined analysis of two different databases i.e., Ensembl, encompassing the peach genome, and miRBase, compiling *P. persica* miRNAs. Note that, for six DEMs, i.e., ppe-MIR7112a and -b, ppe-MIR482a and -c, ppeMIR8127 and ppe-MIR395a, there were two different possible mature sequences that corresponded to the forward (5p) and the complementary (3p) strands, respectively. In these cases, the two sequences were considered potential functional miRNAs for further analyses. In consequence, there were 39 deduced mature miRNAs regulated by PpPep2. The majority (74% of them) were 21 nt long, 22% had 22 nt and the remaining 5% were 20 nt long.

Then, psRNATarget [40] was used to conduct target prediction analysis based on these 39 sequences. Mature miRNAs direct the RISC to target genes through base pairing [41]; hence, target prediction is dependent on sequence homology and secondary structure. The psRNATarget algorithms also consider the fact that, in plants, a single target gene can be targeted by multiple miRNAs and that a single miRNA can have a large number of target genes [42]. Using a p-value threshold of 0.05, we identified a total of 2431 DEM predicted target genes (DEM-Ts). A range of 45 to 117 DEM-Ts were predicted for each DEM, with varying degrees of targeting probability based on miRNA and target complementarity (Table S3).

In order to assist the biological interpretation of miRNA changes in response to PpPep2, DEM-Ts were subjected to functional enrichment analysis. Figure 3 depicts the enriched GO terms associated with predicted targets of miRNAs with differential expression one hour and twenty-four hours after peptide treatment.

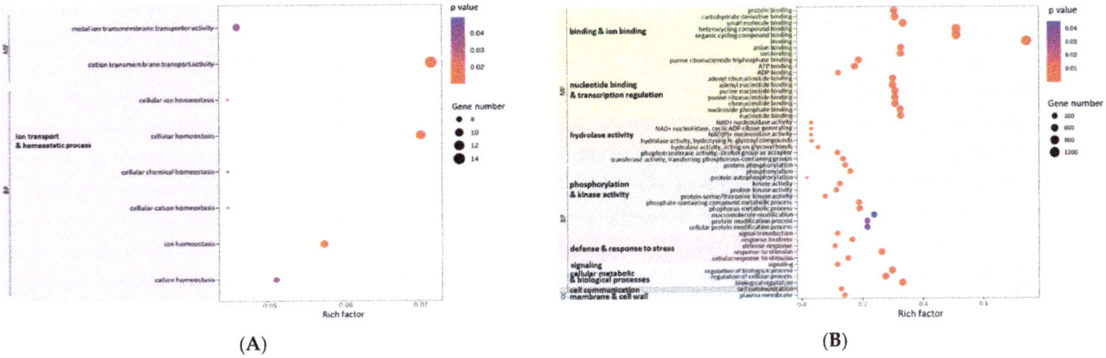

Figure 3. Enriched GO terms associated with DEM-Ts 1 h (**A**) and 24 h (**B**) after PpPep2 treatment of peach leaves. The x-axis indicates the rich factor (ratio of genes in the input list that are annotated to the function); and the y-axis shows the GO terms grouped into upper families shown in rainbow colors. GO terms are also classified into molecular function (MF), biological process (BP), or cellular component (CC) categories. Dot colors indicate p-values, shown in blue (0.05) to red (0.00). Dot radiuses are proportional to gene number (number of genes contributing to each GO term enrichment).

One hour after PpPep2 treatment, there were just four DEMs. They had up to 322 predicted targets, and they primarily belonged to two molecular functions and six biological process terms that were related to ion transport and homeostasis (Figure 3A). "Cation transmembrane transport activity", "cellular homeostasis" and, specifically, "ion homeostasis" and "cation homeostasis" were the terms with the highest gene number and rich factor. Effector perception rapidly triggers ionic flux (mainly affecting Ca^{2+}, H^{+}, K^{+}, Cl^{-} and NO^{3-}), which leads to extracellular alkalinization that triggers the PTI response [43].

Moreover, we previously described that there was a quick and transient regulation of genes related to "calcium transport" and "calcium signaling" terms in peach leaves treated with PpPep2 in an RNA-Seq transcriptome analysis approach [7]. Our miRNA results suggest that miRNAs (particularly ppeMIR156h, ppe-MIR167a, ppe-MIR535a and/or ppe-MIR535b) play a role in these quick initial steps of PTI.

One day after PpPep2 treatment, there was a stronger miRNA response than at the earlier time, i.e., 39 DEMs. Their corresponding DEM-Ts were associated with 47 GO terms (Figure 3B).

The GO terms with the highest gene number and rich factor were related to binding (specifically, the heterocycling compound, organic compound, carbohydrate, protein and ion binding), and "nucleotide binding and transcription regulation". Again, there was the regulation of biological processes, mainly linked to the "defense and response to stress" and "phosphorylation and kinase activity". All these processes are integral to the PTI-like response. Phosphorylation events involving receptors and co-receptors of PAMPs and DAMPs (including PEPR), MAPK cascades and downstream TFs are essential parts of the signal transduction mechanisms. This suggests that miRNAs would have an active role in

PTI regulation at these stages. In coherence, there was also the enrichment of GO terms related to signaling, cell communication and membrane and cell wall modifications, also linked to PTI.

2.3. miRNAs Participate in PpPep2-Driven Regulation of Genes Involved in the Defense Response

Experimental data on the expression patterns of DEM-Ts in response to 1 μM of PpPep2 treatment were used to identify which of these genes are regulated by PpPep2 at the mRNA level. We previously used RNA-Seq to identify peach leaf differentially expressed genes (DEGs) in response to 1 μM of PpPep2 treatment for 1, 24 and 48 h [7]. The intersection of DEGs with all 2431 DEM-Ts retrieved 176 genes (Table 2), corresponding to 7.2% DEM-Ts and 8.9% PpPep2 DEGs. Therefore, both these genes were regulated by PpPep2 at the mRNA level and are predicted targets of PpPep2-regulated miRNAs. Remarkably, there was at least one DEM-T for every DEM. This supports this subset of DEGs are targets of DEMs; and further suggests that the identified miRNAs are involved in the PpPep2-driven regulation of this specific set of genes.

To gain insight into their role, these 176 DEM-Ts were subjected to enrichment analysis (Figure 4). There were 36 enriched GO terms ($p < 0.05$) that were mainly related to protein phosphorylation and cell communication biological processes and to protein kinase and ADP binding activities. We determined the genes that participated in each enriched GO term, as well as their corresponding DEMs (Table S4).

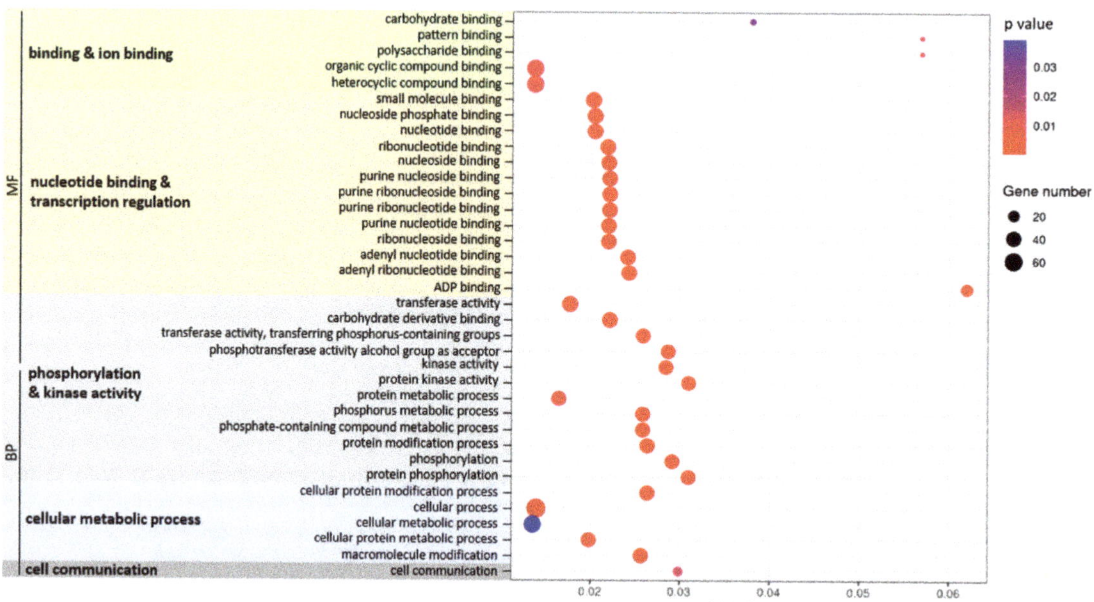

Figure 4. Enriched GO terms associated with DEM-Ts that were differentially expressed after 1 μM of PpPep2 treatment of peach leaves for 1 h, 24 h or 48 h. The *x*-axis indicates the rich factor (ratio of genes in the input list that are annotated to the function), and the *y*-axis shows the GO terms. GO terms are also classified into molecular function (MF) or biological process (BP) categories. Dot color indicate *p*-values, shown in blue (0.05) to red (0.00). Dot radiuses are proportional to the gene number (number of genes contributing to each GO term enrichment).

Table 2. Expression patterns of DEMs and predicted target genes that are differentially expressed in peach leaves in response to PpPep2 treatment. Only those that have opposing expression patterns and up to two predicted target genes per DEM (i.e., those with the lowest *p*-value in target gene prediction) are displayed. t1_t0, t24_t0, and t48_t0 indicate gene expression log2fold-change values at 1, 24, and 48 h after PpPep2 treatment compared to untreated samples, respectively.

DEM				Predicted Target Gene 1					Predicted Target Gene 2					
miRNA	t1_t0	t24_t0	# DEG Targets	Target ID	Target Description	t1_t0	t24_t0	t48_t0	Target ID	Target Description	t1_t0	t24_t0	t48_t0
ppe-MIR156g	6.0	6.8	2	PRUPE.5G208300	Teosinte glume architecture 1	0.7	−0.3	0.3					
ppe-MIR171c	2.1	2.1	4	PRUPE.2G274300	Hexose carrier protein HEX6	1.5	0.3	−0.6					
ppe-miR7122b-3p	1.4	1.7	3										
ppe-miR7122b-5p	1.4	1.7	6	PRUPE.3G227700	Uncharacterized LOC18783989	−0.5	1.1	0.5					
ppe-miR7122a-3p	1.0	1.4	5	PRUPE.6G281500	Receptor-like protein 12	2.0	0.0	0.0	PRUPE.3G039200	Linoleate 13S-lipoxygenase 3-1, chloroplastic	1.9	0.5	0.2
ppe-miR7122a-5p	1.0	1.4	4	PRUPE.1G317500	Lipid phosphate phosphatase 2	0.9	−0.1	−0.2					
ppe-MIR482f	0.5	1.0	11	PRUPE.2G022400	Putative disease resistance RPP13-like protein 1	1.0	−0.6	−0.3					
ppe-miR403	0.5	0.9	2	PRUPE.1G525900	Pentatricopeptide repeat-containing protein At5g27460	1.5	0.1	−0.1	PRUPE.1G385000	Hypothetical protein PRUPE.1G385000	1.7	−0.2	−0.7
ppe-miR482a-3p	0.6	0.9	15	PRUPE.6G295900	Receptor-like protein kinase FERONIA	1.1	−0.5	0.4					
ppe-miR482a-5p	0.6	0.9	10										
ppe-miR396b	0.5	0.9	9	PRUPE.5G013100	Abscisic acid 8′-hydroxylase 1	3.1	−0.7	1.1					
ppe-miR162	0.1	0.7	3	PRUPE.1G385500	Hypothetical protein PRUPE.1G385500	1.0	−0.1	−0.8					
ppe-miR159	0.4	0.6	5	PRUPE.6G073300	Scarecrow-like protein 30	3.4	0.3	−0.8					
ppe-miR482e	0.2	0.6	10	PRUPE.4G284000	Putative disease resistance protein RGA3	0.8	−0.5	−0.4					
ppe-miR171a	0.4	0.6	4	PRUPE.2G274300	Hexose carrier protein HEX6	1.5	0.3	−0.6					
ppe-MIR167a	1.3	0.5	9	PRUPE.2G290200	Protein DETOXIFICATION 49	4.7	0.8	−0.4					
ppe-miR8127-3p	0.0	−0.3	2										
ppe-miR8127-5p	0.0	−0.3	4										
ppe-miR168	−0.3	−0.7	3	PRUPE.7G053500	Glycerol-3-phosphate dehydrogenase [NAD(+)] GPDHC1, cytosolic	−0.3	0.8	0.6					
ppe-MIR395c	−0.8	−1.1	15	PRUPE.6G180002	Sulfate transporter 2.1	0.2	−0.9	0.5	PRUPE.1G023002	ATP sulfurylase 1, chloroplastic	0.2	−0.5	0.9
ppe-MIR482c-3p	−0.7	−1.2	12										
ppe-miR482c-5p	−0.7	−1.2	15	PRUPE.7G243500	SPX domain-containing protein 1	0.7	2.2	1.0					
ppe-MIR535b	−1.4	−1.2	10	PRUPE.4G158000	Nematode-induced LRR-RLK 2 (NILR1)	2.1	−0.7	−0.8					
ppe-MIR172b	0.0	−1.4	12										
ppe-MIR390	−0.5	−1.4	10	PRUPE.4G121900	Receptor-like protein 12	4.2	0.3	0.0					
ppe-MIR395n	−0.3	−1.5	15	PRUPE.6G180009	Sulfate transporter 2.1	0.2	−0.9	0.5	PRUPE.1G023009	ATP sulfurylase 1, chloroplastic	0.2	−0.5	0.9
ppe-MIR166e	−1.2	−1.7	5	PRUPE.7G041600	Two-component response regulator ORR9	−0.9	0.1	0.6					
ppe-MIR395a-3p	−0.8	−1.8	7	PRUPE.6G180000	Sulfate transporter 2.1	0.2	−0.9	0.5	PRUPE.1G023000	ATP sulfurylase 1, chloroplastic	0.2	−0.5	0.9
ppe-miR395a-5p	−0.8	−1.8	4	PRUPE.6G180001	Sulfate transporter 2.1	0.2	−0.9	0.5	PRUPE.1G023001	ATP sulfurylase 1, chloroplastic	0.2	−0.5	0.9
ppe-MIR535a	−2.2	−1.9	8	PRUPE.8G263900	Probable pectinesterase/pectinesterase inhibitor 7	4.7	1.4	1.4					
ppe-MIR156h	−1.3	−1.9	2	PRUPE.4G089000	Cationic amino acid transporter 1	1.6	0.7	0.7					
ppe-MIR395l	−1.1	−2.2	15	PRUPE.6G180007	Sulfate transporter 2.1	0.2	−0.9	0.5	PRUPE.1G023007	ATP sulfurylase 1, chloroplastic	0.2	−0.5	0.9
ppe-MIR169d	−0.8	−2.2	7	PRUPE.7G093100	Nuclear transcription factor Y subunit A-10 (NFYA10)	0.1	0.1	0.3					
ppe-miR530	−1.1	−2.4	1										
ppe-MIR395j	−0.9	−3.0	15	PRUPE.6G180005	Sulfate transporter 2.1	0.2	−0.9	0.5	PRUPE.1G023005	ATP sulfurylase 1, chloroplastic	0.2	−0.5	0.9
ppe-MIR395i	−1.4	−3.1	15	PRUPE.6G180004	Sulfate transporter 2.1	0.2	−0.9	0.5	PRUPE.1G023004	ATP sulfurylase 1, chloroplastic	0.2	−0.5	0.9
ppe-MIR395k	−0.7	−3.4	15	PRUPE.6G180006	Sulfate transporter 2.1	0.2	−0.9	0.5	PRUPE.1G023006	ATP sulfurylase 1, chloroplastic	0.2	−0.5	0.9
ppe-MIR395f	−1.4	−3.8	15	PRUPE.6G180003	Sulfate transporter 2.1	0.2	−0.9	0.5	PRUPE.1G023003	ATP sulfurylase 1, chloroplastic	0.2	−0.5	0.9
ppe-MIR395m	−1.3	−3.9	15	PRUPE.6G180008	Sulfate transporter 2.1	0.2	−0.9	0.5	PRUPE.1G023008	ATP sulfurylase 1, chloroplastic	0.2	−0.5	0.9

Color codes indicate statistically up- (red) and down- (blue) regulated miRNAs and predicted target genes in response to PpPep2. # DEG targets, number of predicted miRNA targets that are differentially expressed upon PpPep2 treatment.

The enrichment pathways (Figure 4) align with the defense-related roles of PpPep2. We previously used RNA-Seq to show that the topical application of PpPep2 onto peach trees is similar to that of Arabidopsis to its specific AtPeps [17] and mimics the natural PTI response [7]. The initial PTI signal transduction steps involve the phosphorylation of different proteins, notably PEPR and co-receptors and MAP kinases that subsequently regulate the transcription of genes participating in the defense response [3,5,43]. MIR390, MIR482a and MIR8127 were predicted to target a number of genes related to protein phosphorylation and kinase activity. These and MIR482c, e and f were anticipated to target genes with nucleoside or nucleotide binding activity. The topic application of PpPep2 quickly regulates biotic stress receptors, signaling through receptor kinases, and RNA transcription. These results suggest that miRNAs, in particular MIR390, MIR482 and MIR8127, are involved in Pep-driven PTI regulation at these initial stages.

The transcriptome response of peaches to PpPep2 involves genes participating in signaling and hormone metabolism. In agreement, here we find the PpPep2-driven regulation of MIR166e, MIR168, MIR172b and MIR482a, c-3p, e and f, which predictably target genes with cell communication roles.

miRNAs modulate gene expression by binding to specific mRNAs and targeting them for cleavage or by directing translational inhibition at the mRNA level [44]. In accordance, for those inducing mRNA cleavage, we might expect the overexpression of a given miRNA to result in the downregulation of target gene mRNA levels and vice versa. We thus compared the expression patterns of DEMs and those of DEM-Ts that are differentially expressed in peach leaves in response to PpPep2 treatment. Most DEMs had one or more DEM-Ts whose expression patterns in response to PpPep2 were diametrically opposed to that of the DEM (Table 2). Although there is no report on the role of these DEMs nor experimental proof of their target genes in peach trees, some have been studied in other species. MIR genes, such as MIR156 [45–47] and MIR390 [45,48–50], are involved in plant development and the response to stress and participate in the growth and defense trade-off. Others, e.g., MIR482 and MIR395, seem to be more specifically involved in the plant defense responses [51,52].

We found that ppe-MIR156g was induced up to six and sevenfold in peach leaves treated with PpPep2 for 1 h and 24 h, respectively, being the most strongly regulated miRNA in these conditions. One of its DEM-Ts is the Squamosa Promoter-Binding-Like (SPL) protein Teosinte Glume Architecture 1 (TGA1) (Table 2), whose mRNA levels were also regulated by PpPep2 in the same peach system [7]. In agreement, ZmMIR156g proved to target TGA1 in maize [47,53]. The evolutionary conserved MIR156 family members are known to target SPLs, whose mRNA levels increase during shoot development as MIR156 expression gradually declines in abundance [45]. MIR156 family members are induced to arrest development under stress conditions, and their miRNAs are suppressed when the plants are returned to favorable conditions, allowing the developmental transition to be accelerated [46].

In peach leaves, we found a negative relation between ppe-MIR390 and its DEM-T, Leucine-Rich Repeat, upon the application of PpPep2 for 1 h and 24 h. A similar relation was described in apple tree, where mdm-miR390a overexpression enhanced the sensitivity to Alternaria infection through the downregulation of MdLRR8 and the LRR-LRK serine/threonine-protein kinase MdRPK2. This way, the fungal induction of apple mdm-miR390a increases the susceptibility of the plant host to infection [48]. Our results suggest that endogenous peptides such as PpPep2 may contribute to pathogen resistance by the downregulation of ppe-MIR390 and the parallel upregulation of LRR. The downregulation of MIR390 in response to stress results in the suppression of growth and yield reduction in rice and *Arabidopsis* [49,54], which might as well be an effect of PpPep2.

The MIR482 family is highly represented among PpPep2-regulated MIRs, with four DEMs out of six annotated miRNAs (MIR482a-3p, MIR482c-3p, MIR482e, MIR482f), representing 66% of the family. Various PpPep2-regulated Resistance (R) genes were among MIR482 DEM-Ts, e.g., putative disease resistance protein RGA3 (ppe-MIR482e target), disease resistance RPP13-like protein 1 (ppe-MIR482f target) and the receptor-like protein kinase FERONIA (ppe-MIR482a-3p target). As expected for miRNA regulation, the expression patterns of these miRNAs and their predicted targets are inversely related (Table 2). Thus, peach MIR482 seems to be related to regulation of the defense response. In agreement, previous studies showed that MIR482 genes target members of different R gene families in apple, cotton, soybean, *Medicago truncatula* and various Solanaceae species, e.g., tomato and potato [48,55–57]. In *Arabidopsis*, MIR482 regulates an NBS-LRR-type R protein that inhibits jasmonic acid (JA) signaling and positively regulates immunity [51].

Up to 27% of DEMs identified in this study were classified in the MIR395 family; nine out of fifteen annotated peach MIR395 members were DEMs (i.e., ppe-MIR395a, -c, -f, -i, -j, -k, -l, -m and -n). All nine ppe-MIR395 family members were downregulated after the PpPep2 application and shared a common mature sequence, which was predicted to target Sulfate Transporter 2.1 (SULTR2) and ATP Sulfurylase (ATPS) transcripts (Table 2). These sulfate-related genes were upregulated one day after PpPep2 treatment. SULTR2 and ATPS are involved in sulfate assimilation into glutathione, providing antioxidant protection against oxygen free radical-mediated damage brought on by pathogens [52,58]. This way, ppe-MIR395 seems to display a role in peach disease resistance, as previously observed in A. thaliana and eggplant [52,59]. In contrast to the Arabidopsis response to the PAMP flg22, MIR395 is among the most strongly regulated miRNAs in response to the exogenous application of the DAMP PpPep2 onto peach.

Taken together, our results enhance the understanding of PpPep2 as a phytosanitary tool to boost peach tree defenses against pathogens, consolidating it as a promising plant disease management tool. PpPep2-driven miRNA regulation is compatible with pathogen defense stimulation and also the developmental arrest associated with stress resistance, which again mimics plant natural defense systems.

3. Materials and Methods

3.1. Plant Material and Peptide Treatments

Peach juvenile plants (*Prunus persica* var. Big Top) were produced using in vitro technology and grown in individual small pots by a professional grower (Agromillora Iberia S.A., Barcelona, Spain). Plant treatments were carried out as previously described [7]. Briefly, plants were acclimatized for two weeks in a glasshouse (21 °C, a 16/8 h light/dark photoperiod and 60% RH) prior to treatments with 1 μM of PpPep2 (YVQRITLRAARPEIST-GSGAQTN, the full-length mature peptide of MW600837 [10,60], chemically synthesized by Caslo ApS (Lyngby, Denmark) with purity above 95% and the identity confirmed by MALDI-TOF. The five youngest fully expanded leaves of each plant were labeled, and PpPep2 was sprayed onto both abaxial and adaxial leaf surfaces. Plants were incubated under standard conditions in the glasshouse for 0, 1 and 24 h. Leaves were detached, and the leaf blades were immediately frozen in liquid nitrogen. The experimental design consisted of three replicates of nine plants per treatment.

3.2. RNA Extraction and Illumina Sequencing

RNA was extracted from a 200 mg aliquot of a ground leaf sample through a TRIzol-based procedure (Invitrogen Life Technologies, Carlsbad, CA, USA) followed by DNAse I (Ambion, Grand Island, NY, USA) digestion and purification with RNeasy MinElute Cleanup Kit (Qiagen, Sollentuna, Sweden). The estimation of RNA concentration was conducted using a NanoDrop ND1000 spectrophotometer (Nanodrop Technologies, Wilmington, DE, USA).

miRNA-Seq was carried out at the Centre Nacional d'Anàlisi Genòmica (CNAG) (Barcelona, Spain). Libraries were prepared using the NEBNext® Small RNA Library Prep Set for Illumina® kit (ref. E7330) (New England Biolabs, Ipswich, MA, USA) according to the manufacturer's protocol. Briefly, RNA was subjected to adaptor 3′and 5′ ligation and first strand cDNA synthesis. Then, DNA fragments with adapter molecules on both ends were selectively enriched by a PCR using NEBNext® Multiplex Oligos for Illumina (Index

Primers Set 1, ref. E7335, Index Primers Set 2, ref. E7500, Index Primers Set 3, ref. E7710 and Index Primers Set 4, ref. E7730) (New England Biolabs, Ipswich, MA, USA). Purification steps were then performed using AgenCourt AMPure XP beads (ref. A63882) (Beckman Coulter, Brea, CA, USA), and final libraries were analyzed using the Agilent Bioanalyzer (ref. 5067-4626) (Agilent Technologies, Santa Clara, CA, USA) to estimate the quantity and check size distribution. A pool was conducted to perform size selection using 6% Novex TBE PAGE Gels (ref. EC6265BOX) (ThermoFisher Scientifics, Waltham, MA, USA), and, then, the final pool was quantified by qPCR using the KAPA Library Quantification Kit (ref. KK4835, KapaBiosystems) (Roche, Basel, Switzerland) prior to amplification with Illumina's cBot. Libraries were sequenced $1 \times 50 + 8$ bp on the Illumina HiSeq2500 device (Illumina, San Diego, CA USA).

3.3. Bioinformatics Analysis

Small RNA-Seq reads were trimmed using Trim Galore version 0.6.6 [61] to remove adapter sequences and low-quality bases. Trimmed reads were mapped against *Prunus persica* reference genome (Prunus_persica_NCBIv2.38) using STAR software version 2.5.3a [62] with ENCODE parameters for small reads. Annotated genes were quantified using release 48 of Ensembl Plants annotation with RSEM version 1.3.0 [63], modifying the default parameter seed length parameter to 16 to improve alignment for small RNA-seq.

Differential expression analysis of annotated pre_miRNA gene biotypes between time points was conducted using the DESeq2 v1.26.0 R package [38]. DESeq2 performs independent filtering by default, using the mean of normalized counts as a filter statistic. Significance testing was performed with the Wald test, and *p*-values were adjusted using the Benjamini–Hochberg method for multiple test correction, and genes were considered statistically significant with an adjusted *p*-value below 0.05.

The miRNA gene target prediction was carried out using psRNATarget [40], which evaluates the complementarity between miRNA and target mRNA sequences and the accessibility of the mRNA based on its secondary structure with a scoring schema. The reliability of psRNATarget predictions is supported by the high recall rates reported in its benchmark analyses. According to the Nucleic Acids Research publication of the tool, psRNATarget achieved a recall rate of 97.3% for Arabidopsis thaliana and 82.7% for rice, demonstrating its capacity to accurately predict miRNA-target interactions across diverse plant species.

Gene Ontology enrichment analysis of miRNA-predicted gene targets was performed using the gprofiler2 v0.1.8 R package [64]. The analysis relies on a cumulative hypergeometric probability (Fisher's one-tail test) to find the over-representation of GO terms from the Ensembl database within the significant gene target list. Principal component analysis (PCA) was carried out on normalized counts using the prcomp R function to assess sample variability. Visualization, including PCA, volcano plots and enrichment dot plots were generated using the R package ggplot2 v3.5.1 [65].

4. Conclusions

Here, we further analyze the mode of action of PpPep2, which causes PTI-like transcriptome reprogramming and enhances resistance to *Xap* in *P. persica* [7,10]. One hour and one day after peptide application, there were expression changes in up to 7% of peach genes. Furthermore, 1 day of pretreatment with chemically synthesized PpPep2 before exposure to massive doses of *Xap* results in about 50% reduction in the disease symptoms.

The miRNA-Seq analysis of the peach tree response to the application of PpPep2 showed that miRNAs participate in the Pep-triggered defense response of plants. About 15% of miRNA genes were regulated by PpPep2 within 24 h. For every DEM, there were some DEM-Ts that were regulated in response to the same PpPep2 treatment. Inversely, about 10% of the PpPep2-regulated protein-coding genes were DEM-Ts. These miRNAs appear directly involved in the PpPep2-driven regulation of their corresponding target genes, which are related to PTI-like processes. A number of experiments carried out in other plant species support the regulatory role of several of these miRNAs towards their predicted target genes in the context of defense and their balance versus growth.

Additionally, our findings indicate that the topical application of Peps activates plant immune systems, involving not only transcriptomic changes but also miRNA regulatory pathways. The parallelism between the plant response to the Pep topic application and natural defense mechanisms indicates its mechanism of action and the lack of safety concerns related to this natural phytosanitary product.

Supplementary Materials: The following supporting information can be downloaded at https://www.mdpi.com/article/10.3390/ijms252313099/s1.

Author Contributions: Conceptualization, A.N. and M.P.; methodology, L.F. and M.P.; formal analysis, L.F., A.E.-C. and B.M.-M.; investigation, L.F.; data curation, L.F., B.M.-M. and A.E.-C.; writing—original draft preparation, L.F., M.P. and A.N.; writing—review and editing, A.N. and M.P.; visualization, L.F., B.M.-M. and A.N. All authors have read and agreed to the published version of the manuscript.

Funding: This research was funded by the Spanish Ministerio de Economía, Industria y Competividad (MINECO), grant number AGL2017–85121-R, and the Spanish Ministerio of Ciencia, Innovación y Universidades (MCIU), grant number PID2021-126505OB-I00. L.F. received a fellowship from Generalitat de Catalunya (2017FI_B_00435).

Institutional Review Board Statement: Not applicable.

Informed Consent Statement: Not applicable.

Data Availability Statement: Data supporting transcriptome analysis are available in the Gene Expression Omnibus (GEO) repository, record GSE214135, enter token slshgkouhfajdyp into the box (https://www.ncbi.nlm.nih.gov/geo/query/acc.cgi?acc=GSE214135; accessed on 1 December 2024).

Acknowledgments: We thank Agromillora Iberia S.A. for the provision of the plant material.

Conflicts of Interest: The authors declare no conflicts of interest.

References

1. Bartels, S.; Boller, T. Quo Vadis, Pep? Plant Elicitor Peptides at the Crossroads of Immunity, Stress, and Development. *J. Exp. Bot.* **2015**, *66*, 5183–5193. [CrossRef]
2. Gust, A.A.; Pruitt, R.; Nürnberger, T. Sensing Danger: Key to Activating Plant Immunity. *Trends Plant Sci.* **2017**, *22*, 779–791. [CrossRef] [PubMed]
3. Couto, D.; Zipfel, C. Regulation of Pattern Recognition Receptor Signalling in Plants. *Nat. Rev. Immunol.* **2016**, *16*, 537–552. [CrossRef]
4. Li, N.; Han, X.; Feng, D.; Yuan, D.; Huang, L.J. Signaling Crosstalk between Salicylic Acid and Ethylene/Jasmonate in Plant Defense: Do We Understand What They Are Whispering? *Int. J. Mol. Sci.* **2019**, *20*, 671. [CrossRef] [PubMed]
5. Li, B.; Meng, X.; Shan, L.; He, P. Transcriptional Regulation of Pattern-Triggered Immunity in Plants. *Cell Host Microbe* **2016**, *19*, 641–650. [CrossRef] [PubMed]
6. Li, P.; Zhao, L.; Qi, F.; Htwe, N.M.P.S.; Li, Q.; Zhang, D.; Lin, F.; Shang-Guan, K.; Liang, Y. The Receptor-like Cytoplasmic Kinase RIPK Regulates Broad-Spectrum ROS Signaling in Multiple Layers of Plant Immune System. *Mol. Plant* **2021**, *14*, 1652–1667. [CrossRef] [PubMed]
7. Foix, L.; Nadal, A.; Zagorščak, M.; Ramšak, Ž.; Esteve-Codina, A.; Gruden, K.; Pla, M. *Prunus persica* Plant Endogenous Peptides PpPep1 and PpPep2 Cause PTI-like Transcriptome Reprogramming in Peach and Enhance Resistance to *Xanthomonas arboricola* pv. *pruni*. *BMC Genom.* **2021**, *22*, 360. [CrossRef]
8. Yamaguchi, Y.; Huffaker, A. Endogenous Peptide Elicitors in Higher Plants. *Curr. Opin. Plant Biol.* **2011**, *14*, 351–357. [CrossRef]
9. Huffaker, A.; Pearce, G.; Veyrat, N.; Erb, M.; Turlings, T.C.J.; Sartor, R.; Shen, Z.; Briggs, S.P.; Vaughan, M.M.; Alborn, H.T.; et al. Plant Elicitor Peptides Are Conserved Signals Regulating Direct and Indirect Antiherbivore Defense. *Proc. Natl. Acad. Sci. USA* **2013**, *110*, 5707–5712. [CrossRef]
10. Ruiz, C.; Nadal, A.; Montesinos, E.; Pla, M. Novel Rosaceae Plant Elicitor Peptides as Sustainable Tools to Control *Xanthomonas arboricola* pv. *pruni* spp. *Mol. Plant Pathol.* **2018**, *19*, 418–431. [CrossRef]
11. Yamada, K.; Yamashita-Yamada, M.; Hirase, T.; Fujiwara, T.; Tsuda, K.; Hiruma, K.; Saijo, Y. Danger Peptide Receptor Signaling in Plants Ensures Basal Immunity upon Pathogen-Induced Depletion of BAK1. *EMBO J.* **2016**, *35*, 46–61. [CrossRef] [PubMed]

12. Bartels, S.; Lori, M.; MBengue, M.; Verk, M.; Klauser, D.; Hander, T.; Böni, R.; Robatzek, S.; Boller, T. The Family of Peps and Their Precursors in *Arabidopsis*: Differential Expression and Localization but Similar Induction of Pattern-Triggered Immune Responses. *J. Exp. Bot.* **2013**, *64*, 5309–5321. [CrossRef] [PubMed]
13. Tang, J.; Han, Z.; Sun, Y.; Zhang, H.; Gong, X.; Chai, J. Structural Basis for Recognition of an Endogenous Peptide by the Plant Receptor Kinase PEPR1. *Cell Res.* **2015**, *25*, 110–120. [CrossRef] [PubMed]
14. Yamaguchi, Y.; Huffaker, A.; Bryan, A.C.; Tax, F.E.; Ryan, C.A. PEPR2 Is a Second Receptor for the Pep1 and Pep2 Peptides and Contributes to Defense Responses in *Arabidopsis*. *Plant Cell* **2010**, *22*, 508–522. [CrossRef]
15. Lori, M.; Van Verk, M.C.; Hander, T.; Schatowitz, H.; Klauser, D.; Flury, P.; Gehring, C.A.; Boller, T.; Bartels, S. Evolutionary Divergence of the Plant Elicitor Peptides (Peps) and Their Receptors: Interfamily Incompatibility of Perception but Compatibility of Downstream Signalling. *J. Exp. Bot.* **2015**, *66*, 5315–5325. [CrossRef]
16. Ruiz, C.; Nadal, A.; Foix, L.; Montesinos, L.; Montesinos, E.; Pla, M. Diversity of Plant Defense Elicitor Peptides within the Rosaceae. *BMC Genet.* **2018**, *19*, 11. [CrossRef]
17. Ross, A.; Yamada, K.; Hiruma, K.; Yamashita-Yamada, M.; Lu, X.; Takano, Y.; Tsuda, K.; Saijo, Y. The *Arabidopsis* PEPR Pathway Couples Local and Systemic Plant Immunity. *EMBO J.* **2014**, *33*, 62–75. [CrossRef]
18. Khraiwesh, B.; Zhu, J.K.; Zhu, J. Role of MiRNAs and SiRNAs in Biotic and Abiotic Stress Responses of Plants. *Biochim. Biophys. Acta Gene Regul. Mech.* **2012**, *1819*, 137–148. [CrossRef]
19. Li, Y.; Zhang, Q.Q.; Zhang, J.; Wu, L.; Qi, Y.; Zhou, J.M. Identification of MicroRNAs Involved in Pathogen-Associated Molecular Pattern-Triggered Plant Innate Immunity. *Plant Physiol.* **2010**, *152*, 2222–2231. [CrossRef]
20. Samarfard, S.; Ghorbani, A.; Karbanowicz, T.P.; Lim, Z.X.; Saedi, M.; Fariborzi, N.; McTaggart, A.R.; Izadpanah, K. Regulatory Non-Coding RNA: The Core Defense Mechanism against Plant Pathogens. *J. Biotechnol.* **2022**, *359*, 82–94. [CrossRef]
21. Song, Y.Y.; Cao, M.; Xie, L.J.; Liang, X.T.; Zeng, R.S.; Su, Y.J.; Huang, J.H.; Wang, R.L.; Luo, S.M. Induction of DIMBOA Accumulation and Systemic Defense Responses as a Mechanism of Enhanced Resistance of Mycorrhizal Corn (*Zea mays* L.) to Sheath Blight. *Mycorrhiza* **2011**, *21*, 721–731. [CrossRef] [PubMed]
22. Bajczyk, M.; Jarmolowski, A.; Jozwiak, M.; Pacak, A.; Pietrykowska, H.; Sierocka, I.; Swida-Barteczka, A.; Szewc, L.; Szweykowska-Kulinska, Z. Recent Insights into Plant MiRNA Biogenesis: Multiple Layers of MiRNA Level Regulation. *Plants* **2023**, *12*, 342. [CrossRef] [PubMed]
23. Iki, T.; Yoshikawa, M.; Nishikiori, M.; Jaudal, M.C.; Matsumoto-Yokoyama, E.; Mitsuhara, I.; Meshi, T.; Ishikawa, M. In Vitro Assembly of Plant RNA-Induced Silencing Complexes Facilitated by Molecular Chaperone HSP90. *Mol. Cell* **2010**, *39*, 282–291. [CrossRef] [PubMed]
24. Li, M.; Yu, B. Recent Advances in the Regulation of Plant MiRNA Biogenesis. *RNA Biol.* **2021**, *18*, 2087–2096. [CrossRef]
25. Tomassi, A.H.; Re, D.A.; Romani, F.; Cambiagno, D.A.; Gonzalo, L.; Moreno, J.E.; Arce, A.L.; Manavella, P.A. The Intrinsically Disordered Protein CARP9 Bridges HYL1 to AGO1 in the Nucleus to Promote MicroRNA Activity. *Plant Physiol.* **2020**, *184*, 316–329. [CrossRef]
26. Vaucheret, H.; Vazquez, F.; Crété, P.; Bartel, D.P. The Action of ARGONAUTE1 in the MiRNA Pathway and Its Regulation by the MiRNA Pathway Are Crucial for Plant Development. *Genes. Dev.* **2004**, *18*, 1187–1197. [CrossRef]
27. Yu, Y.; Jia, T.; Chen, X. The 'how' and 'where' of plant microRNAs. *New Phytol.* **2017**, *216*, 1002–1017. [CrossRef]
28. Song, X.; Li, Y.; Cao, X.; Qi, Y. MicroRNAs and Their Regulatory Roles in Plant-Environment Interactions. *Annu. Rev. Plant Biol.* **2019**, *70*, 489–525. [CrossRef]
29. Liu, W.W.; Meng, J.; Cui, J.; Luan, Y.S. Characterization and Function of MicroRNA∗s in Plants. *Front. Plant Sci.* **2017**, *8*, 2200. [CrossRef]
30. Baumberger, N.; Baulcombe, D.C. *Arabidopsis* ARGONAUTE1 Is an RNA Slicer That Selectively Recruits MicroRNAs and Short Interfering RNAs. *Proc. Natl. Acad. Sci. USA* **2005**, *102*, 11928–11933. [CrossRef]
31. Feng, H.; Wang, X.; Zhang, Q.; Fu, Y.; Feng, C.; Wang, B.; Huang, L.; Kang, Z. Monodehydroascorbate Reductase Gene, Regulated by the Wheat PN-2013 MiRNA, Contributes to Adult Wheat Plant Resistance to Stripe Rust through ROS Metabolism. *Biochim. Biophys. Acta* **2014**, *1839*, 1–12. [CrossRef] [PubMed]
32. Hou, S.; Liu, Z.; Shen, H.; Wu, D. Damage-Associated Molecular Pattern-Triggered Immunity in Plants. *Front. Plant Sci.* **2019**, *10*, 646. [CrossRef]
33. Liu, S.; Wu, L.; Qi, H.; Xu, M. LncRNA/CircRNA–MiRNA–MRNA Networks Regulate the Development of Root and Shoot Meristems of Populus. *Ind. Crops Prod.* **2019**, *133*, 333–347. [CrossRef]
34. Wei, F.L.; Wang, J.H.; Ding, G.; Yang, S.Y.; Li, Y.; Hu, Y.J.; Wang, S.L. Mechanical Force-Induced Specific MicroRNA Expression in Human Periodontal Ligament Stem Cells. *Cells Tissues Organs* **2014**, *199*, 353–363. [CrossRef] [PubMed]
35. Xu, W.; Meng, Y.; Wise, R.P. Mla- and Rom1-Mediated Control of MicroRNA398 and Chloroplast Copper/Zinc Superoxide Dismutase Regulates Cell Death in Response to the Barley Powdery Mildew Fungus. *New Phytol.* **2014**, *201*, 1396–1412. [CrossRef] [PubMed]
36. Kumar, R. Role of MicroRNAs in Biotic and Abiotic Stress Responses in Crop Plants. *Appl. Biochem. Biotechnol.* **2014**, *174*, 93–115. [CrossRef] [PubMed]
37. Navarro, L.; Dunoyer, P.; Jay, F.; Arnold, B.; Dharmasiri, N.; Estelle, M.; Voinnet, O.; Jones, J.D.G. A Plant MiRNA Contributes to Antibacterial Resistance by Repressing Auxin Signaling. *Science* **2006**, *312*, 436–439. [CrossRef]

38. Love, M.I.; Huber, W.; Anders, S. Moderated Estimation of Fold Change and Dispersion for RNA-Seq Data with DESeq2. *Genome Biol.* **2014**, *15*, 550. [CrossRef]
39. Jiang, C.H.; Li, Z.J.; Zheng, L.Y.; Yu, Y.Y.; Niu, D.D. Small RNAs: Efficient and Miraculous Effectors That Play Key Roles in Plant–Microbe Interactions. *Mol. Plant Pathol.* **2023**, *24*, 999–1013. [CrossRef]
40. Dai, X.; Zhuang, Z.; Zhao, P.X. PsRNATarget: A Plant Small RNA Target Analysis Server (2017 Release). *Nucleic Acids Res.* **2018**, *46*, W49–W54. [CrossRef]
41. Tyagi, S.; Sharma, S.; Ganie, S.A.; Tahir, M.; Mir, R.R.; Pandey, R. Plant MicroRNAs: Biogenesis, Gene Silencing, Web-Based Analysis Tools and Their Use as Molecular Markers. *3 Biotech.* **2019**, *9*, 1–12. [CrossRef] [PubMed]
42. Zhang, C.; Zhang, B.; Ma, R.; Yu, M.; Guo, S.; Guo, L.; Korir, N.K. Identification of Known and Novel MicroRNAs and Their Targets in Peach (*Prunus persica*) Fruit by High-Throughput Sequencing. *PLoS ONE* **2016**, *11*, e0159253. [CrossRef] [PubMed]
43. Bigeard, J.; Colcombet, J.; Hirt, H. Signaling Mechanisms in Pattern-Triggered Immunity (PTI). *Mol. Plant* **2015**, *8*, 521–539. [CrossRef] [PubMed]
44. Bartel, D.P. MicroRNAs: Genomics, Biogenesis, Mechanism, and Function. *Cell* **2004**, *116*, 281–297. [CrossRef] [PubMed]
45. Spanudakis, E.; Jackson, S. The Role of MicroRNAs in the Control of Flowering Time. *J. Exp. Bot.* **2014**, *65*, 365–380. [CrossRef]
46. Xu, Y.; Zhang, L.; Wu, G. Epigenetic Regulation of Juvenile-to-Adult Transition in Plants. *Front. Plant Sci.* **2018**, *9*, 1048. [CrossRef]
47. Chuck, G.; Cigan, A.M.; Saeteurn, K.; Hake, S. The Heterochronic Maize Mutant Corngrass1 Results from Overexpression of a Tandem MicroRNA. *Nat. Genet.* **2007**, *39*, 544–549. [CrossRef]
48. Qin, L.; Zhao, L.; Wu, C.; Qu, S.; Wang, S. Identification of MicroRNA Transcriptome in Apple Response to Alternaria Alternata Infection and Evidence That MiR390 Is Negative Regulator of Defense Response. *Sci. Hortic.* **2021**, *289*, 110435. [CrossRef]
49. Lu, Y.; Feng, Z.; Liu, X.; Bian, L.; Xie, H.; Zhang, C.; Mysore, K.S.; Liang, J. MiR393 and MiR390 Synergistically Regulate Lateral Root Growth in Rice under Different Conditions. *BMC Plant Biol.* **2018**, *18*, 261. [CrossRef]
50. Zhou, Z.; Schenke, D.; Shen, E.; Fan, L.; Cai, D. MicroRNAs Constitute an Additional Layer in Plant Response to Simultaneous Bio- and Abiotic Stresses as Exemplified by UV-B Radiation and Flg22-Treatment on *Arabidopsis Thaliana*. *Plant Cell Environ.* **2024**, *47*, 14773. [CrossRef]
51. Guo, H.; Nolan, T.M.; Song, G.; Liu, S.; Xie, Z.; Chen, J.; Schnable, P.S.; Walley, J.W.; Yin, Y. FERONIA Receptor Kinase Contributes to Plant Immunity by Suppressing Jasmonic Acid Signaling in *Arabidopsis thaliana*. *Curr. Biol.* **2018**, *28*, 3316–3324.e6. [CrossRef] [PubMed]
52. Mu, X.Y.; Liu, X.R.; Cai, J.H.; Zhu, W.J.; Wang, Z.; Yang, Q.; You, X. MiR395 Overexpression Increases Eggplant Sensibility to *Verticillium dahliae* Infection. *Russ. J. Plant Physiol.* **2018**, *65*, 203–210. [CrossRef]
53. Wei, H.; Zhao, Y.; Xie, Y.; Wang, H. Exploiting SPL Genes to Improve Maize Plant Architecture Tailored for High-Density Planting. *J. Exp. Bot.* **2018**, *69*, 4675–4688. [CrossRef] [PubMed]
54. Luo, P.; Di, D.W.; Wu, L.; Yang, J.; Lu, Y.; Shi, W. MicroRNAs Are Involved in Regulating Plant Development and Stress Response through Fine-Tuning of TIR1/AFB-Dependent Auxin Signaling. *Int. J. Mol. Sci.* **2022**, *23*, 510. [CrossRef] [PubMed]
55. Zhu, Q.H.; Fan, L.; Liu, Y.; Xu, H.; Llewellyn, D.; Wilson, I. MiR482 Regulation of NBS-LRR Defense Genes during Fungal Pathogen Infection in Cotton. *PLoS ONE* **2013**, *8*, e84390. [CrossRef]
56. Eckardt, N.A. A MicroRNA Cascade in Plant Defense. *Plant Cell* **2012**, *24*, 840. [CrossRef]
57. Shivaprasad, P.V.; Chen, H.M.; Patel, K.; Bond, D.M.; Santos, B.A.C.M.; Baulcombe, D.C. A MicroRNA Superfamily Regulates Nucleotide Binding Site–Leucine-Rich Repeats and Other MRNAs. *Plant Cell* **2012**, *24*, 859. [CrossRef]
58. Liu, Z.; Sun, Z.; Zeng, C.; Dong, X.; Li, M.; Liu, Z.; Yan, M. The Elemental Defense Effect of Cadmium on Alternaria Brassicicola in Brassica Juncea. *BMC Plant Biol.* **2022**, *22*, 17. [CrossRef]
59. Matthewman, C.A.; Kawashima, C.G.; Húska, D.; Csorba, T.; Dalmay, T.; Kopriva, S. MiR395 Is a General Component of the Sulfate Assimilation Regulatory Network in *Arabidopsis*. *FEBS Lett.* **2012**, *586*, 3242–3248. [CrossRef]
60. Trivilin, A.P.; Hartke, S.; Moraes, M.G. Components of Different Signalling Pathways Regulated by a New Orthologue of AtPROPEP1 in Tomato Following Infection by Pathogens. *Plant Pathol.* **2014**, *63*, 1110–1118. [CrossRef]
61. Krueger, F. Trim Galore. Available online: https://github.com/FelixKrueger/TrimGalore (accessed on 30 September 2024).
62. Dobin, A.; Davis, C.A.; Schlesinger, F.; Drenkow, J.; Zaleski, C.; Jha, S.; Batut, P.; Chaisson, M.; Gingeras, T.R. STAR: Ultrafast Universal RNA-Seq Aligner. *Bioinformatics* **2013**, *29*, 15–21. [CrossRef] [PubMed]
63. Li, B.; Dewey, C.N. RSEM: Accurate Transcript Quantification from RNA-Seq Data with or without a Reference Genome. *BMC Bioinform.* **2011**, *12*, 323. [CrossRef] [PubMed]
64. Peterson, H.; Kolberg, L.; Raudvere, U.; Kuzmin, I.; Vilo, J. Gprofiler2—An R Package for Gene List Functional Enrichment Analysis and Namespace Conversion Toolset g: Profiler. *F1000Res* **2020**, *9*, 709. [CrossRef]
65. Wickham, H. *Ggplot2: Elegant Graphics for Data Analysis*, 1st ed.; Springer: New York, NY, USA, 2009; ISBN 978-0-387-98141-3.

International Journal of
Molecular Sciences

MDPI

Article

The Transcriptional Regulator TfmR Directly Regulates Two Pathogenic Pathways in *Xanthomonas oryzae* pv. *oryzicola*

Zheng Chang [1,2], Zengfeng Ma [3], Qian Su [1], Xinqi Xia [1], Wenxin Ye [1], Ruifang Li [2,*] and Guangtao Lu [1,*]

1 State Key Laboratory for Conservation and Utilization of Subtropical Agro-Bioresources, College of Life Science and Technology, Guangxi University, Nanning 530004, China; 2008401017@st.gxu.edu.cn (Z.C.); 2208401009@st.gxu.edu.cn (Q.S.); 2208391038@st.gxu.edu.cn (X.X.); 1808301061@st.gxu.edu.cn (W.Y.)
2 Plant Protection Research Institute, Guangxi Academy of Agricultural Science/Key Laboratory of Green Prevention and Control on Fruits and Vegetables in South China Ministry of Agriculture and Rural Affairs/Guangxi Key Laboratory of Biology for Crop Diseases and Insect Pests, Nanning 530007, China
3 Rice Research Institute, Guangxi Academy of Agricultural Sciences, Nanning 530007, China; mazengfeng@gxaas.net
* Correspondence: ruifangli@gxaas.net (R.L.); lugt@gxu.edu.cn (G.L.)

Abstract: *Xanthomonas oryzae* pv. *oryzicola* (*Xoc*) is a notorious plant pathogen. Like most bacterial pathogens, *Xoc* has evolved a complex regulatory network to modulate the expression of various genes related to pathogenicity. Here, we have identified TfmR, a transcriptional regulator belonging to the TetR family, as a key player in the virulence mechanisms of this phytopathogenic bacterium. We have demonstrated genetically that *tfmR* is involved in the hypersensitive response (HR), pathogenicity, motility and extracellular polysaccharide production of this phytopathogenic bacterium. Our investigations extended to exploring TfmR's interaction with RpfG and HrpX, two prominent virulence regulators in *Xanthomonas* species. We found that TfmR directly binds to the promoter region of RpfG, thereby positively regulating its expression. Notably, constitutive expression of RpfG partly reinstates the pathogenicity compromised by TfmR-deletion mutants. Furthermore, our studies revealed that TfmR also exerts direct positive regulation on the expression of the T3SS regulator HrpX. Similar to RpfG, sustained expression of HrpX partially restores the pathogenicity of TfmR-deletion mutants. These findings underscore TfmR's multifaceted role as a central regulator governing key virulence pathways in *Xoc*. Importantly, our research sheds light on the intricate molecular mechanisms underlying the regulation of pathogenicity in this plant pathogen.

Keywords: *Xanthomonas*; transcriptional regulator; virulence; RpfG; HrpX

1. Introduction

The genus *Xanthomonas* comprises an important group of Gram-negative phytopathogenic bacteria that cause severe diseases in many important crops [1,2]. Among the most prominent of these pathogens is the species *Xanthomonas oryzae* pv. *oryzicola* (*Xoc*), the causal agent of bacterial leaf streak in rice (BLS), a bacterial disease that is a serious threat to rice productivity [3].

In recent years a large number of *Xoc* virulence factors have been identified and characterized. *Xoc* has been shown to produce and secrete a number of factors to promote its pathogenicity, including the extracellular polysaccharide (EPS), adhesion, diffusible signal factor (DSF) cell–cell signals, extracellular enzymes, and type III effectors (T3SEs) secreted through the type III secretion system (T3SS) et al. [3–12]. However, the molecular mechanisms that underpin the regulation of these virulence factors in *Xoc* are less explored and understood. Currently, the main pathogenic pathogenesis-related genes that have been identified in *Xoc* include the *rpf* gene and the T3SS gene, among others [5,6].

The *rpf* gene cluster in *Xoc* mainly encodes the proteins RpfF, RpfC and RpfG. In the genus *Xanthomonas* they are associated with a regulatory system involving the diffusible signaling factor (DSF) [7]. RpfF is involved in the synthesis of DSF, which is important for intercellular signaling [7–9]. RpfC and RpfG constitute a two-component regulatory system, and they are responsible for the perception of DSF signaling molecules and for signaling [7]. When RpfC senses DSF signaling, it undergoes self-phosphorylation and passes the phosphate group to the RpfG [7]. RpfG has a two-component system of a receptive domain and an HD-GYP structural domain [7,10]. Phosphorylation of RpfG activates its HD-GYP structural domain, which possesses the phosphodiesterase activity that hydrolyzes cyclic di-GMP, leading to changes in the intracellular level of cyclic di-GMP [7,11]. In *Xoc*, RpfG positively regulates pathogenicity and EPS production, and negatively regulates T3SS gene expression [6]. Despite the above studies on the function of RpfG, few transcription factors that directly regulate RpfG have been studied in *Xoc*.

T3SS plays a crucial role in both plant and animal pathogens successfully infected into the host [12,13]. As with many other bacterial pathogens, T3SS, which is encoded by a cluster of over 20 hypersensitive response and pathogenicity (*hrp*) genes, is an essential virulence determinant of *X. oryzae* [14,15]. Mutations in the T3SS gene in *X. oryzae* result in loss of the ability to elicit its hypersensitive response (HR) in the nonhost plant *Nicotiana benthamiana*. And the T3SS gene in *X. oryzae* is essential for the pathogenicity of the host rice plant [14]. In basic culture (XOM2), the T3SS gene is activated [16,17]. Notably the T3SS gene is directly activated by the AraC family transcriptional regulator HrpX in *X. oryzae* [18]. And the OmpR family regulator HrpG positively regulates HrpX transcription directly[17,19]. Although there are regulatory proteins that directly positively regulate HrpG/HrpX in *Xanthomonas* spp., there are also other transcription factors that have not yet been identified [20,21].

Several other regulatory proteins have been shown to control the expression of virulence factor genes in other *Xanthomonas* species. TfmR is a transcriptional regulator belonging to the TetR family of transcriptional regulators (TFRs). In *Xanthomonas citri* subsp. *citri* (*Xcci*) strain 306, TfmR indirectly regulates the expression of T3SS genes through the direct regulation of fatty acid metabolism genes [22]. TfmR plays a positive role in the pathogenicity, HR, motility, and production of EPS in the *Xcci* strain 306 [22]. TFRs are usually used as transcriptional repressors to regulate the expression of target gene expression [23]. Their regulatory functions vary widely and are now found to be related to bacterial drug resistance, community sensing, catabolism, antibiotic production, and host response [24,25].

Despite the above advances in TfmR in the *Xcci 306* strain, little is known about the function of TfmR in *Xoc*. Too investigate its role in *Xoc*, we knocked out TfmR in the *Xoc* GX01 strain. We found that the pathogenicity of the EPS production, motility, HR, and proliferation in the host plant of the deletion mutant Δ*tfmR* were significantly attenuated.

Interestingly, the expression of *rpfG* in the Δ*tfmR* mutant restores motility and EPS yield, and partially restores the pathogenicity and proliferative capacity of the mutant in host plants. In addition, through a series of experiments we showed that TfmR can directly bind to the promoter region of *rpfG* and positively regulate the expression of the *rpfG* gene. Our experimental evidence indicates that TfmR can directly bind to the *hrpX* promoter region and activate its gene expression, and that HrpX restores the mutant's HR and partially restores pathogenicity and proliferative capacity in the host plant. Based on our evidence, we can show that in *Xoc*, TfmR regulates a series of cellular processes through the regulation of RpfG and HrpX, respectively, which in turn regulates pathogenicity in the host plant.

2. Results

2.1. TfmR Plays a Role in Xoc Full Virulence

To explore the role of TetR family transcriptional regulators in *Xoc*, we conducted sequence comparison and protein sequence-identity analysis of all nine TetR family transcriptional regulators in *Xoc*. The analysis showed low sequence identity of these proteins. (Figure S1A,B), suggesting potential functional diversity in *Xoc*.

Subsequently, we generated deletion mutants for each of the nine TetR family transcriptional regulators using the homology double-swap method. These mutants were then inoculated into rice (*Oryza sativa* L. ssp. *Japonica* cultivar *Nipponbare*) using the leaf-infiltration method (Figure S1C). Our results indicated a significant reduction in pathogenicity, specifically in the mutant strain ΔXOCgx_1556. Remarkably, the amino acid sequence encoded by *XOC*gx_1556 (accession number QEO96549.1) shared a high similarity (98.5%) with the *Xcci* TfmR (strain 306) [22]. Consequently, we designated the protein encoded by XOCgx_1556 as TfmR.

To further elucidate the function of TfmR in *Xoc*, we utilized the pXUK [26] plasmid to construct the complementary strain CΔ*tfmR*. Pathogenicity assays were performed on wild-type strain GX01, deletion mutant Δ*tfmR*, and complementation strain CΔ*tfmR*. Lesions induced by Δ*tfmR* were significantly shorter compared to the wild type (Figure 1A). Notably, the complemented strain CΔ*tfmR* exhibited virulence symptoms (lesion length) similar to those of the wild type (Figures 1A and S3A).

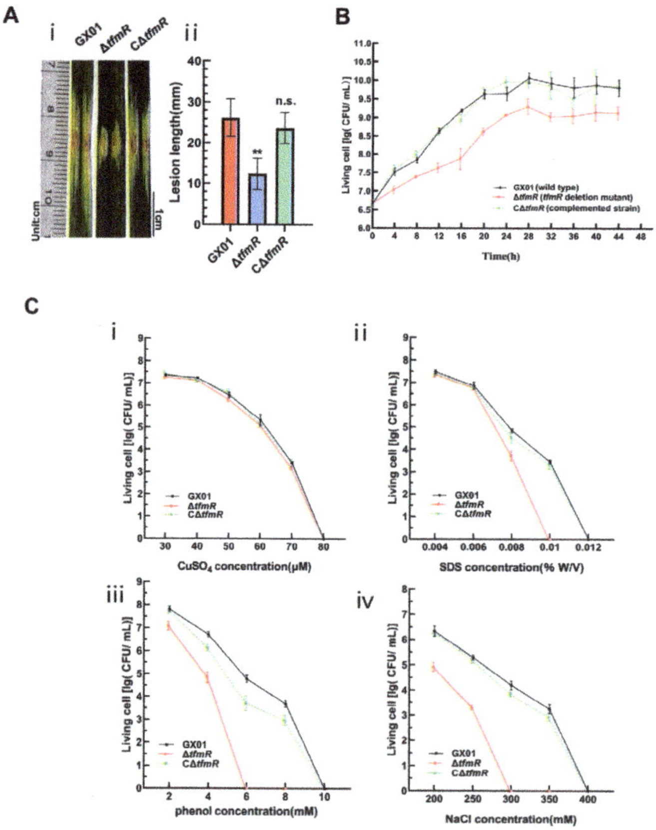

Figure 1. TfmR is required for full virulence of *Xoc*. (**A**) *Xoc* strain pathogenicity assay. (**i**) Phenotypic chart of lesion length. (**ii**) Statistical chart of lesion length, values are means ± SD ($n = 30$). Significance was determined by ANOVA and Dunnett's post hoc test to compare with the wild type. ** $p < 0.01$; n.s., not significant. The experiment was repeated three times with similar results. (**B**) Growth of *Xoc* strains in complex medium (NB). The strains were inoculated into NB medium at a final density of 0.01 (OD_{600}). The growth of the strains was recorded every 4 h. (**C**) Stress tolerance test of *Xoc* strains. Survival experiments performed by subculturing strains overnight on fresh NA plates supplemented with different concentrations of $CuSO_4$ (**i**), SDS (**ii**), phenol (**iii**), and NaCl (**iv**). Values given are the means ± SD ($n = 3$ biological repeats).

To investigate the specific functions of TfmR in manipulating *Xoc* pathogenesis, we conducted a series of phenotypic tests. Compared to the wild type, the growth rate of the deletion mutant Δ*tfmR* in nutrient-rich liquid medium (NB) significantly lagged behind during the early exponential phase (Figure 1B). Notably, the growth status of the complementation strain CΔ*tfmR* in NB medium remained consistent with that of the wild type (Figure 1B), suggesting an indispensable role of TfmR in *Xoc* growth. Furthermore, our results demonstrated that the deletion mutant Δ*tfmR* exhibited a lower survival rate under stress conditions induced by sodium dodecyl sulfate (SDS), phenol, and NaCl, while the complemented strain responded similarly to the wild type (Figure 1C). Additionally, the secretion of protease and amylase was higher in the deletion mutant Δ*tfmR* compared to the wild type, and in the complemented strain, the enhanced enzyme activities were restored to wild-type levels (Figure S4).

These findings suggest that in *Xoc*, TfmR has a key role in bacterial growth and confers resistance to specific stressors, but appears to negatively regulate extracellular enzyme production. However, despite these findings, the regulatory mechanisms of TfmR in *Xoc* pathogenicity remain unclear.

2.2. TfmR Is a Global Regulator That Regulates the Expression of a Large Number of Genes Involved in Various Cellular Processes

To further elucidate the regulatory role of TfmR in *Xoc*, we conducted RNA sequencing (RNA-Seq) to determine the transcriptome of the TfmR-deletion mutant strain Δ*tfmR*. Both the mutant strain and the wild-type strain GX01 were cultured in NB medium until reaching the mid-exponential phase (OD_{600} = 0.8). Subsequently, bacterial RNA was extracted, library preparation was performed, and the generated data were analyzed for differential gene expression.

Analysis of the transcriptome data revealed differential expression of 446 genes in the mutant Δ*tfmR*, with 191 genes significantly down-regulated and 255 genes significantly up-regulated (Table S1). To validate the transcriptome data, we randomly selected 16 genes and assessed their relative expression levels using reverse transcription quantitative real-time PCR (RT-qPCR). The results showed broadly consistent expression patterns with those observed in the transcriptome analysis (Figure 2A).

Functional clustering analysis of the differentially expressed genes, based on the genome annotation of *Xoc* strain GX01, revealed that they could be categorized into 19 functional categories. Among these, 108 genes were classified as belonging to the 'poorly characterized or undefined category'. The majority of the remaining genes were associated with 'mobile genetic elements (mobilomes)', 'pathogenicity, virulence, invasion and adaptation', 'cell structure', 'cellular processes and community', 'degradation of small molecules', 'amino acid biosynthesis', and 'fatty acid and phosphatidic acid biosynthesis'. Notably, 72 genes were categorized under 'pathogenicity, virulence, invasion and adaptation' (Figure 2B). These findings provide valuable insights into the regulatory mechanisms orchestrated by TfmR in *Xoc*.

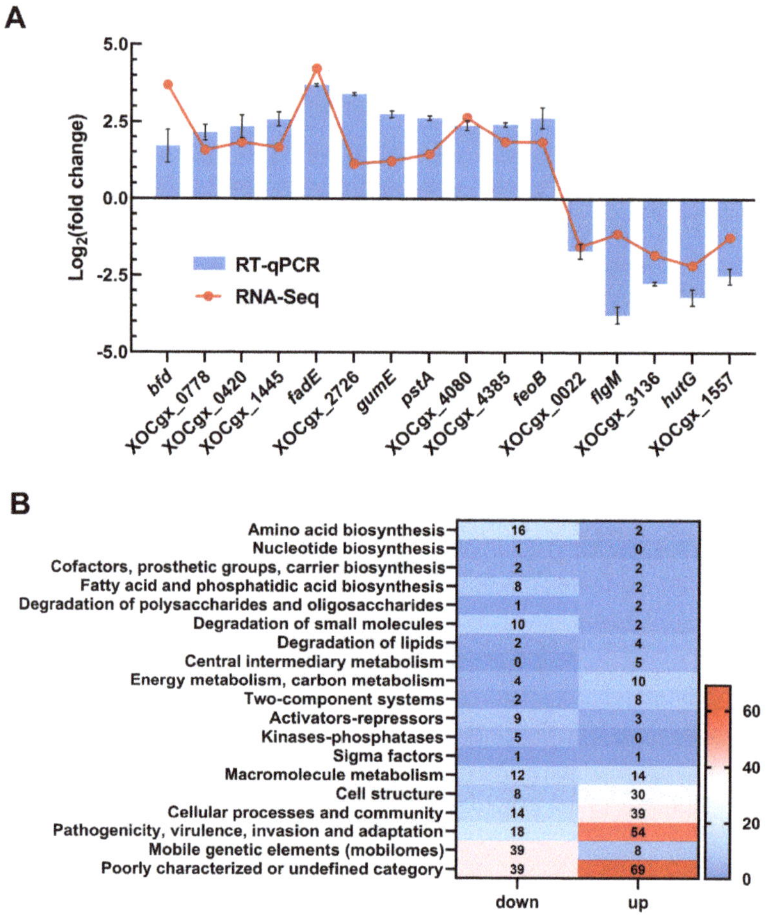

Figure 2. TfmR is a global regulatory protein that affects the expression of many genes. (**A**) A total of 16 genes from the transcriptome data were selected to verify the accuracy of the transcriptome by RT–qPCR. Values are the means ± SD ($n = 3$ biological replicates). (**B**) Functional categories of differentially expressed genes (DEGs) ($|\log 2(\text{fold change})| \geq 1$) in Δ*tfmR* mutants. The transcriptomes of *Xoc* strains cultured in NB medium were investigated by RNA–Seq. In the 446 DEGs of Δ*tfmR* mutants, 191 genes were significantly down–regulated and 255 genes significantly up–regulated.

2.3. *Xoc TfmR Directly Binds to the Promoter of rpfG and Activates Its Transcription*

Although our transcriptomic data showed that 191 genes were significantly downregulated in the mutant Δ*tfmR*, there were few genes with respect to pathogenicity among these, since the mutant Δ*tfmR* has a significant decrease in pathogenicity compared to the wild type, and the production of extracellular polysaccharides as well as motility is reduced (Figures 1A, S3A and 3A,B). This is very similar to the phenotype after deletion of the *rpf* gene in *Xanthomonas* [6,7,27–30]. We guessed that TfmR is likely to regulate the expression of the *rpf* gene. Therefore, we performed an electrophoretic mobility shift assay (EMSA) on the promoter regions of *rpfF*, *rpfC* and *rpfG*, and found that TfmR did bind to the promoter region of *rpfG* (Figures 3C and S6A), but not to the promoter regions of *rpfF* and *rpfC* (Figure S5). After that, we investigated in depth the regulatory mechanism of TfmR on RpfG.

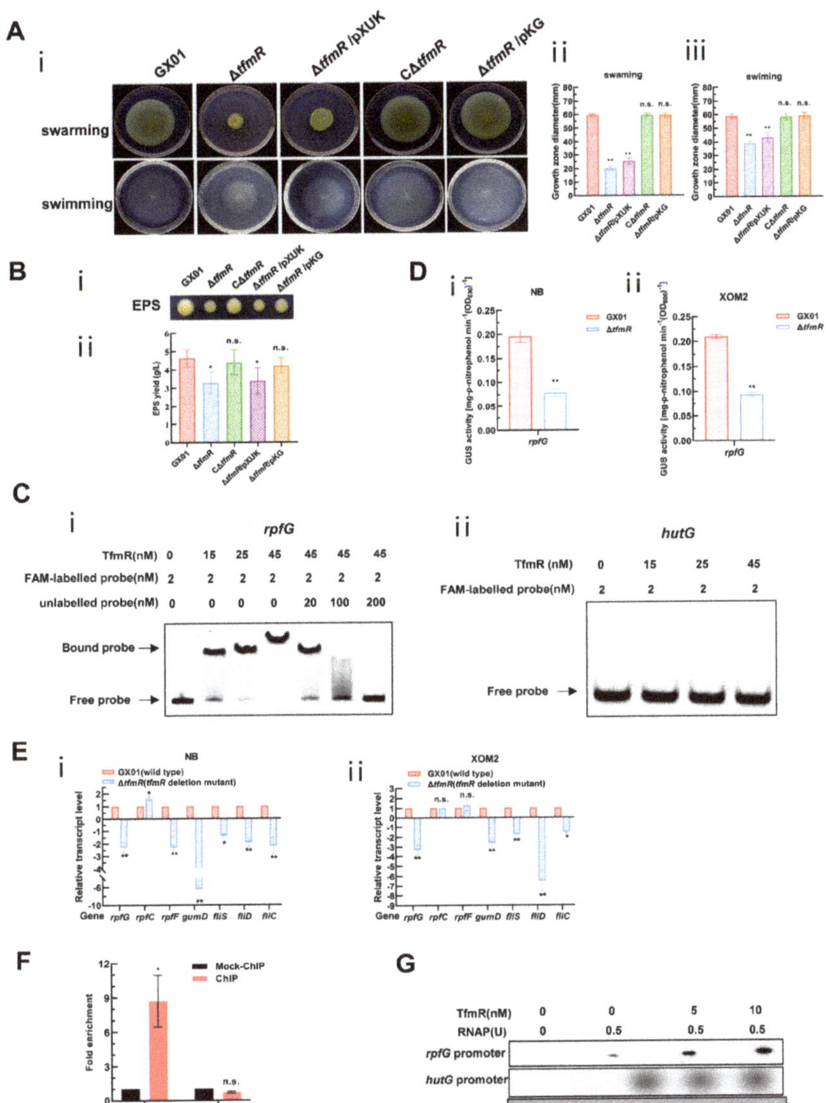

Figure 3. Constitutive expression of *rpfG* restores motility and EPS production of the mutant Δ*tfmR*. And *Xoc* TfmR binds directly to the promoter of RpfG and activates its transcription. (**A**) Constitutive expression of *rpfG* restores motility of the mutant Δ*tfmR*. (**i**) Example photo of a bacterial strain. (**ii**) Mean measurements of colony diameter for each strain on "swarming" plates. (**iii**) Mean measurements of colony diameter for each strain on "swimming" plates. Data shown are the mean ± SD ($n = 10$). Significance was determined by ANOVA and Dunnett's post hoc test for comparison with to the wild type. ** $p < 0.01$; n.s., not significant. (**B**) Constitutive expression of *rpfG* restores the yield of the mutant Δ*tfmR* EPS. (**i**) *Xoc* strains were grown on NA plates supplemented with 2% sucrose for 3 days. (**ii**) *Xoc* strains were cultured in NB medium supplemented with 2% sucrose for 3 days and EPS was precipitated from the culture supernatant. Values given are the means ± SD of triplicate measurements from a representative experiment, and significance was determined by analysis of variance (ANOVA) and Dunnett's post hoc test for comparison with the wild type. * $p < 0.05$; n.s., not significant. Similar results

were obtained in two other independent experiments. (**C**) Electrophoretic mobility shift and competition assays of TfmR with the promoter region of *rpfG* (**i**) and *hutG* (**ii**) (negative control); the bound– and free–DNA fragments are marked with the words Bound probe and Free probe, respectively, and the concentrations are indicated at the top of each lane. (**D**) ß–Glucuronidase (GUS) activity of the *gusA* reporter of the *rpfG* gene promoter in the Δ*tfmR* mutant and the wild type in NB medium (**i**), or in XOM2 medium (**ii**). The data shown are the mean and standard deviation of three measurements. The experiment was repeated three times and similar results were obtained. Differences were evaluated by Student's *t*-test (** $p < 0.01$; * $p < 0.05$; n.s., no significance at $p \leq 0.05$). (**E**) Detection of Δ*tfmR* mutant and wild–type expression of *rpf* genes in NB medium (**i**), or XOM2 medium (**ii**), revealed by RT–qPCR analysis. Values are the means ± SD ($n = 3$ biological replicates). Differences were evaluated by Student's *t*-test (** $p < 0.01$; * $p < 0.05$; n.s., no significance at $p \leq 0.05$). (**F**) Fold enrichment of the promoter region of *rpfG* in the GX01/TfmR::3 × Flag–ChIP samples compared with the Mock–ChIP samples (with anti–HA antibody), as measured by ChIP–qPCR using *hutG* as the negative control. Data are presented as means ± SD ($n = 3$). Differences were evaluated using Student's *t*-test (* $p < 0.05$; n.s., no significance at $p \leq 0.05$). (**G**) In vitro transcription experiments showing TfmR activates the transcription of *rpfG*. RNA was produced from a 323 bp template DNA fragment containing the *rpfG* promoter using *E. coli* RNA polymerase (RNAP) holoenzyme. A 334 bp template DNA fragment containing the *hutG* promoter and a 150 bp template DNA fragment of the *rpfG* coding sequence were used as controls. Lane 1, template DNA alone; lane 2, template DNA with RANP; lanes 3–4, template DNA with RANP and 5 and 10 nM TrxA–TfmR.

To further investigate, we engineered a reporter plasmid for RpfG, where its promoter sequence was fused with the *gusA* gene, enhancing *gusA* activity. The introduction of this reporter plasmid into both the wild-type and mutant-Δ*tfmR* strains, followed by cultivation in nutrient-rich (NB) and plant-mimicking (XOM2) media, revealed significantly lower β-glucuronidase (GUS) activity in the mutant Δ*tfmR* compared to the wild type in both conditions (Figure 3D).

Furthermore, analysis of *rpfG*, *rpfC*, and *rpfF* transcript levels via reverse transcription quantitative real-time PCR (RT-qPCR) corroborated diminished *rpfG* expression in the mutant Δ*tfmR* across both media types. Hence, we posit that TfmR activates *rpfG* expression in *Xoc*. Consistent with previous findings [6,30], reduced expression of genes *gumD*, *fliS*, *fliD*, and *fliC* was observed in the deletion mutant of *tfmR*, akin to the effects of *rpfG* deletion mutants, possibly due to decreased *rpfG* expression (Figure 3E).

Based on our previous EMSA experiments using TrxA-tagged TfmR proteins, it was indeed shown that there are mobility shifts in the DNA probes spanning the *rpfG* promoter region in vitro, which intensified with increasing TrxA-TfmR protein concentration. Notably, no shifted bands were observed in the presence of TrxA protein alone or with a control DNA fragment (Figures 3C and S6A). Chromatin immunoprecipitation-qPCR (ChIP-qPCR) analysis further confirmed direct binding of TfmR to the *rpfG* promoter in vivo (Figures 3F and S7), while in vitro transcription experiments demonstrated that TfmR induced *rpfG* transcription in the presence of RNA polymerase allosteric enzyme (RNAP) (Figure 3G). Collectively, these findings strongly support the hypothesis that TfmR regulates *rpfG* expression by directly binding to its promoter region, thus elucidating a crucial regulatory pathway in *Xoc* pathogenesis.

2.4. Constitutive Expression of rpfG Restores Mutant ΔtfmR Motility, EPS Production, Pathogenicity, and In Vivo Plant Growth

The significantly reduced pathogenicity of the *Xoc* GX01 mutant strain Δ*tfmR* implies that TfmR may be involved in other cellular processes required for full pathogenicity of *Xoc*, such as cell motility, and extracellular polysaccharide (EPS) production. To confirm that the effect of TfmR on these cellular processes is due to the regulation of *rpfG* expression, we constitutively expressed RpfG in the TfmR-deficient mutant Δ*tfmR* and determined its phenotype. To accomplish this, we utilized the pXUK plasmid to construct the cross-complementary strain Δ*tfmR*/pKG (Table S2), while we introduced the pXUK null plasmid into the mutant strain Δ*tfmR*, named Δ*tfmR*/pXUK (Table S2). The results showed that both the swarming and swimming abilities of the mutant strain Δ*tfmR* were significantly reduced, relative to that of the wild-type strain GX01 under the tested conditions, and that the complementary strain CΔ*tfmR* could completely restore the motility of mutant strain Δ*tfmR* (Figure 3A). Whereas the Δ*tfmR*/pXUK strain had similar motility to the mutant Δ*tfmR*, the Δ*tfmR*/pKG strain had similar motility to GX01 and the complementary strain CΔ*tfmR* (Figure 3A). Under the tested conditions, the yield of EPS of the deletion-mutant Δ*tfmR* was significantly lower than that of the wild-type strain GX01, and the complementary strain CΔ*tfmR* could completely restore its phenotype; the yield of EPS of the Δ*tfmR*/pXUK strain was similar to that of Δ*tfmR*, and the yield of EPS of the Δ*tfmR*/pKG strain was also similar to that of GX01 and CΔ*tfmR* (Figure 3B). The above results suggest that constitutive expression of *rpfG* can restore motility and EPS production in Δ*tfmR* strains.

To test the significant reduction in pathogenicity of the mutant strain Δ*tfmR* in relation to RpfG, we inoculated the wild-type strain GX01, mutant strain Δ*tfmR*, complementary strain CΔ*tfmR*, constitutively expressed strain Δ*tfmR*/pKG and Δ*tfmR*/pXUK strain, respectively, into the host rice plant (*Oryza sativa* L. ssp. *Japonica* cultivar *Nipponbare*) by the leaf-infiltrating method. The results showed that the length of diseased spots was significantly lower than that of the wild-type GX01 in both the mutant strain Δ*tfmR* and the Δ*tfmR*/pKG strains; both had similar spot lengths that were significantly lower than that of the wild-type GX01, whereas the complementary strain CΔ*tfmR* could fully recover to wild-type levels and the Δ*tfmR*/pKG strain produced spots that could mostly recover to wild-type levels (Figures 4A and S3B). In our subsequent experiments to detect the proliferation of the pathogen in host tissues, we could observe that the ability of the Δ*tfmR*/pKG strain to proliferate in the host plant was also mostly restored to that of the wild type (Figure 4B). These results suggest that constitutive expression of *rpfG* can largely restore the pathogenicity and growth of the Δ*tfmR* strain in host plants.

2.5. *Xoc TfmR Directly Binds to the Promoter of hrpX and Activates Its Transcription*

According to a previous report by Sun et al., RpfG negatively regulates the expression of *hrpG* and *hrpX* genes [6]. Given that TfmR positively regulates RpfG expression, we constructed two GUS reporter plasmids for two key *hrp*-regulated genes, *hrpG* and *hrpX*, as per the previous method, and introduced them into both the wild-type and mutant-Δ*tfmR* strains.

As anticipated, the GUS enzyme activity of the HrpG reporter in the mutant Δ*tfmR* was significantly higher than that of the wild type in both NB and XOM2 medium (Figure 5A). Interestingly, however, the GUS enzyme activity of the HrpX reporter in mutant Δ*tfmR* was significantly lower than that of the wild type in both media (Figure 5A). To validate this result, we conducted RT-qPCR to analyze the expression changes of a set of T3SS or T3SS-related genes in both the wild type and mutant Δ*tfmR*. The results revealed that while the expression of *hrpG* in both media of mutant Δ*tfmR* was significantly higher than that of the wild type, the expression of *hrpX* and most T3SS or T3SS-related genes in mutant Δ*tfmR* was significantly lower than that of the wild type in both the NB and XOM2 medium (Figure 5B). These findings are consistent with the results of the GUS reporter system, suggesting a significant decrease in HrpX expression in the mutant Δ*tfmR*, potentially indicating direct regulation of HrpX expression by TfmR.

To investigate whether TfmR directly binds to the HrpX promoter to activate its expression, we conducted EMSA experiments in vitro. The results demonstrated that TrxA-TfmR protein could bind to the promoter region of HrpX under experimental conditions, while TrxA protein did not exhibit binding under the same conditions (Figures 5C and S6B). Notably, under the same conditions, the TrxA-TfmR protein also failed to bind to the promoter region of HutG (negative control) (Figure 5C). Subsequently, ChIP-qPCR experiments conducted in vivo in bacteria revealed significant enrichment of the HrpX promoter region in the TfmR-ChIP samples compared to mock ChIP control samples, confirming the direct binding of TfmR to the HrpX promoter region in vivo. Conversely, no binding was detected in the negative control *hutG* promoter region (Figures 5D and S7). These experiments conclusively demonstrated that TfmR directly binds to the promoter of HrpX.

Following this, we conducted in vitro transcription experiments, which showed that the transcription products of *hrpX* increased with the concentration of TfmR protein, while the transcription products of *hutG* remained unchanged (Figure 5E). In summary, these experiments confirm that TfmR directly binds to the promoter of HrpX and activates its transcription.

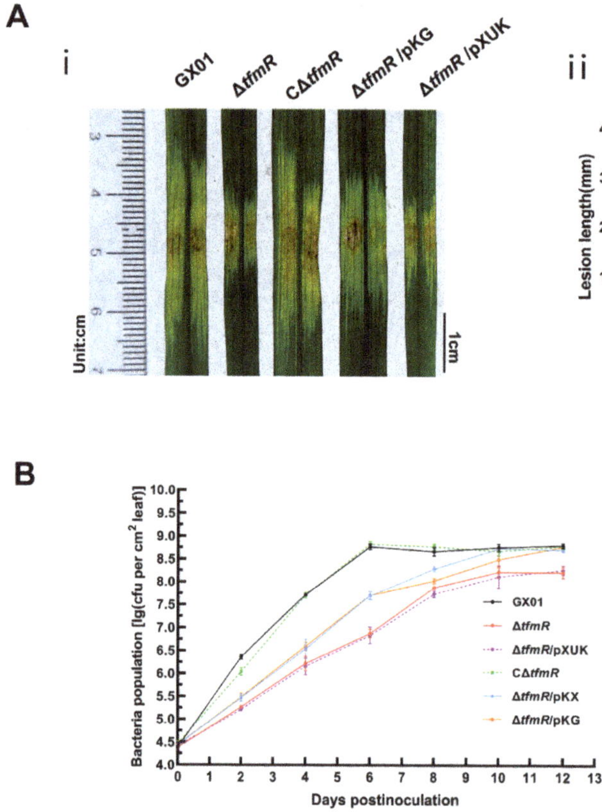

Figure 4. Constitutive expression of *rpfG* restores restores Δ*tfmR* mutant pathogenicity and proliferation in host plants. (**A**) Constitutive expression of *rpfG* can partially characterize the pathogenicity of the mutant Δ*tfmR*. (**i**) Phenotypic chart of lesion length. (**ii**) Statistical chart of lesion length, values are means ± SD ($n = 30$). Significance was determined by ANOVA and Dunnett's post hoc test to compare with the wild type. ** $p < 0.01$; n.s., not significant. The experiment was repeated three times, with similar results. (**B**) Both constitutive expression of *hrpX* and constitutive expression of *rpfG* partially restored the proliferative capacity of the mutant Δ*tfmR* in the host rice plant. Data are shown as the mean ± SD ($n = 3$ biological repeats).

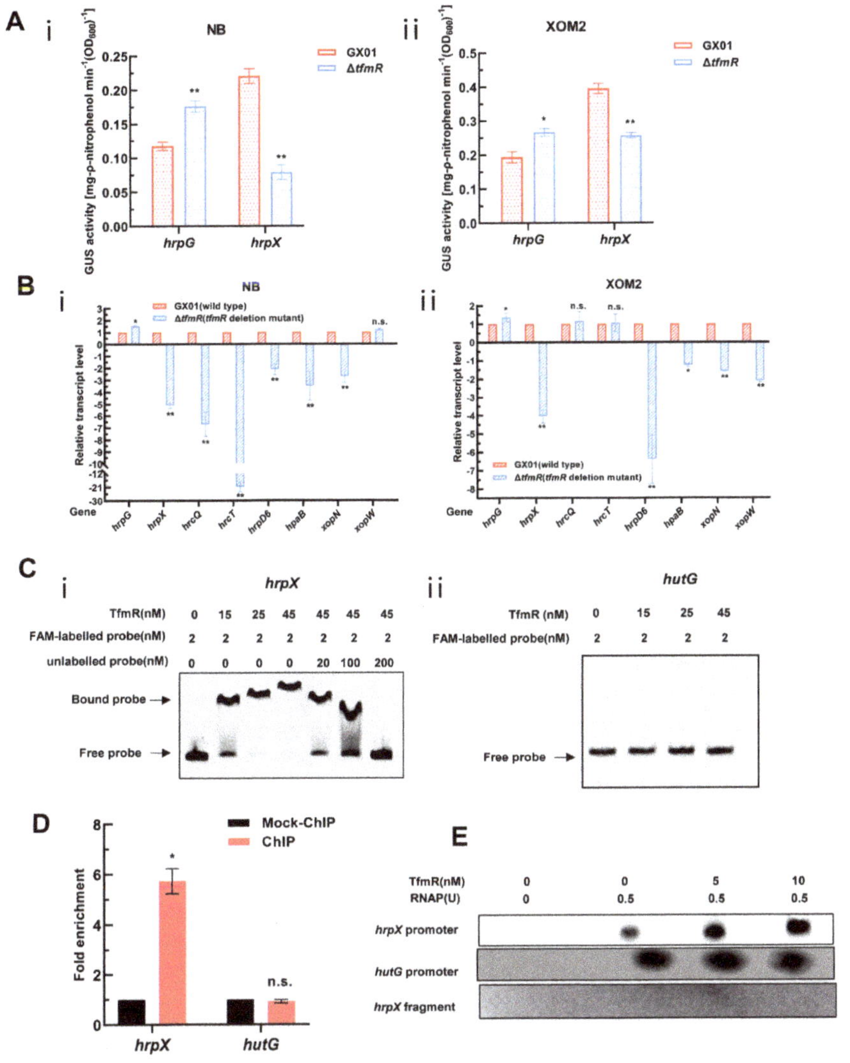

Figure 5. *Xoc* TfmR binds directly to the promoter of HrpX and activates its transcription. (**A**) ß–Glucuronidase (GUS) activity of the *gusA* reporter of the *hrpG* and *hrpX* gene promoter in the Δ*tfmR* mutant and the wild type in NB medium (**i**), or in XOM2 medium (**ii**). The data shown are the mean and standard deviation of three measurements. The experiment was repeated three times and similar results were obtained. Differences were evaluated by Student's *t*-test (** $p < 0.01$; * $p < 0.05$; n.s., no significance at $p \leq 0.05$). (**B**) Detection of Δ*tfmR*–mutant and wild-type expression of T3SS genes in NB medium (**i**), or XOM2 medium (**ii**) revealed by RT–qPCR analysis. Values are the means ± SD (n = 3 biological replicates). Differences were evaluated by Student's *t*-test (** $p < 0.01$; * $p < 0.05$; n.s., no significance at $p \leq 0.05$). (**C**) Electrophoretic mobility shift and competition assays of TfmR with the promoter region of *hrpX* (**i**) and *hutG* (**ii**) (negative control); the bound– and free–DNA fragments are marked with the words Bound probe and Free probe, respectively, and the concentrations are indicated at the top of each lane. (**D**) Fold enrichment of the promoter region of *hrpX* in the GX01/TfmR::3 × Flag–ChIP samples compared with the Mock–ChIP samples (with anti–HA antibody), as measured by ChIP–qPCR using *hutG* as the negative control. Data are presented as means ± SD (n = 3). Differences were evaluated using Student's *t*–test (* $p < 0.05$; n.s., no significance at $p \leq 0.05$). (**E**) In vitro transcription experiments

showing TfmR activates the transcription of *hrpX*. RNA was produced from a 371 bp template DNA fragment containing the *hrpX* promoter using *E. coli* RNA polymerase (RNAP) holoenzyme. A 334 bp template DNA fragment containing the *hutG* promoter and a 161 bp template DNA fragment of the *hrpX* coding sequence were used as controls. Lane 1, template DNA alone; lane 2, template DNA with RANP; lanes 3–4, template DNA with RANP and 5 and 10 nM TrxA–TfmR.

2.6. Constitutive Expression of hrpX Restores ΔtfmR Mutant HR, Pathogenicity, and Proliferation in Host Plants

The impact of TfmR mutation on the expression of the T3SS gene prompted an evaluation of its effect on HR induction using an osmotic assay. While the wild-type strain and the complementary strain C*ΔtfmR* exhibited pronounced HR symptoms at 24 h, the mutant strain *ΔtfmR* showed observable HR symptoms only at 36 h, indicating delayed and attenuated HR (Figure 6A). To confirm that TfmR's effect on plant T3SS is mediated by the modulation of *hrpX* expression, we tested a TfmR mutant constitutively expressing *hrpX* in various HR-induced assays.

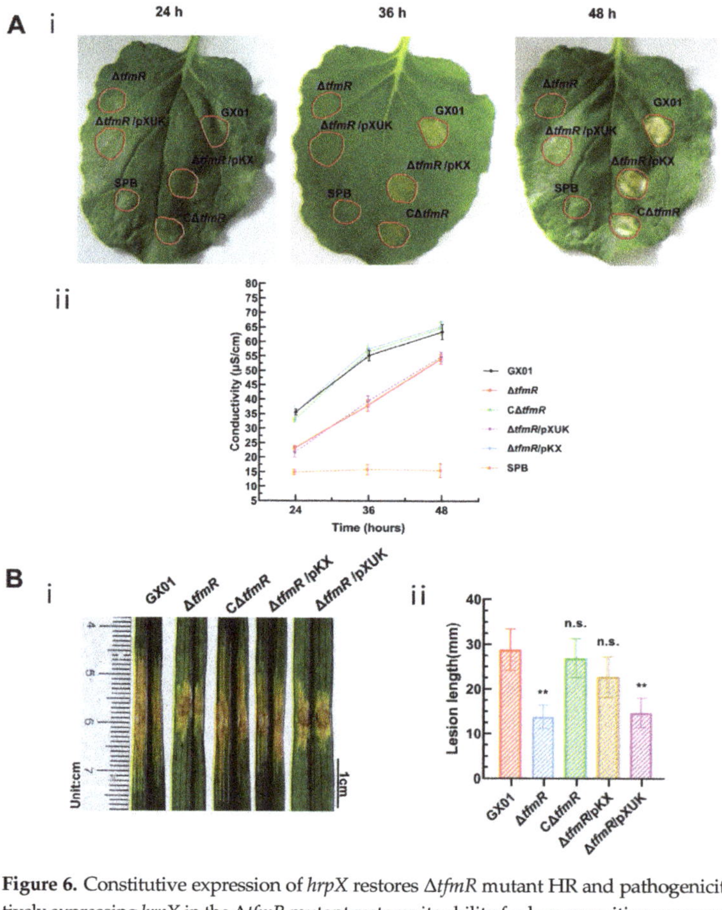

Figure 6. Constitutive expression of *hrpX* restores *ΔtfmR* mutant HR and pathogenicity. (**A**) Constitutively expressing *hrpX* in the *ΔtfmR* mutant restores its ability for hypersensitive response (HR) induction in nonhost plant *N. benthamiana*. (**i**) The HR symptoms recorded at 24, 36, and 48 hours post-inoculation (hpi). Three replications were performed in each experiment, and the experiment was repeated three times. The results presented are from a representative experiment, and similar results were obtained in all other independent experiments. (**ii**) The electrolyte leakage from *N. benthamiana* leaves inoculated with *Xoc* strains. Three samples were taken for each measurement in each experiment. Data are shown

as mean and standard deviation. (**B**) Constitutive expression of *hrpX* partially restores the pathogenicity of mutant Δ*tfmR*. (**i**) Phenotypic chart of lesion length. (**ii**) Statistical chart of lesion length, values are means ± SD ($n = 30$). Significance was determined by ANOVA and Dunnett's post hoc test to compare with the wild type. ** $p < 0.01$; n.s., not significant. The experiment was repeated three times with similar results.

A plasmid constitutively expressing *hrpX* was constructed and introduced into the mutant Δ*tfmR*, designated as Δ*tfmR*/pKX (Table S2). This strain exhibited HR comparable to the wild-type strain GX01. Quantitative assessment of HR induction using an electrolyte leakage assay revealed that both the mutant-Δ*tfmR* and Δ*tfmR*/pXUK strains had significantly lower HR than GX01 at 24 h, 36 h, and 48 h. Conversely, both the complementary strain CΔ*tfmR* and the constitutively expressed strain Δ*tfmR*/pKX showed HR levels similar to the wild-type strain (Figure 6A).

To examine the relationship between the decrease in pathogenicity of mutant Δ*tfmR* and the T3SS gene, pathogenicity tests were conducted. Results mirrored those of the HR experiments, with both mutant strains Δ*tfmR* and Δ*tfmR*/pXUK exhibiting significantly reduced pathogenicity compared to GX01. However, the complementary strain CΔ*tfmR* fully restored pathogenicity, while the constitutively expressed strain Δ*tfmR*/pKX partially restored it (Figures 6B and S3C). Additionally, the in-plant growth curve demonstrated that the mutant strain Δ*tfmR* exhibited significantly weaker growth, while the constitutively expressed strain Δ*tfmR*/pKX partially restored proliferation in plants (Figure 4B).

These results collectively indicate that constitutive expression of *hrpX* can restore HR, pathogenicity, and proliferation of the Δ*tfmR* strain in plants, suggesting that TfmR regulates the T3SS gene by modulating *hrpX* expression.

3. Discussion

The TetR family of transcriptional regulators governs various bacterial processes [22], and our study has identified TfmR as a significant regulator in the *Xoc* GX01 strain. Through our investigations, we have demonstrated that *Xoc* TfmR acts as a global regulator, influencing diverse cellular processes including motility, extracellular enzymes, EPS production, bacterial proliferation, and T3SS gene expression. Specifically, we have identified TfmR as a key regulator of two pathogenesis-related pathways in *Xoc*. Deletion of TfmR resulted in reduced expression of *rpfG* and T3SS genes. Our findings indicate that TfmR directly promotes the expression of *rpfG* and *hrpX* genes, underscoring its crucial role in positively regulating pathogenicity in *Xoc* (Figure 7).

The *rpf* genes cluster in the *Xanthomonas* spp. plays a pivotal role in pathogenicity, community sensing, and various cellular processes [7]. In the *Xanthomonas* spp., the two-component system comprising RpfG and the sensor kinase RpfC responds to the cell–cell signaling molecule DSF to regulate the synthesis of virulence factors [7,11,29]. RpfG, with its HD-GYP cyclic di-GMP phosphodiesterase domain, is a well-studied regulator within the *rpf* gene cluster. Studies across different *Xanthomonas* species have highlighted its importance in pathogenicity, motility, and virulence-factor production [6,7,27,28,31–35]. RpfG was critical for pathogenicity and colonization of rice in the *Xoc* strain BLS256, and deletion of the *rpfG* resulted in a significant decrease in the yield of EPS in the mutant [6]. In *Xanthomonas oryzae* pv. *oryzae* (*Xoo*), RpfG was positively correlated with pathogenicity and proliferation in the host plant, but did not affect *Xoo* growth in enriched medium [31,32]. In *Xanthomonas campestris* pv. *campestris* (*Xcc*), knockout of *rpfG* resulted in a significant reduction in pathogenicity, reduced cellulase, reduced EPS production, and reduced motility [7,27,28]. In *Xanthomonas albilineans* (*Xal*), single deletion of the *rpfG* and *rpfC* genes resulted in mutants with no change in virulence or swimming ability, but double deletion of the *rpfG* and *rpfC* genes resulted in a slight decrease in virulence and severely impaired swimming ability [35]. Since the contribution of the *rpfG* gene to pathogenicity in *Xanthomonas* is particularly important, a number of genes have also been identified that can regulate the expression of the *rpfG* gene. In the *Xoc* RS105 strain, deletion of the *zwf* gene leads to an increase in RpfG at the transcriptional level [34]. The expression of two genes, *rpfC* and *rpfG*, was down-regulated in *Xoo* with the deletion of the *thiG* gene [33]. Importantly, to the best of our knowledge, no transcriptional regulator that can directly regulate RpfG in *Xanthomonas* has been reported yet. Our experimental results for EMSA and ChIP-qPCR indicate that TfmR can directly bind the promoter region of RpfG (Figure 3). And, according to our in vitro transcription, RT-qPCR and GUS enzyme activity experiments, TfmR is a positive transcriptional regulator for the *rpfG* gene in the *Xoc* GX01 strain (Figure 3). However, the regulation of the *rpfG* gene is very complex. Therefore, the next direction of research should focus on the regulatory network of regulatory factors for the *rpfG* gene.

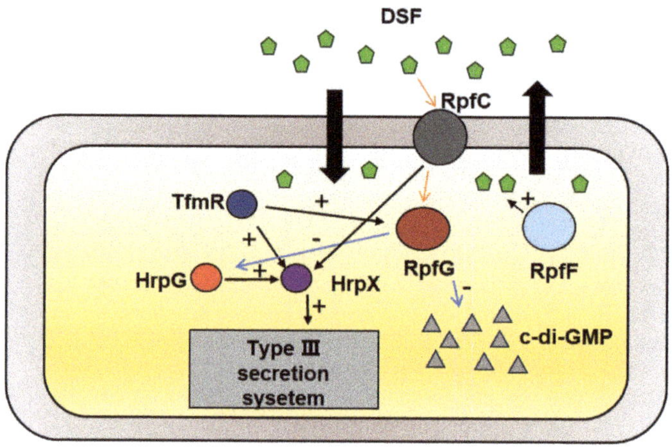

Figure 7. Schematic representation of the genetic regulation between TfmR and its target genes in *Xoc*. TfmR is involved in activating the expression of RpfG and HrpX.

Moreover, our findings regarding TfmR's regulation of *hrpX* expression elucidate a previously unexplored aspect of *Xoc* pathogenesis. T3SS is essential for Gram-negative bacterial pathogens [12,13]. In *Xanthomonas*, virulence is dependent on the secretion and transport of T3SS effectors [12]. These effectors are regulated by two master transcription regulators, HrpG and HrpX [17]. Great efforts have been made to reveal the regulatory network of the HrpG/HrpX regulon [17]. The GntR-family transcriptional regulators HpaR1/YtrA from *Xcc* and *Xcci* negatively regulate *hrpG* expression in plant minimal medium, and positively regulate gene expression in plants [36,37]. In *Xcc*, HpaS is a membrane-bound histidine kinase sensor that forms a two-component signal transduction system with HrpG and activates HrpG activity by phosphorylation [38]. In the *Xoc* RS105 strain, there is a cross-talk between CheA/VemR and HpaS/HrpG-mediated signaling events that coordinate the *hrp* gene expression [39]. In addition to HrpG, regulatory factors have been identified in other *Xanthomonas* spp. that can directly regulate HrpX. In *Xoo*, the Sigma factor 70 RpoD directly binds to the promoters of *hrpG* and *hrpX* to activate T3SS gene expression [40]. And the GntR-family transcriptional regulator Sar can positively regulate the expression of the T3SS gene by directing the promoter of *hrpG* and *hrpX* in *Xoo* [20]. In *Xcc*, the *fis*-type transcriptional regulator Flp interacts with the *hrpX* promoter and positively regulates its expression [41]. It is noteworthy that no proteins (except for HrpG) have been reported in *Xoc* that directly positively regulate the HrpX. Through EMSA, ChIP-qPCR, and in vitro transcription experiments, we have confirmed in the *Xoc* GX01 strain TfmR's direct binding to the *hrpX* promoter, thereby activating its transcription (Figure 5). Although in previous studies TfmR could indirectly regulate the expression of T3SS genes in *Xcci 306* by regulating fatty acid metabolism, TfmR could not directly bind to the promoters of *hrpX* and *hrpG* in the *Xcci* 306 strain [22]. Further studies are needed regarding the differences between the two proteins, *Xoc* TfmR and *Xcci* TfmR, in regulating T3SS gene expression.

In addition, differences in culture environments lead to changes in the expression of *hrpG/hrpX* [17]. According to previous studies, in *Xanthomonas*, *hrpG/hrpX* gene expression was repressed in nutrient-rich medium but induced in plants and minimal medium [16,17]. Although many proteins regulating *hrpG/hrpX* have been identified, the exact mechanism is currently unknown. We performed the transcriptome analysis in nutrient-rich medium, which may be the reason why we did not find significant changes in *hrpX* expression in our transcriptome. According to Sun et al., in *Xoc*, RpfG negatively regulates *hrpG* expression [6], whereas TfmR positively regulates RpfG expression. The slightly higher expression of *hrpG* in the mutant compared to the wild type in minimal culture is likely to be caused by the decreased expression of RpfG. Our results demonstrate that TfmR positively regulates *hrpX* and T3SS gene expression while concurrently activating *rpfG* gene expression. This discovery underscores the intricate regulatory mechanisms governing *Xoc* virulence.

In conclusion, our work highlights the multifaceted signaling pathways underlying *Xoc* virulence and describes TfmR as a pivotal regulator orchestrating T3SS gene expression, HR induction, and pathogenesis. By elucidating the regulatory mechanisms involving TfmR, RpfG, and HrpX, we provide valuable insights for developing strategies to combat plant pathogens such as *Xoc*.

4. Materials and Methods

4.1. Bacterial Strains, Plasmids, and Growth Conditions

The bacterial strains and plasmids used in this work are enumerated in Table S2. *E. coli* (*Escherichia coli*) strains were cultured in LB (Luria Bertani) medium (containing 5 g yeast extract, 10 g NaCl, and 10 g tryptone per liter) or on LB agar plates (LB supplemented with 30 g agar per liter) at 37 °C. *Xoc* strains were cultured in NB medium (comprising 1 g yeast extract, 3 g beef extract, 5 g polypeptone, and 10 g sucrose per liter), NA (NB with 30 g agar per liter), or the minimal medium XOM2 [42]. Antibiotics were used, according to the concentrations as required: kanamycin at 25 μg/mL, rifampicin at 50 μg/mL, ampicillin at 100 μg/mL, spectinomycin at 50 μg/mL and Tet at 5 μg/mL for *Xoc* strains and 15 μg/mL for *E. coli* strains.

4.2. Nucleic Acid Manipulations

The procedure for nucleic acid manipulation was conducted according to the method outlined by Sambrook et al. [43]. Conjugation between *Xoc* and *E. coli* strains was executed as detailed by Turner et al. [37]. The necessary enzymes—restriction endonuclease, T4 DNA ligase, and Pfu polymerase—were supplied by Promega (Shanghai, China). Total RNA was extracted from *Xoc* strain cultures utilizing the Total RNA Extraction Kit (Invitrogen, Carlsbad, CA, USA), while cDNA synthesis was carried out using the cDNA Synthesis Kit (Invitrogen). For RT-qPCR, the obtained cDNA was diluted and utilized as a template with selected primers targeting specific genes. RT-qPCR was performed in the qPCR thermocycler (Analytik jena qTOWER2.0, Jena, Germany). Each reaction comprised ChamQ universal SYBR qPCR master mix (Vazyme Nanjing, China), appropriate primers, and cDNA. The relative mRNA levels were normalized relative to those of the target genes in the wild-type strain GX01 (equal to 1). The expression levels of the 16S rRNA genes were used as an internal standard. Triplicate RT-qPCR assays were conducted, and all primer sequences applied in this study are listed in Table S3.

4.3. Construction of Mutant Strains

The deletion mutant in *Xoc* was constructed using the previously employed homology double-swap method. As an example, the mutant Δ*tfmR* (XOCgx_1556) was constructed as follows [4]. PCR was used to amplify 425 bp of the upstream sequence and 309 bp of the downstream sequence of the *tfmR* gene, utilizing the appropriate primers. The two fragments were fused with the suicide plasmid pK18mobsacB [44] and transformed into *E. coli* DH5α. The recombinant plasmid obtained was introduced into *Xoc* using triparental conjugation with the help of plasmid pRK2073 [45]. Screening for triparental conjugates was carried out using NA plates containing Rif with Kan antibiotics. The single colonies that could grow on the double-resistant plates were all those in which the recombinant plasmid was integrated into the *Xoc* genome by the first homologous exchange, called a single exchanger. Since the *sacB* gene in the pK18mobsacB plasmid is a sucrose-lethal gene, the high concentrations of sucrose are utilized to force a second homologous exchange of the single exchanger. This resulted in the deletion mutant Δ*tfmR*.

For complementation of the Δ*tfmR*, the full length (606 bp) of *tfmR* was amplified by PCR from the total DNA of the *Xoc* strain GX01 and inserted into the pXUK vector [26] (adapted from an endogenous plasmid in the *Xoc* strain GX01), creating the plasmid pKC*tfmR*. For constitutive expression of RpfG or HrpX in Δ*tfmR*, we inserted the full-length (1137 bp) of *rpfG* or the full length (1431 bp) of *hrpX* into the pXUK vector, named pKG or pKX, respectively. These plasmids, including the pXUK null vector, were introduced into the Δ*tfmR* by triparental mating, respectively.

Xoc strains chromosomally encoding the proteins fused with a 3 × Flag-tag at the C-terminus were constructed using the method previously described [4]. Using the genomic DNA of strain GX01 as a template and the corresponding primers, a 1069 bp DNA fragment consisting of a 425 bp sequence upstream of the *tfmR* start codon, a 606 bp coding sequence for *tfmR*, and a 38 bp coding sequence for Flag was obtained by PCR amplification. Meanwhile, PCR amplification using the corresponding primers resulted in a 362 bp DNA fragment comprising 50 bp of the Flag coding sequence, 3 bp of the stop codon of *tfmR* and 309 bp downstream of the *tfmR* stop codon. The two fragments were jointed using overlap extension PCR, and the resulting recombinant fragment was cloned into the suicide plasmid pK18mobsacB. The plasmid was further introduced into the *Xoc* strain GX01 by conjugation. The transconjugants were screened on selective agar plates and were confirmed by DNA sequencing. This constructed variant strain was named as GX01/TfmR::3 × Flag.

4.4. Pathogenicity Tests, HR Assays, Leakage Assays and in-Plant Growth Curve

To assess the pathogenicity of *Xoc* in the host rice plants (*Oryza sativa* L. ssp. *Japonica* cultivar *Nipponbare*) the infiltration method was employed, as previously outlined [4]. Rice seedlings were cultivated in the greenhouse for 6 weeks before being inoculated with *Xoc* strains on their leaves. The *Xoc* strains, collected from overnight cultures, were washed and adjusted to a consistent final density (OD_{600} of 0.3, approximately 1×10^8 CFU/mL). These resuspended bacterial cells were inoculated into 6-week-old rice leaves under relevant conditions, and the resulting lesions and symptoms were evaluated 14 days post-inoculation.

HR was tested on *Nicotiana benthamiana* leaves, as previously described [5]. Following a similar procedure, *Xoc* cells from cultures were washed and resuspended in sodium phosphate buffer (SPB, 5.8 mM Na_2HPO_4 and 4.2 mM NaH_2PO_4, pH 7.0) to an OD_{600} of 0.5 (5×10^8 CFU/mL). Resuspended bacterial cells were infiltrated into the *N. benthamiana* leaves. Symptoms were monitored, and conductivity measurements were taken at 24, 36, and 48 h post-inoculation. For conductivity measurements, samples (four 0.4 cm^2 leaf disks) were collected using a punch. These obtained leaf discs were immersed in 10 mL of ultrapure water and shaken at 200 rpm for 30 min. Then the leaf disk was removed and the conductivity of the liquid was measured with a DDS-307A conductivity meter.

Quantification of bacterial growth in plants was carried out as per prior procedures [4]. To elaborate, the *Xoc* bacterial solution adjusted to a certain concentration was inoculated into the rice leaves, and one infiltrated leaf per group of inoculated plants was sampled and homogenized every 48 h. The homogenates were serially diluted with NB medium and then coated onto selective NA plates with appropriate antibiotics, and after a 3-day incubation period, colony-forming units (CFUs) were counted.

4.5. Stress Tolerance Assay

To investigate the susceptibility of the *Xoc* strain to various environmental stresses, such as heavy metal salts ($CuSO_4$), sodium dodecyl sulfate (SDS), phenol, and hyperosmotic challenge (NaCl), the minimum inhibitory concentration (MIC) method was employed [4]. In brief, the *Xoc* strain was cultured overnight and subsequently diluted to an OD_{600} of 1, and the diluted cultures were further diluted and incubated onto corresponding NA plates supplemented with different reagents. After 3 days of incubation at 28 °C, the number of surviving colonies was counted.

4.6. Extracellular-Polysaccharide and -Enzyme Assays

Extracellular-polysaccharide (EPS) and extracellular-enzyme assays were conducted, following previously established protocols [4,5]. For EPS production analysis, the *Xoc* strain was spotted on NA plates supplemented with 2% sucrose and incubated for 3 days. To quantify EPS production, the *Xoc* strain was cultured in NB medium containing 2% sucrose at 28 °C with shaking at 220 rpm for 3 days. The EPS was then precipitated from the culture supernatant using anhydrous ethanol.

To assess protease and amylase activities, *Xoc* strains were spotted on NA plates containing skim milk (for protease) or soluble starch (for amylase) and incubated for 2 days. For quantification of enzymes, bacterial cells were cultured in NB medium for 24 h and adjusted to the same concentration.

4.7. Motility Assay

The method for detecting cell motility followed the previously described protocol [4]. To begin, the *Xoc* strain culture that had been incubated overnight was rewashed and resuspended to achieve the same concentration (OD_{600} of 0.5). Subsequently, 3 μL of the resuspended bacterial suspension was stabbed into 0.28% agar plates containing 0.03% bacterial peptone and 0.03% yeast extract for swimming, or spotted onto NA plates consisting of 2% sucrose and 0.6% agar for swarming. After a resting incubation at 28 °C for 5 days, the diameter of the area occupied by the bacterial cells was measured.

4.8. GUS Activity Assays

The measurement of GUS activity in *Xoc* strains followed the previously described protocol [41]. Both wild-type and mutant strains carrying the reporter plasmid were cultured at 28 °C in the appropriate medium. Bacterial cells were then collected and resuspended in 375 μL 1 mM p-nitrophenyl-β-d-glucuronide extraction buffer (comprising 50 mM sodium dihydrogen phosphate, 0.1% Triton X-100, and 10 mM β-mercaptoethanol, pH 7.2), and incubated for 10 min at 37 °C. This was followed by incubation with 200 mL of 2.5 M 2-amino-2-methyl-1,3-propylene glycol to terminate the reaction. GUS activity assays were performed in triplicate.

4.9. Overexpression and Purification of TrxA-TfmR Protein

To obtain purified TrxA-TfmR protein, the *tfmR* gene was first cloned and introduced into the expression vector pET32a. Subsequently, the recombinant plasmid was transformed into *E. coli* strain BL21. The protein expression purification process followed a well-established method, involving induction of the recombinant strain with isopropyl β-D-thiogalactopyranoside (IPTG) and purification of the fusion protein using Ni-NTA resin (Qiagen, Alameda, CA, USA).

4.10. Electrophoretic Mobility Shift Assay (EMSA)

The EMSA procedure followed a previously described protocol [4]. Briefly, DNA fragments labeled with FAM at the 5′ terminal were amplified using the appropriate primers. The purified protein was then incubated with the FAM-tagged DNA fragment in binding buffer [20 mm Tris-HCl, 10 mm NaCl, 1 mm ethylenediminetetraacetic acid (EDTA) and 1 mm dithiothreitol, pH 8.0] containing 1 μg of sonicated salmon sperm DNA and 3 μg of bovine serum albumin, incubated at 28 °C for 30 min, Samples were then loaded onto a 6% polyacrylamide-Tris-borate-EDTA (TBE) gel, and visualized.

4.11. In Vitro Transcription Assays

The in vitro transcription assays were conducted following a previously established protocol [4]. First, the corresponding DNA fragments (including the promoter region and a portion of the coding region of the target gene) were amplified using the appropriate primers. The TrxA-TfmR protein and DNA fragments were incubated in transcription buffer at 28 °C for 30 min. Then, the NTP mixture (250 μM each of ATP, CTP, and GTP; 250 μM biotin 16-UTP) and 0.5 U of *E. coli* RNA polymerase holoenzyme (New England BioLabs, Ipswich, MA, USA) were added and the reaction was carried out for 1 h at 37 °C. Transcription products were analyzed by electrophoresis at the end of the reaction. The obtained transcripts were visualized using a fluorescence imager screen (GE AI600, Boston, MA, USA).

4.12. Chromatin Immunoprecipitation–Quantitative PCR (ChIP-qPCR)

We performed ChIP-qPCR experiments as previously described, with only minor modifications [4]. Briefly, GX01/TfmR::3 × Flag reporter strains were cultured in medium to the corresponding OD_{600} values, collected and crosslinked using formaldehyde for cross-linking. Then sonication was performed and, for each ChIP sample, 50 μL of Flag antibody (agarose-conjugated) was added to 3.5 mL of bacterial lysate and incubated at 4 °C overnight. Unbound-DNA fragments were washed and bound-DNA fragments and proteins were eluted with 0.25 M glycine (pH 2.5). RNA was removed from the DNA samples by incubation with RNaseA for 2 h at 37 °C, and then proteins were removed by incubation with proteinase K for 2 h at 55 °C. The DNA samples were subsequently purified using a PCR purification kit (Qiagen, Alameda, CA, USA). To measure the enrichment of RpfG, HrpX, and HutG (negative control) in the ChIP DNA samples, relative-abundance quantitative PCR (qPCR) was performed, using the ChamQ Universal SYBR-qPCR Master Mix (Vazyme) and a real-time PCR thermocycler (Analytik jena qTOWER2.0; jena). The quantity of IP DNA (eluted DNA) was calculated as the percentage of the DNA present in the input DNA (10 μL sample taken prior to IP) using the ΔΔCt method to calculate the relative enrichment as the fold change [46]. The relative IP was calculated by normalizing IP to a mock ChIP using an anti-HA-tag monoclonal antibody (Solarbio, Beijing, China).

4.13. Transcriptome Analysis of the TfmR Mutant

Transcriptome analysis was conducted following the previously described protocol [5]. In brief, RNA was extracted from cell cultures with an OD_{600} of 0.8. To eliminate any contamination from genomic DNA, RNase-free DNase I was used. After quantification and quality assessment, the total-RNA samples were sent to Novogene (Beijing, China) for library construction and strand-specific RNA sequencing. The resulting clean reads were mapped to the genome of the *Xoc* strain GX01 (assembly ID 4726371), and gene expression levels were calculated using the RPKM (reads-per-kilobase per million-mapped-reads) method. Differentially expressed genes (DEGs) were identified based on a false discovery rate (FDR) of ≤0.05 and a | log2FC | (log2 of fold change) ≥ 1 threshold. The details of the DEGs can be found in Table S1.

4.14. Western Blotting

Western blotting was performed using the method previously used by Li et al. [47]. First, bacterial proteins separated by SDS-PAGE gels were electro-transferred onto polyvinylidene difluoride (PVDF) membranes (Millipore, Billerica, MA, USA). The membrane was then blocked with 1% milk for 1 h at room temperature, after which it was incubated with a 1:2500 dilution of anti-Flag-tagged mouse monoclonal antibody (Abmart) as the primary antibody, and then washed with Tris-buffer saline and Tween buffer (Tris 20 mM, NaCl 0.3 M, Tween 20 0.08%). Goat anti-mouse immunoglobulin G (IgG) (Beyotime Biotechnology) coupled with diluted 1:2500 horseradish peroxidase (HRP) was used as secondary antibody. Finally, the luminescent signal was detected according to the manufacturer's instructions.

Supplementary Materials: The following supporting information can be downloaded at: https://www.mdpi.com/article/10.3390/ijms25115887/s1.

Author Contributions: G.L. and R.L. conceived and designed the project. Z.C. carried out the experiments. Z.M., Q.S., X.X. and W.Y. provided the resources and analyzed the data. Z.C. and R.L. wrote the manuscript. All authors have read and agreed to the published version of the manuscript.

Funding: This research was funded by the National Natural Science Foundation of China, grant number 32360649, 32160617 and 32260501.

Institutional Review Board Statement: Not applicable.

Informed Consent Statement: Not applicable.

Data Availability Statement: The RNA sequencing data generated in this study are available in the NCBI SRA database under the accession codes PRJNA1098699. Other data are presented within the manuscript and Supplementary Materials.

Conflicts of Interest: The authors declare no conflicts of interest.

References

1. Mansfield, J.; Genin, S.; Magori, S.; Citovsky, V.; Sriariyanum, M.; Ronald, P.; Dow, M.A.X.; Verdier, V.; Beer, S.V.; Machado, M.A.; et al. Top 10 plant pathogenic bacteria in molecular plant pathology. *Mol. Plant Pathol.* **2012**, *13*, 614–629. [CrossRef] [PubMed]
2. Ryan, R.P.; Vorhölter, F.-J.; Potnis, N.; Jones, J.B.; Van Sluys, M.-A.; Bogdanove, A.J.; Dow, J.M. Pathogenomics of *Xanthomonas*: Understanding bacterium–plant interactions. *Nat. Rev. Microbiol.* **2011**, *9*, 344–355. [CrossRef] [PubMed]
3. NiÑO-Liu, D.O.; Ronald, P.C.; Bogdanove, A.J. *Xanthomonas* oryzae pathovars: Model pathogens of a model crop. *Mol. Plant Pathol.* **2006**, *7*, 303–324. [CrossRef] [PubMed]
4. Peng, J.-L.; Shi, J.-F.; Ma, Z.-F.; Zhou, X.-L.; Ye, W.-X.; Su, Q.; Zhu, G.-N.; Tang, J.-L.; Li, R.-F.; Lu, G.-T. XrvB regulates the type III secretion system by directly repressing *hrpG* transcription in *Xanthomonas oryzae* pv. *oryzicola*. *Phytopathol. Res.* **2023**, *5*, 43. [CrossRef]
5. Li, R.-F.; Ren, P.-D.; Zhang, D.-P.; Cui, P.; Zhu, G.-N.; Xian, X.-Y.; Tang, J.-L.; Lu, G.-T. HpaP divergently regulates the expression of *hrp* genes in *Xanthomonas oryzae* pathovars *oryzae* and *oryzicola*. *Mol. Plant Pathol.* **2023**, *24*, 44–58. [CrossRef] [PubMed]
6. Zhang, Y.; Wei, C.; Jiang, W.; Wang, L.; Li, C.; Wang, Y.; Dow, J.M.; Sun, W. The HD-GYP domain protein RpfG of *Xanthomonas oryzae* pv. *oryzicola* regulates synthesis of extracellular polysaccharides that contribute to biofilm formation and virulence on rice. *PLoS ONE* **2013**, *8*, e59428. [CrossRef] [PubMed]
7. Ryan, R.P.; McCarthy, Y.; Andrade, M.; Farah, C.S.; Armitage, J.P.; Dow, J.M. Cell–cell signal-dependent dynamic interactions between HD-GYP and GGDEF domain proteins mediate virulence in *Xanthomonas campestris*. *Proc. Natl. Acad. Sci. USA* **2010**, *107*, 5989–5994. [CrossRef] [PubMed]
8. Dow, J.M.; Crossman, L.; Findlay, K.; He, Y.-Q.; Feng, J.-X.; Tang, J.-L. Biofilm dispersal in *Xanthomonas campestris* is controlled by cell–cell signaling and is required for full virulence to plants. *Proc. Natl. Acad. Sci. USA* **2003**, *100*, 10995–11000. [CrossRef] [PubMed]
9. Barber, C.E.; Tang, J.L.; Feng, J.X.; Pan, M.Q.; Wilson, T.J.G.; Slater, H.; Dow, J.M.; Williams, P.; Daniels, M.J. A novel regulatory system required for pathogenicity of *Xanthomonas campestris* is mediated by a small diffusible signal molecule. *Mol. Microbiol.* **1997**, *24*, 555–566. [CrossRef]
10. Dow, J.M.; Fouhy, Y.; Lucey, J.F.; Ryan, R.P. The HD-GYP domain, cyclic di-GMP signaling, and bacterial virulence to plants. *Mol. Plant-Microbe Interact.* **2006**, *19*, 1378–1384. [CrossRef]
11. Ryan, R.P.; Fouhy, Y.; Lucey, J.F.; Jiang, B.-L.; He, Y.-Q.; Feng, J.-X.; Tang, J.-L.; Dow, J.M. Cyclic di-GMP signalling in the virulence and environmental adaptation of *Xanthomonas campestris*. *Mol. Microbiol.* **2007**, *63*, 429–442. [CrossRef] [PubMed]
12. Deng, W.; Marshall, N.C.; Rowland, J.L.; McCoy, J.M.; Worrall, L.J.; Santos, A.S.; Strynadka, N.C.J.; Finlay, B.B. Assembly, structure, function and regulation of type III secretion systems. *Nat. Rev. Microbiol.* **2017**, *15*, 323–337. [CrossRef]
13. McCann, H.C.; Guttman, D.S. Evolution of the type III secretion system and its effectors in plant–microbe interactions. *New Phytol.* **2008**, *177*, 33–47. [CrossRef] [PubMed]
14. Zou, L.-F.; Wang, X.-P.; Xiang, Y.; Zhang, B.; Li, Y.-R.; Xiao, Y.-L.; Wang, J.-S.; Walmsley Adrian, R.; Chen, G.-Y. Elucidation of the *hrp* clusters of *Xanthomonas oryzae* pv. *oryzicola* that control the hypersensitive response in nonhost tobacco and pathogenicity in susceptible host rice. *Appl. Environ. Microbiol.* **2006**, *72*, 6212–6224. [CrossRef] [PubMed]
15. Niu, X.-N.; Li, Y.; Carpenter, S.C.D.; Dan, X.; Li, T.; Wu, Q.; Wang, L.; Jiang, W.; Huang, S.; Tang, J.-L.; et al. Complete genome resource of *Xanthomonas oryzae* pv. *oryzicola* GX01 isolated in south China. *Mol. Plant-Microbe Interact.* **2022**, *35*, 357–359. [CrossRef] [PubMed]

16. Büttner, D.; Bonas, U. Regulation and secretion of *Xanthomonas* virulence factors. *FEMS Microbiol. Rev.* **2010**, *34*, 107–133. [CrossRef] [PubMed]
17. Teper, D.; Pandey, S.S.; Wang, N. The HrpG/HrpX Regulon of *Xanthomonads*—An insight to the complexity of regulation of virulence traits in phytopathogenic bacteria. *Microorganisms* **2021**, *9*, 187. [CrossRef] [PubMed]
18. Koebnik, R.; Krüger, A.; Thieme, F.; Urban, A.; Bonas, U. Specific binding of the *Xanthomonas campestris* pv. *vesicatoria* AraC-type transcriptional activator HrpX to plant-inducible promoter boxes. *J. Bacteriol.* **2006**, *188*, 7652–7660. [CrossRef] [PubMed]
19. Guo, Y.; Figueiredo, F.; Jones, J.; Wang, N. HrpG and HrpX play global roles in coordinating different virulence traits of *Xanthomonas axonopodis* pv. *citri*. *Mol. Plant-Microbe Interact.* **2011**, *24*, 649–661. [CrossRef]
20. Shao, Y.; Tang, G.; Huang, Y.; Ke, W.; Wang, S.; Zheng, D.; Ruan, L. Transcriptional regulator Sar regulates the multiple secretion systems in *Xanthomonas oryzae*. *Mol. Plant Pathol.* **2023**, *24*, 16–27. [CrossRef]
21. Huang, D.-L.; Tang, D.-J.; Liao, Q.; Li, H.-C.; Chen, Q.; He, Y.-Q.; Feng, J.-X.; Jiang, B.-L.; Lu, G.-T.; Chen, B.; et al. The Zur of *Xanthomonas campestris* functions as a repressor and an activator of putative zinc homeostasis genes via recognizing two distinct sequences within its target promoters. *Nucleic Acids Res.* **2008**, *36*, 4295–4309. [CrossRef] [PubMed]
22. Teper, D.; Zhang, Y.; Wang, N. TfmR, a novel TetR-family transcriptional regulator, modulates the virulence of *Xanthomonas citri* in response to fatty acids. *Mol. Plant Pathol.* **2019**, *20*, 701–715. [CrossRef] [PubMed]
23. Cuthbertson, L.; Nodwell Justin, R. The TetR family of regulators. *Microbiol. Mol. Biol. Rev.* **2013**, *77*, 440–475. [CrossRef] [PubMed]
24. Ramos Juan, L.; Martínez-Bueno, M.; Molina-Henares Antonio, J.; Terán, W.; Watanabe, K.; Zhang, X.; Gallegos María, T.; Brennan, R.; Tobes, R. The TetR family of transcriptional repressors. *Microbiol. Mol. Biol. Rev.* **2005**, *69*, 326–356. [CrossRef] [PubMed]
25. Deng, W.; Li, C.; Xie, J. The underling mechanism of bacterial TetR/AcrR family transcriptional repressors. *Cell. Signal.* **2013**, *25*, 1608–1613. [CrossRef]
26. Zhu, P.-C.; Li, Y.-M.; Yang, X.; Zou, H.-F.; Zhu, X.-L.; Niu, X.-N.; Xu, L.-H.; Jiang, W.; Huang, S.; Tang, J.-L.; et al. Type VI secretion system is not required for virulence on rice but for inter-bacterial competition in *Xanthomonas oryzae* pv. *oryzicola*. *Res. Microbiol.* **2020**, *171*, 64–73. [CrossRef] [PubMed]
27. Ryan, R.P.; Dow, J.M. Intermolecular interactions between HD-GYP and GGDEF domain proteins mediate virulence-related signal transduction in *Xanthomonas campestris*. *Virulence* **2010**, *1*, 404–408. [CrossRef] [PubMed]
28. An, S.-Q.; Febrer, M.; McCarthy, Y.; Tang, D.-J.; Clissold, L.; Kaithakottil, G.; Swarbreck, D.; Tang, J.-L.; Rogers, J.; Dow, J.M.; et al. High-resolution transcriptional analysis of the regulatory influence of cell-to-cell signalling reveals novel genes that contribute to *Xanthomonas* phytopathogenesis. *Mol. Microbiol.* **2013**, *88*, 1058–1069. [CrossRef]
29. He, Y.-W.; Ng, A.Y.-J.; Xu, M.; Lin, K.; Wang, L.-H.; Dong, Y.-H.; Zhang, L.-H. *Xanthomonas campestris* cell–cell communication involves a putative nucleotide receptor protein Clp and a hierarchical signalling network. *Mol. Microbiol.* **2007**, *64*, 281–292. [CrossRef]
30. Guo, Y.; Zhang, Y.; Li, J.-L.; Wang, N. Diffusible Signal Factor-Mediated Quorum sensing plays a central role in coordinating gene expression of *Xanthomonas citri* subsp. *citri*. *Mol. Plant-Microbe Interact.* **2011**, *25*, 165–179. [CrossRef]
31. Cho, J.H.; Yoon, J.M.; Lee, S.W.; Noh, Y.H.; Cha, J.S. *Xanthomonas oryzae* pv. *oryzae* RpfE regulates virulence and carbon source utilization without change of the DSF production. *Plant Pathol. J.* **2013**, *29*, 364–373. [CrossRef] [PubMed]
32. Han, S.W.; Lee, M.A.; Yoo, Y.; Cho, M.H.; Lee, S.W. Genome-wide screening to identify responsive regulators involved in the virulence of *Xanthomonas oryzae* pv. *oryzae*. *Plant Pathol. J.* **2019**, *35*, 84–89. [CrossRef] [PubMed]
33. Yu, X.; Liang, X.; Liu, K.; Dong, W.; Wang, J.; Zhou, M.-G. The *thiG* gene is required for full virulence of *Xanthomonas oryzae* pv. *oryzae* by preventing cell aggregation. *PLoS ONE* **2015**, *10*, e0134237. [CrossRef]
34. Guo, W.; Zou, L.-F.; Cai, L.-L.; Chen, G.-Y. Glucose-6-phosphate dehydrogenase is required for extracellular polysaccharide production, cell motility and the full virulence of *Xanthomonas oryzae* pv. *oryzicola*. *Microb. Pathog.* **2015**, *78*, 87–94. [CrossRef] [PubMed]
35. Rott, P.; Fleites, L.A.; Mensi, I.; Sheppard, L.; Daugrois, J.-H.; Dow, J.M.; Gabriel, D.W. The RpfCG two-component system negatively regulates the colonization of sugar cane stalks by *Xanthomonas albilineans*. *Microbiology* **2013**, *159*, 1149–1159. [CrossRef] [PubMed]
36. An, S.-Q.; Lu, G.-T.; Su, H.-Z.; Li, R.-F.; He, Y.-Q.; Jiang, B.-L.; Tang, D.-J.; Tang, J.-L. Systematic mutagenesis of all predicted *gntR* genes in *Xanthomonas campestris* pv. *campestris* reveals a GntR family transcriptional regulator controlling hypersensitive response and virulence. *Mol. Plant-Microbe Interact.* **2011**, *24*, 1027–1039. [CrossRef]
37. Zhou, X.; Yan, Q.; Wang, N. Deciphering the regulon of a GntR family regulator via transcriptome and ChIP-exo analyses and its contribution to virulence in *Xanthomonas citri*. *Mol. Plant Pathol.* **2017**, *18*, 249–262. [CrossRef] [PubMed]
38. Li, R.-F.; Lu, G.-T.; Li, L.; Su, H.-Z.; Feng, G.-f.; Chen, Y.; He, Y.-Q.; Jiang, B.-L.; Tang, D.-J.; Tang, J.-L. Identification of a putative cognate sensor kinase for the two-component response regulator HrpG, a key regulator controlling the expression of the *hrp* genes in *Xanthomonas campestris* pv. *campestris*. *Environ. Microbiol.* **2014**, *16*, 2053–2071. [CrossRef] [PubMed]
39. Cai, L.; Ma, W.; Zou, L.; Xu, X.; Xu, Z.; Deng, C.; Qian, W.; Chen, X.; Chen, G. *Xanthomonas oryzae* pv. *oryzicola* response regulator VemR is co-opted by the sensor kinase CheA for phosphorylation of multiple pathogenicity-related targets. *Front. Microbiol.* **2022**, *13*, 928551. [CrossRef]

40. Xu, Z.-Z.; Wu, G.-C.; Wang, B.; Guo, B.-D.; Sheng, C.; Zhao, Y.-Y.; Tang, B.; Zhao, Y.-C.; Liu, F.-Q. Sigma factor 70 RpoD contributes to virulence by regulating cell motility, oxidative stress tolerance, and manipulating the expression of *hrpG* and *hrpX* in *Xanthomonas oryzae* pv. *oryzae*. *J. Integr. Agric.* **2023**. [CrossRef]
41. Leng, M.; Lu, Z.-J.; Qin, Z.-S.; Qi, Y.-H.; Lu, G.-T.; Tang, J.-L. Flp, a Fis-like protein, contributes to the regulation of type III secretion and virulence processes in the phytopathogen *Xanthomonas campestris* pv. *campestris*. *Mol. Plant Pathol.* **2019**, *20*, 1119–1133. [CrossRef]
42. Tsuge, S.; Furutani, A.; Fukunaka, R.; Oku, T.; Tsuno, K.; Ochiai, H.; Inoue, Y.; Kaku, H.; Kubo, Y. Expression of *Xanthomonas oryzae* pv. *oryzae* hrp genes in XOM2, a novel synthetic medium. *J. Gen. Plant Pathol.* **2002**, *68*, 363–371. [CrossRef]
43. Sambrook, J.; Maniatis, T.; Fritsch, E. *Molecular Cloning: A Laboratory Manual*; Cold Spring Harbor Laboratories: Cold Spring Harbor, NY, USA, 1989; Volume 1.
44. Schäfer, A.; Tauch, A.; Jäger, W.; Kalinowski, J.; Thierbach, G.; Pühler, A. Small mobilizable multi-purpose cloning vectors derived from the *Escherichia coli* plasmids pK18 and pK19: Selection of defined deletions in the chromosome of *Corynebacterium glutamicum*. *Gene* **1994**, *145*, 69–73. [CrossRef]
45. Leong, S.A.; Ditta, G.S.; Helinski, D.R. Heme biosynthesis in Rhizobium. Identification of a cloned gene coding for delta-aminolevulinic acid synthetase from *Rhizobium meliloti*. *J. Biol. Chem.* **1982**, *257*, 8724–8730. [CrossRef]
46. Livak, K.J.; Schmittgen, T.D. Analysis of relative gene expression data using Real-Time quantitative PCR and the $2^{-\Delta\Delta CT}$ method. *Methods* **2001**, *25*, 402–408. [CrossRef] [PubMed]
47. Li, R.-F.; Peng, J.-L.; Liu, Q.-Q.; Chang, Z.; Huang, Y.-X.; Tang, J.-L.; Lu, G.-T. *Xanthomonas campestris* VemR enhances the transcription of the T3SS key regulator HrpX via physical interaction with HrpG. *Mol. Plant Pathol.* **2023**, *24*, 232–247. [CrossRef]

International Journal of
Molecular Sciences

MDPI

Article

FgGET3, an ATPase of the GET Pathway, Is Important for the Development and Virulence of *Fusarium graminearum*

Caihong Liu [1,2,3,†], Lu Lei [2,†], Jing Zhu [2], Lirun Chen [2], Shijing Peng [2], Mi Zhang [2], Ziyi Zhang [2], Jie Tang [2], Qing Chen [2], Li Kong [2], Youliang Zheng [2], Maria Ladera-Carmona [3], Karl-Heinz Kogel [3,4], Yuming Wei [2,*] and Pengfei Qi [2,*]

[1] State Key Laboratory of Crop Gene Exploration and Utilization in Southwest China, Sichuan Agricultural University, Chengdu 611130, China; sicaulch@outlook.com
[2] Triticeae Research Institute, Sichuan Agricultural University, Chengdu 611130, China; leilu@stu.sicau.edu.cn (L.L.); zhujing@stu.sicau.edu.cn (J.Z.); lirunchen88@stu.sicau.edu.cn (L.C.); pengshijing@stu.sicau.edu.cn (S.P.); zhangmi@stu.sicau.edu.cn (M.Z.); zhangjiajia@stu.sicau.edu.cn (Z.Z.); tangjie@stu.sicau.edu.cn (J.T.); qingchen83@sicau.edu.cn (Q.C.); kongli@sicau.edu.cn (L.K.); ylzheng@sicau.edu.cn (Y.Z.)
[3] Institute of Phytopathology, Research Centre for BioSystems, Land Use and Nutrition, Justus Liebig University Giessen, Heinrich-Buff-Ring 26, 35392 Giessen, Germany; maria.ladera-carmona@agrar.uni-giessen.de (M.L.-C.); karl-heinz.kogel@agrar.uni-giessen.de (K.-H.K.)
[4] Institut de Biologie Moléculaire des Plantes, CNRS, Université de Strasbourg, 12 rue du Général Zimmer, 67084 Strasbourg, France
* Correspondence: ymwei@sicau.edu.cn (Y.W.); pengfeiqi@sicau.edu.cn (P.Q.); Tel.: +86-28-82650337 (P.Q.); Fax: +86-28-82650350 (P.Q.)
† These authors contributed equally to this work.

Abstract: GET3 is an ATPase protein that plays a pivotal role in the guided entry of the tail-anchored (GET) pathway. The protein facilitates the targeting and inserting of tail-anchored (TA) proteins into the endoplasmic reticulum (ER) by interacting with a receptor protein complex on the ER. The role of GET3 in various biological processes has been established in yeast, plants, and mammals but not in filamentous fungi. *Fusarium graminearum* is the major causal agent of Fusarium head blight (FHB), posing a threat to the yield and quality of wheat. In this study, we found that FgGET3 exhibits a high degree of sequence and structural conservation with its homologs across a wide range of organisms. Ectopic expression of *FgGET3* in yeast restored the growth defects of the *Saccharomyces cerevisiae ScGET3* knock-out mutant. Furthermore, FgGET3 was found to dimerize and localize to the cytoplasm, similar to its homologs in other species. Deletion of *FgGET3* in *F. graminearum* results in decreased fungal growth, fragmented vacuoles, altered abiotic stress responses, reduced conidia production, delayed conidial germination, weakened virulence on wheat spikes and reduced DON production. Collectively, these findings underscore the critical role of *FgGET3* in regulating diverse cellular and biological functions essential for the growth and virulence of *F. graminearum*.

Keywords: *Fusarium graminearum*; Fusarium head blight; GET pathway; FgGET3; stress response; virulence

1. Introduction

Tail-anchored (TA) proteins are a distinct class of membrane-integrated proteins [1] that are involved in diverse cellular processes, including protein translocation across organelle membranes, vesicular trafficking, apoptosis, and protein quality control [2,3]. In yeast, the GET pathway is the most extensively studied pathway for targeting and inserting endoplasmic reticulum (ER)-destined TA proteins [4]. This pathway starts with the transfer of TA to a GET4/GET5/SGT2 sorting complex. GET3, an ATPase, can bind to GET4 thereby facilitating the loading of TA to GET3. Subsequently, GET3 chaperones TA proteins to the ER membrane by interacting with the ER-membrane localized GET1-GET2 receptor proteins [5,6].

GET3 is involved in multiple biological processes in different organisms, including metal homeostasis, signal transduction, and the ubiquitin-proteasome system [7]. In *Saccharomyces cerevisiae*, the knock-out (KO) of *ScGET3* leads to several seemingly unrelated phenotypes, including sensitivity to hygromycin, hydroxyurea, metal, H_2O_2, and heat stresses [5,8,9]. In *Arabidopsis thaliana*, there are three GET3 homologs; however, only AtGET3a has the conserved GET1 binding domain and is localized in the cytosol [10]. *AtGET3a* KO mutant flowers earlier than the wild-type (WT) [11]. In mammals, the deletion of *GET3* leads to embryonic lethality in mice [12], and upregulation of *GET3* has been observed in several types of human cancers [7]. In the protozoan parasite *Plasmodium falciparum*, the function of *GET3* is required in the intraerythrocytic developmental stage [13]. Homologs of GET3 are well conserved across organisms, but the functional studies of *GET3* remain poorly understood in filamentous fungi.

Fusarium graminearum, which is included in the top ten top 10 economically important fungal pathogens, is the predominant causal agent of Fusarium head blight (FHB) in wheat and other cereal crops [14]. The main consequence of FHB is the reduction in grained yield and quality, which affects the overall harvest and economy [15,16], and a serious threat to animal and human health by contaminating grains with mycotoxins, including the trichothecene deoxynivalenol (DON), a potent inhibitor of protein synthesis in eukaryotes [17,18]. So far, the control of FHB remains problematic. Due to the unavailability of resistant wheat cultivars, the major approach for controlling disease and limiting mycotoxin accumulation is still the use of chemical fungicides, which partly exhibit negative traits like fungicide tolerance and hazardous effects on the environment [19,20]. Therefore, there is an urgency to develop effective control strategies for FHB. Previous studies have shown that growth, development, and wheat infection of *F. graminearum* are regulated by various genes involved in different regulatory pathways [21–23]; however, the GET pathway associated with growth and virulence remains poorly understood. In *F. graminearum*, GET pathway genes have not been identified yet. In this study, we provide evidence that *FGSG_09891* encodes a GET3 homolog, designated as FgGET3, and we uncovered the genetic and biological functions of *FgGET3* using cellular, genetic, and biochemical approaches.

2. Results

2.1. Identification of FgGET3 in F. graminearum

To search for the corresponding homolog of GET3 protein in *F. graminearum*, we used the amino acid (aa) sequence of ScGET3 from *S. cerevisiae* (NP_010183.1) and queried against the *F. graminearum* database by using Basic Local Alignment Search Tool (BLAST) in NCBI database. The top hit in all queries was FGSG_09891 (XP_011318797.1), which we designated as FgGET3. FgGET3 is a 341 aa protein, sharing 47.2% identity with *S. cerevisiae* ScGET3, 50.4% identity with *A. thaliana* AtGET3a (NP_563640.1), 49.9% identity with *Homo sapiens* HsGET3 (NP_004308.1), and 42.8% identity with *P. falciparum* PfGET3 (XP_001351457.1). Domain prediction reveals that FgGET3 exhibits conserved protein motifs that are characteristic of other GET3 proteins (Figure 1), including the presence of (i) the ATPase activity-related motifs [24]; among them, the P-loop, Switch I and Switch II show conserved sequences across various species, while the B-loop and A-loop aa motif in FgGET3 is relatively conserved compared to other GET3 homologs but not entirely identical; (ii) a motif that interacts with both GET1 and GET2 [25,26]; the 'DELYED' motif in yeast ScGET3, which can bind to ScGET1, and the RERR motif in ScGET2, which corresponds to 'EELYAED' in FgGET3; (iii) other aa residues that facilitate interaction with GET1; specific residues in ScGET3, including F246, L249, Y250, Y298 and L305 [26], are all conserved in FgGET3; (iv) the 'CXXC' motif, which promotes the formation of GET3 homodimers [27], corresponds to 'CDQC' in FgGET3; (v) aa residues involved in binding with ScGET4; in ScGET3 residues D253, K293, K297, D300, D303, E304, E307 and D308 interact with ScGET4 [28]; in FgGET3, these residues are conserved except for the substitution of Asp303 with Glu; and (vi) the 'CVC' motif, associated with oxidative stress sensitivity [9], which is consistently conserved across different species.

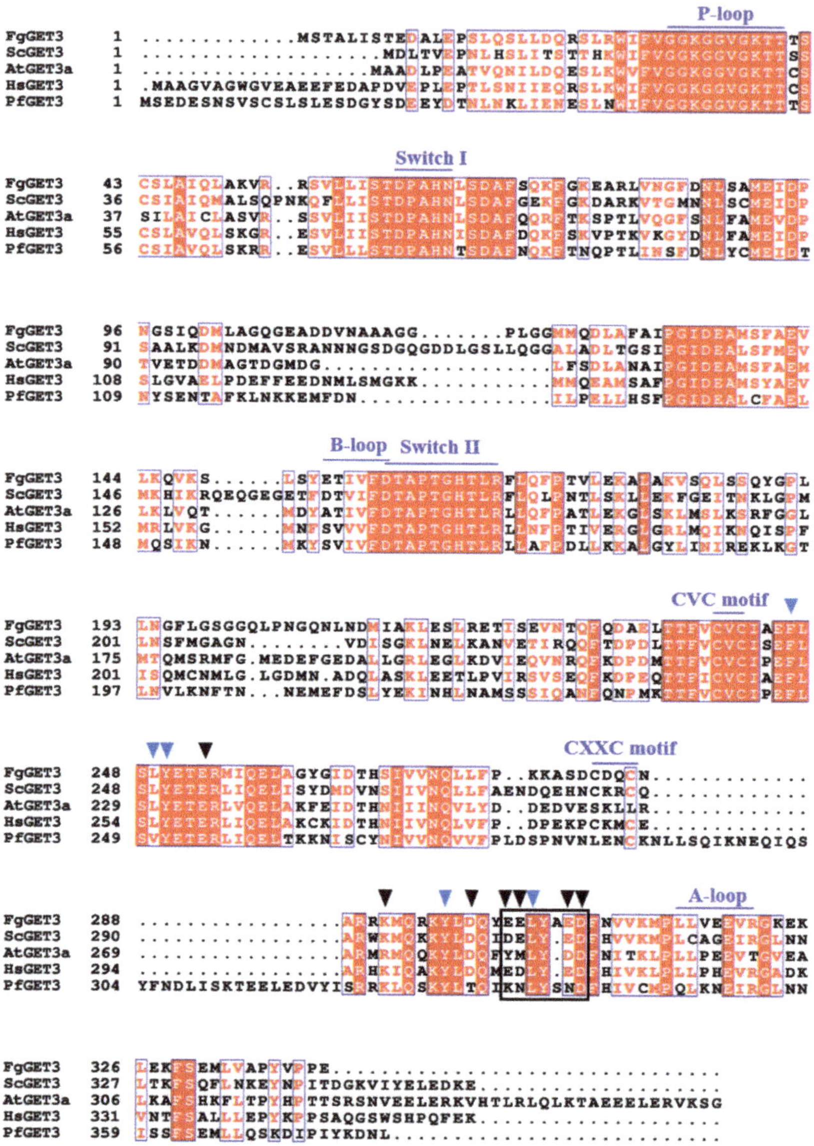

Figure 1. Identification of the FgGET3 protein in *Fusarium graminearum*. The aa sequence of FgGET3 (XP_011318797.1), ScGET3 (NP_010183.1) of *Saccharomyces cerevisiae*, AtGET3a (NP_563640.1) of *Arabidopsis thaliana*, HsGET3 (NP_004308.1) of *Homo sapiens* and PfGET3 (XP_001351457.1) of *Plasmodium falciparum* was aligned using Clustal W2. ESPript 3.0 was used to highlight identical (white font, highlighted in red with red), well-conserved (red font, boxed in blue) residues. The conserved motifs are labeled above with blue lines and font. The sequence involved in both the GET1-GET3 and GET2-GET3 interactions is enclosed in a black frame. The other aa involved in the GET1-GET3 interaction are indicated with blue inverted triangles above. The aa involved in the GET4-GET3 interaction are indicated with black inverted triangles.

2.2. *FgGET3* Restores the Growth Defect of the Yeast *ScGET3* KO Mutant

In *S. cerevisiae*, *ScGET3* KO mutant (Δ*get3*) showed clear phenotypic changes under various stress conditions [5]. To investigate comparable function between *FgGET3* and *ScGET3*, we transformed Δ*get3* with either the empty pYES2 vector or the expression vector pYES2-FgGET3 containing the full-length cDNA of *FgGET3* and assessed their growth. As shown in Figure 2, there was no apparent difference in growth between the *S. cerevisiae* WT *BY4741* and Δ*get3* in the absence of stresses. However, when exposed to stresses such as 3 mM $CuSO_4$ at 37 °C or 200 mM hygromycin at 30 °C, the Δ*get3* mutant and Δ*get3* mutant carrying the empty pYES2 vector consistently exhibited sensitive growth phenotypes. Those phenotypes were successfully rescued by the *in trans* expression of *FgGET3* to a similar level as seen in the WT *BY4741* cells. These results confirmed that *FgGET3* is a homolog of *ScGET3*.

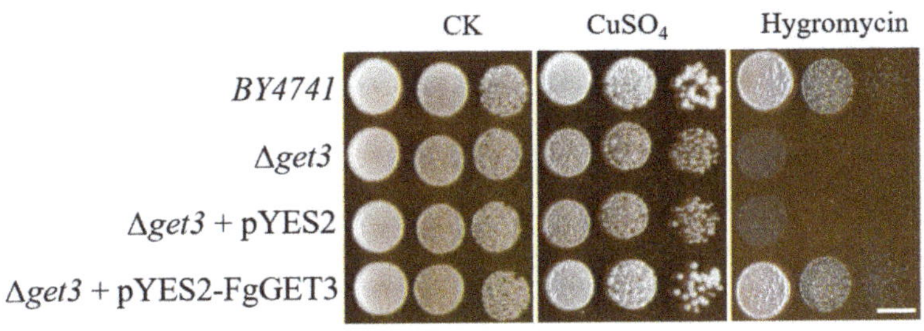

Figure 2. Complementation of *S. cerevisiae* strain Δ*get3* with *FgGET3*. The plasmid pYES2-FgGET3 was transformed into Δ*get3*. The cell growth (ten-fold dilutions of a starting concentration of OD_{600} = 1) without (CK) and in the presence of 3 mM $CuSO_4$ at 37 °C or 200 mM hygromycin at 30 °C on YPD plates is shown. The empty pYES2 vector was transformed into Δ*get3* as a negative control. The growth of each strain was examined after 3 days of incubation. Scale bar = 5 mm.

2.3. *FgGET3* Forms a Homodimer

In *S. cerevisiae*, ScGET3 interacts with itself to form a homodimer [3,6]. To test whether a self-interaction exists in its homologs FgGET3, a yeast two-hybrid (Y2H) assay was performed. The BD-FgGET3 bait and AD-FgGET3 prey vectors were constructed and co-transformed into *Y2HGold* strains. The result showed that FgGET3 interacts with itself (Figure 3).

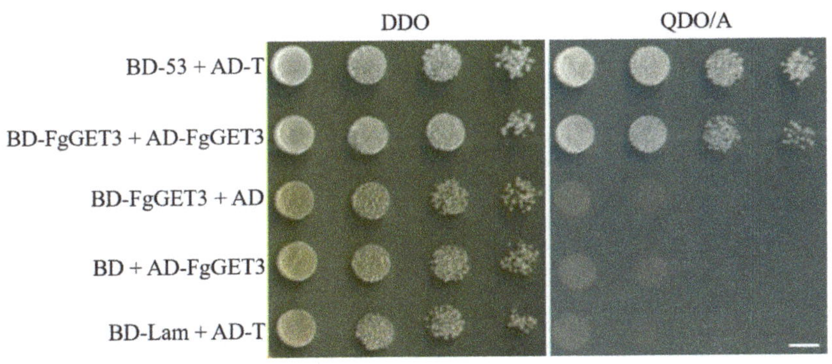

Figure 3. Detection of FgGET3 homodimerization using the yeast two-hybrid assay. Y2HGold strains transformed with the pair of bait and prey vectors were grown on DDO and QDO/A media (ten-fold dilutions of a starting concentration of OD_{600} = 1). The interactions of BD-53/AD-T and BD-Lam/AD-T were the positive and negative controls, respectively. AD: pGADT7; BD: pGBKT7; DDO: Double dropout medium (SD-Leu-Trp); QDO/A: Quadruple dropout medium (SD-Ade-His-Leu-Trp) supplemented with 70 ng/mL Aureobasidin A. Scale bar = 5 mm.

2.4. *FgGET3 Localizes in the Cytoplasm*

Previous studies have identified that GET3 from other species localize to the cytoplasm [5,10]. Consistent with this, FgGET3 was predicted to lack a transmembrane helix (Figure S1). To verify that FgGET3 is indeed a soluble protein as predicted, we tagged the C-terminus of FgGET3 with a green fluorescent protein (eGFP) and expressed it in the WT *F. graminearum* strain. We found that FgGET3-eGFP localized to the cytoplasm in both ungerminated (Figure 4A) and germinated conidia (Figure 4B), consistent with the subcellular localization of GET3 in other species.

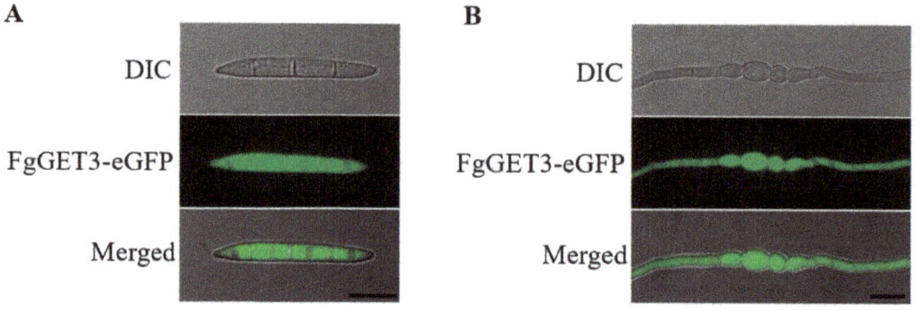

Figure 4. Subcellular localization of FgGET3 in (**A**) ungerminated conidia and (**B**) conidia germinating for 8 h. Scale bar = 10 μm; DIC, differential interference contrast.

2.5. *Generation of FgGET3 Mutants*

To elucidate the biological function of *FgGET3* in *F. graminearum*, the KO mutant Δ*Fgget3* was created. The fragments flanking upstream and downstream of *FgGET3* were cloned into the pRF-HU2 vector, which harbored the hygromycin B phosphotransferase (*HPH*) gene. The constructed FgGET3-pRF-HU2 vector was transformed into the WT strain, and the coding sequence of *FgGET3* was replaced with the *HPH* gene through homologous recombination (Figure S2A). Transformants exhibiting resistance to hygromycin were verified via PCR analysis (Figure S2B).

The complemented strains Δ*Fgget3*-C were generated by introducing the neomycin-resistance (NEO) vector FgGET3-JM45, which encompasses a 2.1-kb region containing the gene and the promoter region, into the Δ*Fgget3* mutant. The expression of *HPH* and *NEO* genes in Δ*Fgget3*-C was detected by reverse transcription PCR (RT-PCR), demonstrating the presence of the KO and complementary vectors. Additionally, the transcript levels of *FgGET3* in the WT and Δ*Fgget3*-C strains were found to be similar, but completely absent in Δ*Fgget3* (Figure S2C). These results demonstrate successful generation of KO and complemented strains.

2.6. *FgGET3 Is Involved in Hyphal Growth and Vacuole Fusion*

To investigate the requirement of *FgGET3* in hyphal growth, WT, Δ*Fgget3*, and Δ*Fgget3*-C strains were cultured on both nutrient-poor medium mSNA and nutrient-rich medium PDA. Compared with the WT and Δ*Fgget3*-C, the Δ*Fgget3* mutant showed a reduced vegetative growth rate on both types of media (Figure 5A,B). Further microscopic observation revealed that Δ*Fgget3* grew abnormally producing multiple hyphal branches (Figure 5C).

In addition, by staining the hyphae with the vacuole tracker 7-amino-4-chloromethylcoumarin (CAMC), we observed that Δ*Fgget3* hyphae produced fragmented small vacuoles, whereas WT cells exhibited a single large vacuole (Figure 5D). Transmission electron microscopy further confirmed the presence of fragmented vacuoles with numerous irregular spheroids in Δ*Fgget3* hyphae but not in WT (Figure 5E). These findings indicate that *FgGET3* is crucial for both hyphal growth and vacuole fusion in *F. graminearum*.

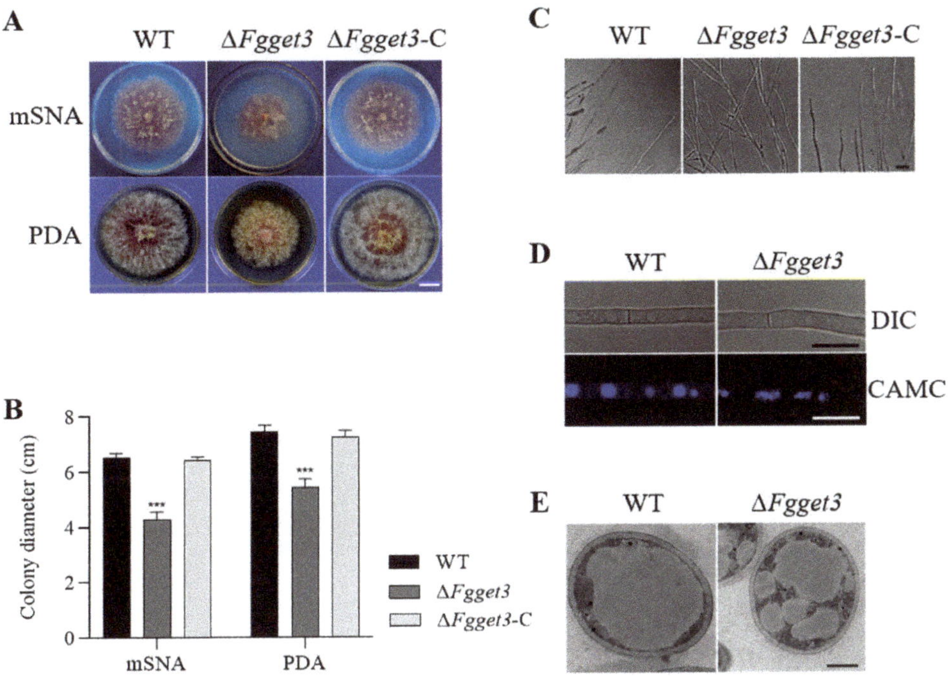

Figure 5. *FgGET3* in involved in vegetative growth and vacuole morphology. (**A**) Colony morphology of WT, Δ*Fgget3*, and Δ*Fgget3*-C on PDA and mSNA plates. Photos were taken after incubation at 25 °C for 4 days. Scale bar = 2 cm. (**B**) Colony diameters of indicated strains on mSNA and PDA media. Means and standard deviations were calculated from three replicates; the asterisks indicate significant differences from the WT group (Student's *t*-test, *** $p < 0.001$). (**C**) Microscopic observation of hyphal branching patterns of WT, and Δ*Fgget3* grown on PDA medium for 2 days. Scale bar = 20 μm. (**D**) Fluorescence microscopy images of WT and Δ*Fgget3* hyphae stained with the vacuole tracker dye CMAC. DIC: differential interference contrast. Scale bar = 20 μm. (**E**) Vacuole structures in hyphae of WT and Δ*Fgget3* observed using transmission electron microscopy. Scale bar = 1 μm.

2.7. *FgGET3* Is Involved in Fungal Responses to Various Stresses

To investigate the function of *FgGET3* in stress responses, WT, Δ*Fgget3*, and Δ*Fgget3*-C were cultured on a PDA medium supplemented with different stress-inducing agents. Firstly, when the mycelia of strains were treated with environmental stresses, we observed that compared with WT and Δ*Fgget3*-C, Δ*Fgget3* displayed a decreased inhibition rate when treated with osmotic stress agents (Sorbitol and NaCl) and oxidative stress agents (H_2O_2), but was more sensitive to cell wall-perturbing agent Congo Red (CR), but not to SDS (Figure 6A,D). In addition, Δ*Fgget3* was more sensitive to high-temperature stress at 33 °C. Secondly, we determined the sensitivity of Δ*Fgget3* to metal ions, however, Δ*Fgget3* did not exhibit changed sensitivity to heavy metal ions Mg^{2+}, Cu^{2+}, Zn^{2+}, and Fe^{2+} (Figure S3). Thirdly, in drug sensitivity assays, Δ*Fgget3* exhibited increased sensitivity to pyraclostrobin, but not to carbendazim and tebuconazole (Figure 6B,E). Fourthly, compared to that of the WT and Δ*Fgget3*-C, the Δ*Fgget3* mutant showed increased sensitivity to dithiothreitol (DTT) induced ER stress (Figure 6C,F).

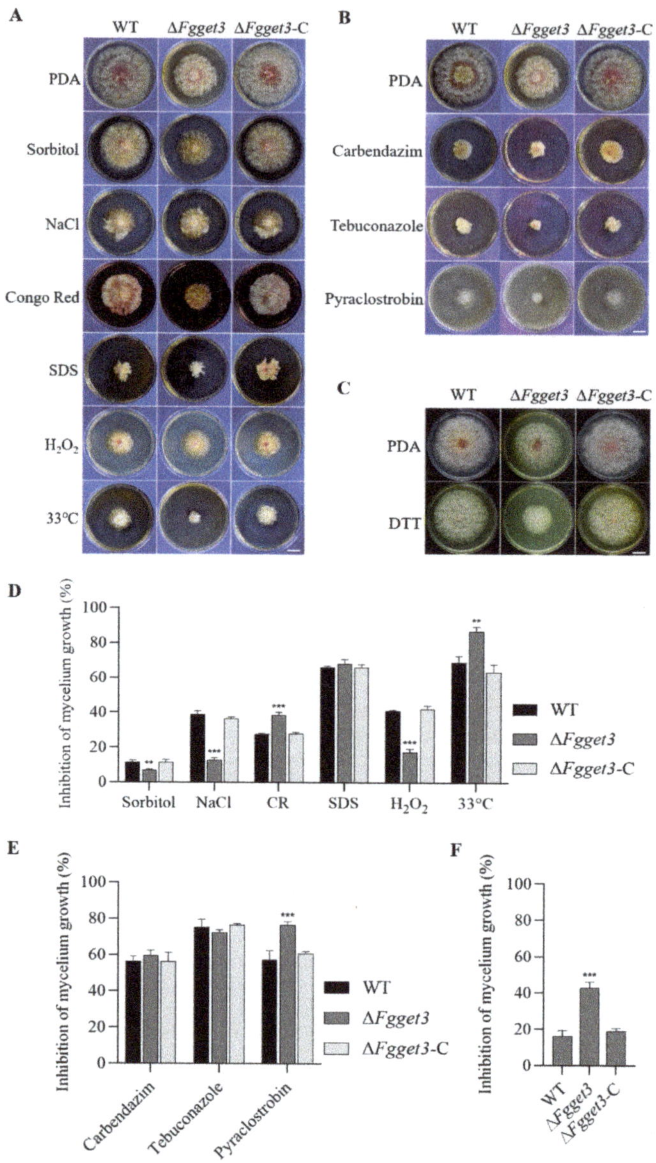

Figure 6. *FgGET3* impacts the responses of *F. graminearum* to various stresses. (**A**) Colony morphology of WT, Δ*Fgget3*, and Δ*Fgget3*-C in response to environmental stresses including 1M Sorbitol, 1 M NaCl, 0.5 mM Congo Red (CR), 0.025% SDS, 0.05% H_2O_2 and high temperature at 33 °C. Scale bar = 2 cm. (**B**) Colony morphology of indicated strains in response to fungicides including 0.4 μg/mL carbendazim, 5 μg/mL tebuconazole, and 0.6 μg/mL pyraclostrobin. Scale bar = 2 cm. (**C**) Colony morphology of WT, Δ*Fgget3*, and Δ*Fgget3*-C in response to 5 mM DTT. Scale bar = 2 cm. (**D**) Percentage of mycelium growth inhibition by environmental stresses. (**E**) Percentage of mycelium growth inhibition by fungicides. (**F**) Percentage of mycelium growth inhibition by DTT. Means and standard deviations were calculated from three replicates; the asterisks indicate significant differences from the WT group (Student's *t*-test, ** $p < 0.01$, *** $p < 0.001$).

2.8. FgGET3 Plays a Critical Role in Asexual Development

To investigate whether *FgGET3* is involved in conidiation, equal amounts of fresh mycelium of WT, Δ*Fgget3*, and Δ*Fgget3*-C were inoculated into a liquid CMC medium. After four days, conidia of WT and Δ*Fgget3*-C were observed forming on phialides. In contrast, these structures were rarely seen in Δ*Fgget3* (Figure 7A). Compared to those of WT and Δ*Fgget3*-C, the amount of conidia produced by Δ*Fgget3* was decreased by over 95% (Figure 7B). Microscopic examination revealed that the conidia of Δ*Fgget3* displayed morphological defects with a 40% reduction in length compared to WT and Δ*Fgget3*-C (Figure 7C). Additionally, most of the conidia of Δ*Fgget3* had two septa or fewer (82%). In contrast, conidia of WT and Δ*Fgget3*-C typically had three or more septa (91%) (Figure 7D). To explore the involvement of *FgGET3* in conidial germination, conidia were inoculated into a YEPD liquid medium. Microscopic examination revealed that, although the conidial germination of Δ*Fgget3* was not blocked completely, a significant delay was observed (Figure 7E). After 6 h, only 11% of Δ*Fgget3* conidia produced three or more germ tubes, compared to over 43% of WT and Δ*Fgget3*-C conidia. After 9 h, 20% of Δ*Fgget3* conidia produced three or more germ tubes, compared to over 81% of WT and Δ*Fgget3*-C conidia (Figure 7F). These results underscored the critical roles of *FgGET3* in conidiation, conidial morphology, and germination in *F. graminearum*.

2.9. FgGET3 Is Required for Virulence and DON Production

To determine the role of *FgGET3* in the virulence of *F. graminearum*, conidia of WT, Δ*Fgget3*, and Δ*Fgget3*-C were point inoculated into flowering wheat spikes. At 2 days post-inoculation (dpi), fungal quantification in the inoculated spikes showed a 55% decrease in Δ*Fgget3* biomass compared to WT and Δ*Fgget3*-C (Figure 8A). At 14 dpi, WT and Δ*Fgget3*-C caused typical head blight symptoms in the inoculated and nearby spikelets with dark brown rachises and shriven seeds, while Δ*Fgget3* only caused limited discoloration at the inoculation site and failed to spread to adjacent rachis nodes (Figure 8B). Moreover, as DON is the most characterized virulence factor in fungal infection on wheat [29], spikelets inoculated with conidial suspensions were harvested at 8 dpi and used to assess the DON production. As the result showed (Figure 8C), compared to WT and Δ*Fgget3*-C, the DON production on infected spikes was reduced by about 68% when inoculating with Δ*Fgget3* (Figure 8D). Consistent with the wheat spikes infection results, Δ*Fgget3* also produced a much lower level (65%) of DON than WT and Δ*Fgget3*-C in a liquid medium (Figure 8E). These findings indicated that *FgGET3* contributes to the virulence and DON production of *F. graminearum*.

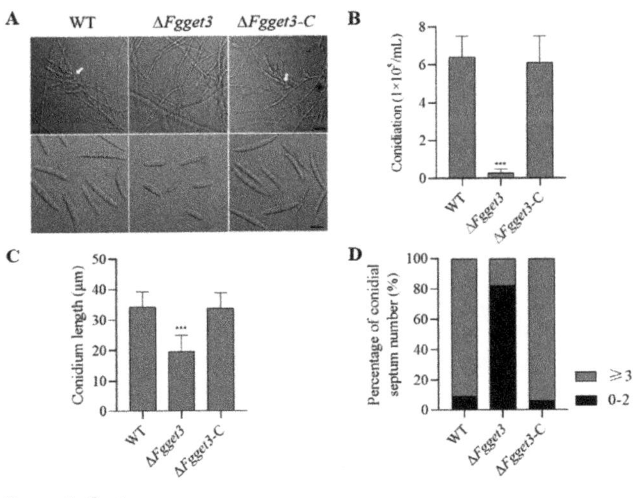

Figure 7. *Cont.*

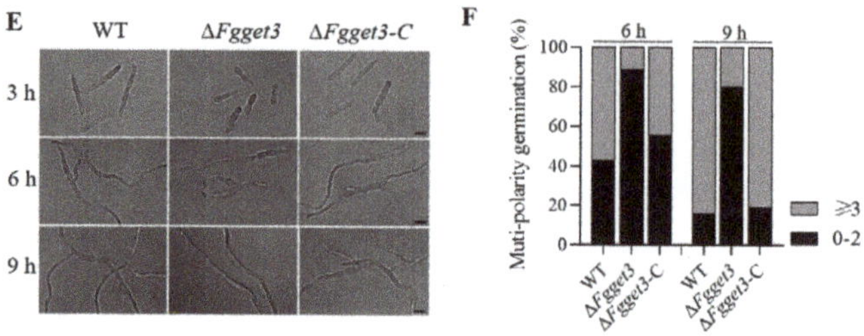

Figure 7. *FgGET3* is crucial for conidiation, conidial morphology, and germination of *F. graminearum*. (**A**) Morphology of phialide (white arrows) and conidia of WT, Δ*Fgget3*, and Δ*Fgget3*-C strains. Scale bar = 20 μm. (**B**) Conidiation capacity of the indicated strains in CMC liquid medium. (**C**) Conidial length of the indicated strains. Means and standard deviations were calculated from three replicates, at least 100 conidia were observed in each replicate, and the asterisks indicate significant differences from the WT group. (Student's *t*-test, *** $p < 0.001$). (**D**) Percentages of conidia with different numbers of septa in the indicated strains. At least 300 conidia were observed from three replicates. (**E**) Morphology of germinated conidia at 3 h, 6 h, and 9 h in liquid YEPD medium. Scale bar = 20 μm. (**F**) Germination rates of the indicated strains under a microscope after 6 h and 9 h incubation in YEPD liquid medium. At least 100 conidia were randomly observed at each time point.

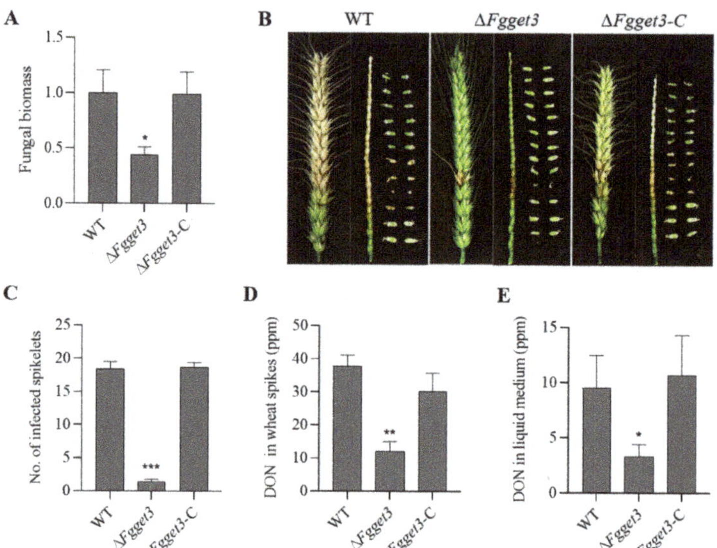

Figure 8. *FgGET3* contributes to the virulence and DON production of *F. graminearum*. (**A**) Fungal biomass in the inoculated spikelets was quantified using quantitative PCR (qPCR) at 2 dpi. The relative amount of fungal DNA to spikelet DNA was determined by comparing the *F. graminearum FgTUB2* gene to the wheat *TaGAPDH* gene using qPCR. The measurement was referred to as the value 1.0 obtained for the WT treatment. (**B**) The symptoms in spikelets, seeds, and rachises of infected wheat spikes infected with a conidial suspension of WT, Δ*Fgget3*, and Δ*Fgget3*-C were observed at 14 dpi. (**C**) The numbers of infected and bleached spikelets were counted at 14 dpi. (**D**) DON production in wheat spikes infected with conidia suspensions of indicated strains was quantified at 8 dpi. (**E**) The DON concentration of indicated strains was determined in a liquid medium. Means and standard deviations were calculated from three replicates; the asterisks indicate significant differences from the WT group (Student's *t*-test, * $p < 0.01$, ** $p < 0.005$, *** $p < 0.001$).

3. Discussion

Within the GET pathway, GET3 homodimeric ATPase is the central targeting factor that delivers TA proteins to the ER by interacting with the transmembrane complex [4,30]. In the present study, we identified a single GET3 homolog in *F. graminearum*, encoded by *FGSG_09891* and named FgGET3. FgGET3 exhibits high aa sequence identity with GET3 homologs from other eukaryotes and possesses a conserved protein sequence, including ATPase activity-related motifs, interaction sites with GET1, GET2, GET4, and conserved aa sites for dimerization (Figure 1). Functional complementation in *S. cerevisiae* demonstrated that *FgGET3* can restore the function of *ScGET3* in yeast (Figure 2), indicating its conserved gene function. Yeast two-hybrid assays confirmed that FgGET3 forms homodimers (Figure 3). Moreover, FgGET3 is predicted to lack transmembrane domains (Figure S1) and further confirmed to localize to the cytoplasm in conidia and hyphae as shown by eGFP fusion expression analysis (Figure 4), consistent with GET3 homologs in other species. Our results confirmed that FgGET3 is indeed a GET3 functional homolog in *F. graminearum*.

In *S. cerevisiae*, the deletion of *ScGET3* had no significant impact on vegetative growth [5]. In contrast, the Δ*Fgget3* mutant exhibited impaired vegetative growth on both nutrient-poor and nutrient-rich media, accompanied by increased hyphal branching (Figure 5A). Additionally, Δ*Fgget3* exhibited defects in conidial reproduction, morphology, and germination (Figure 7). *ScGET3* plays a role in mediating the responses of *S. cerevisiae* to different environmental stress conditions; its KO mutant Δ*get3* showed increased sensitivity to cell-wall stress [31], oxidative stress [9], elevated temperature stress, and metal stress [5]. Similarly, Δ*Fgget3* was sensitive to cell wall stress and elevated temperature (Figure 6A). However, Δ*Fgget3* showed enhanced tolerance to osmotic stress (Figure 6A) and no impact on metal stress (Figure S2). Furthermore, Δ*Fgget3* showed increased tolerance to osmotic stress (Figure 6A). Given the observed defect in vacuole morphology (Figure 5D,E), it is reasonable to speculate that Δ*Fgget3* exhibited an altered response to various environmental stresses. These discrepancies suggest that GET3 proteins may have partly distinct roles among different fungi.

In *F. graminearum*, virulence is a complex mechanism regulated by multiple factors [21,22]. On flowering wheat spikes, Δ*Fgget3* produced less fungal biomass in the early stages of infection and failed to spread through the rachis (Figure 8), possibly due to defects in both mycelial and conidial development stages that could potentially influence virulence. In addition, DON is considered another key pathogenicity factor in *F. graminearum* [32,33]. The total amount of DON in Δ*Fgget3* was reduced both on infected spikes and in liquid medium (Figure 8), which could be an additional factor contributing to the impaired infection capability of *F. graminearum*. Currently, the application of chemical fungicides is still one main approach for controlling FHB. Interestingly, Δ*Fgget3* showed higher sensitivity to pyraclostrobin (Figure 6), suggesting that *FgGET3* may be a potential target for this fungicide.

Deletion of *ScGET3* in yeast leads to various growth defects, possibly due to the mislocalization or aggregation of TA proteins near the ER [5]. In *A. thaliana*, the KO of *AtGET3* resulted in increased sensitivity of seedlings to the ER stressor DTT [11]. Similarly, the KO of *FgGET3* in *F. graminearum* also exhibited increased sensitivity to DTT (Figure 6), which may indicate aggregation of TA proteins in the cytoplasm upon gene deletion. Hence, the multifaceted defects caused by the loss of *FgGET3* function in *F. graminearum* may also be attributed to the impaired functionality of TA proteins and thus require further validation.

4. Materials and Methods

4.1. Fungal Strains and Growth Conditions

F. graminearum strain DAOM180378 (Canadian Fungal Culture Collection, AAFC, 270 Ottawa, ON, Canada) was used as the parental WT. The colony characteristics and growth rate of both the WT and transformants were examined on Potato Dextrose Agar (PDA) plates (Aobox, Beijing, China) and modified Synthetischer Nährstoffarmer Agar (mSNA) medium, and further cultured at 25 °C for 4 days under dark conditions. The radial growth of mycelium was measured as previously described [34]. To assess the sensitivity of mycelium to stresses, 5 mm mycelial plugs of each strain were inoculated onto PDA plates containing various stressors at concentrations indicated in the figure legends. The percentage of inhibition of mycelial radial growth was calculated as described by [35]. Each treatment was carried out in at least five Petri dishes and the experiments were repeated three times.

For the evaluation of conidiation, 20 mg of fresh mycelium taken from the periphery of a 3-day-old colony was cultured in a 150 mL flask containing 50 mL of carboxymethyl-cellulose (CMC) medium [36]. The cultures were shaken at 180 rpm at 25 °C for 4 days and the number of conidia in each flask was counted using a hemocytometer (Sigma-Aldric, St. Louis, MO, USA). The conidial morphology and the number of septa were observed using an Olympus-BX63 fluorescence microscope (DP80; Olympus, Tokyo, Japan), and the colonial length and width were measured using the Olympus CellSens Dimension software ver.1.18 (Olympus, Tokyo, Japan). Germination of conidia was determined by culturing 1×10^5 conidia of each strain in 5 mL yeast extract peptone dextrose (YEPD) liquid medium [37] and incubating at 25 °C on a rotary shaker. The germ tubes per 100 total conidia were counted at 3 h, 6 h, and 9 h after inoculation.

4.2. Sequence Analysis

All the sequence data used in this study can be found in the National Center for Biotechnology Information (NCBI, https://www.ncbi.nlm.nih.gov/) (accessed on 20 August 2024). The detailed accession numbers for FgGET3 and its homologs from various species used in this study are as follows: *F. graminearum* (XP_011318797.1), *S. cerevisiae* ScGET3 (NP_010183.1), *A. thaliana* AtGET3a (NP_563640.1), *H. sapiens* HsGET3 (NP_004308.1), and *P. falciparum* PfGET3 (XP_001351457.1). Clustal W2 was used to generate amino acid sequence comparisons.

4.3. Fungal Transformation

The KO mutant of FgGET3 was generated using a homologous recombination strategy to replace *FgGET3* with a hygromycin B phosphotransferase gene (*HPH*) gene. For this purpose, a gene replacement vector was first constructed. 748 bp upstream (UP) and 550 bp downstream (Down) flanking sequences of *FgGET3* were amplified by PCR using the genomic DNA of WT, and the resulting amplicons were fused to the pRF-HU2 vector [38] with ClonExpress II One Step Cloning Kit (Vazyme, Nanjing, China). The constructed FgGET3-pRF-HU2 vector was transformed into WT by the *Agrobacterium tumefaciens*-mediated transformation method [39]. Putative KO transformants were selected with 100 μg/mL hygromycin, and confirmed by PCR assays. To complement the Δ*Fgget3* mutation, a PCR product, including the native promoter and open reading fragment of the gene, was cloned into the vector JM45 that contains a neomycin resistance gene. The constructed vector was transformed into the protoplast of the KO mutant by the polyethylene glycol (PEG)-mediated fungal transformation method [40]. Putative complementary transformants were selected by 100 μg/mL neomycin and further identified by PCR and RT-PCR assay. For generating the FgGET3-eGFP fusion construct, the coding sequence of *FgGET3* was amplified and inserted into the vector pRFHUE-eGFP [41], and the resulting construct pRFHUE-eGFP-FgGET3 was transformed into the WT strain by *A. tumefaciens*-mediated transformation. FgGET3-GFP transformants were selected with hygromycin and identified by PCR. All primers used are listed in Table S1.

4.4. Virulence and DON Production Assays

The wheat (*Triticum aestivum*) cultivar SM482, which is highly susceptible to *F. graminearum*, was used for the virulence tests. The plants were grown as previously described [36]. Conidia of WT, Δ*Fgget3*, and Δ*Fgget3*-C were collected from a CMC medium and adjusted to a concentration of 1×10^5 conidia in 1 mL of sterile water. The floral cavity of the outer florets of the 5th spikelet from the bottom of the wheat spike was inoculated with 10 μL of conidial suspension at each point. After inoculation, the spikes were covered with a plastic bag for 48 h to maintain high humidity. The infected florets from five spikes were harvested at 48 h post-inoculation to estimate the relative fungal biomass within the plants. Following the removal of the plastic bags, the wheat spikelets were cultivated for an additional 12 days to evaluate typical FHB symptoms. The experiments were repeated three times with each repetition including at least ten replicates for each fungal strain.

To measure DON production in wheat spikes, conidial suspensions of WT, Δ*Fgget3*, and Δ*Fgget3*-C strains were injected into each floret of the whole spikes. There were at least five spikes for each strain. The infected spikes were collected at 8 dpi and ground to a fine powder in liquid nitrogen. To test the DON production in the liquid medium, a two-stage protocol was employed as previously described [32]. The amount of DON was quantified by using a competitive ELISA-based DON detection kit (Mlbio, Shanghai, China) and a Multiskan Spectrum instrument (Thermo Fisher Scientific, Waltham, MA, USA).

4.5. Nucleic Acid Extraction and PCR Assay

The genomic DNA from mycelium and plants was extracted using the DNA extraction kit (Biofit, Chengdu, China). Total RNA was extracted with Trizol Reagent (Thermo Fisher Scientific), and further used to perform reverse transcription to generate First-strand cDNA using HiScript III 1st Strand cDNA Synthesis Kit (Vazyme, Nanjing, China).

To measure the relative amount of fungal biomass in infected wheat spikelets, total DNA was extracted, and the β-tubulin gene *FgTUB2* (*FGSG_09530*) of *F. graminearum* was quantified relative to the *TaGAPDH* gene of *Triticum aestivum* using quantitative PCR (qPCR) [42]. The qPCR was run on the MyiQ Real-Time PCR Detection System (Bio-Rad, Hercules, CA, USA) with the Taq Pro Universal SYBR qPCR Master Mix (Vazyme, Nanjing, China). The relative quantification was calculated by the $2^{-\Delta\Delta CT}$ method [43] with three biological replicates for each treatment. The primers used throughout the paper are listed in Table S1.

4.6. Staining and Microscopic Examination

For observing the hyphae morphology, fresh mycelia of WT and Δ*Fgget3* taken from 2-day-old colonies on PDA plates were fixed with 2.5% (*v*/*v*) glutaraldehyde. Subsequently, the mycelia were dehydrated in a graded series of ethanol and embedded in Epon812 (Sigma-Aldric, St. Louis, MO, USA). After cutting the ultrathin section with an ultramicrotome EM UC7 (Leica, Weztlar, Germany), specimens were stained with uranyl acetate and lead citrate, and then the hyphal morphology was observed with an HT7700 120 KV transmission electron microscopy (Hitachi, Tokyo, Japan). For vacuole morphology examination, hyphae of WT and Δ*Fgget3* were stained with 7-amino-4-chloromethylcoumarin (CAMC) [44]. To observe the expression of *FgGET3*, conidia of the *FgGET3-eGFP* strain were cultured in YEPD medium for 8 h, and the GFP fluorescence was observed using an Olympus-BX63 fluorescence microscope (DP80; Olympus, Tokyo, Japan) with the GFP filter set (excitation 488 nm, emission 520 nm).

4.7. Yeast Complementation Assay

S. cerevisiae strains, WT strain *BY4741*, and Δ*get3* mutant (YDL100C) were obtained from HORIZON (Dublin, Ireland). The full length of *FgGET3* was amplified from the cDNA of WT and subsequently cloned into the pYES2 vector (Thermo Fisher Scientific) to construct the pYES2-FgGET3 vector. The pYES2-*FgGET3* vector was transformed into Δ*get3* using the lithium acetate method according to the manufacturer's instructions (Thermo Fisher Scientific). At the same time, the empty pYES2 vector was transformed into the *BY4741* and Δ*get3* strains as controls. The uracil-deficient selective medium was used to select the transformants (Clontech, Mountain View, CA, USA). Transformants exhibiting uracil prototrophy were then diluted in distilled water, and aliquots of 10 μL were plated in 10-fold serial dilutions on YPD medium (10 g yeast extract, 20 g peptone, 20 g glucose,

10 g agar per liter) supplemented with chemicals including hydroxyurea, hygromycin and $CuSO_4$ at concentrations as indicated in figure legends.

4.8. Yeast Two-Hybrid Assay

To verify whether FgGET3 forms a homodimer, the cDNA of the *FgGET3* gene was inserted into both the prey vector pGADT7 (AD) and bait vector pGBKT7 (BD) (Clontech, Mountain View, CA, USA). The pair of AD-FgGET3 and BD-FgGET3 vectors were co-transformed into *S. cerevisiae* strain *Y2HGold* according to the manufacturer's instructions (Clontech, Mountain View, CA, USA). The growth of the transformants was determined on both DDO (Double dropout medium: SD-Leu/Trp) and QDO/X/A (Quadruple dropout medium: SD/-Ade/-His/-Leu/-Trp supplemented with Aureobasidin A) medium for 3 days. The AD-T/BD-53 pair and AD-T/BD-Lam pair were used as the positive and negative control, respectively.

4.9. Statistical Analyses

All data analysis was performed using GraphPad 8.0 and graphs are shown as means $\pm$ standard deviation (SD). The significant differences between experimental and control groups were determined using the two-tailed Student's *t*-test, and all experiments in this study were performed with at least three replicated measurements.

5. Conclusions

Our results indicate that *FgGET3* plays a pleiotropic role in the *F. graminearum* disease cycle, involving mycelium growth, stress response, asexual development, and plant infection. Understanding the biological roles of *FgGET3* in impacting the plant disease cycle provides fundamental new knowledge for the development of novel methods of disease intervention methods. Further insight into the role of *FgGET3* in TA protein targeting and inserting could help elucidate its cellular function and how it regulates.

Supplementary Materials: The following supporting information can be downloaded at https://www.mdpi.com/article/10.3390/ijms252212172/s1.

Author Contributions: Conceptualization, C.L., P.Q. and Y.W.; methodology, C.L., L.L., J.Z. and L.C.; software, S.P. and Z.Z.; validation, M.Z. and J.T.; formal analysis, C.L. and L.K.; investigation, Q.C.; resources, C.L. and L.K.; data curation, Q.C. and Y.Z; writing—original draft preparation, C.L., L.L. and P.Q.; writing—review and editing, C.L., P.Q., M.L.-C. and K.-H.K.; visualization, C.L. and J.Z.; supervision, Y.W. and P.Q.; project administration, Y.Z.; funding acquisition, Y.W. and P.Q. All authors have read and agreed to the published version of the manuscript.

Funding: This work was supported by the National Key Research and Development Program of China (2023YFD1200404) and the National Natural Science Foundation of China (31901961).

Institutional Review Board Statement: Not applicable.

Informed Consent Statement: Not applicable.

Data Availability Statement: The data that support the findings of this study are available from the corresponding author upon reasonable request.

Conflicts of Interest: The authors declare no conflicts of interest.

References

1. Kutay, U.; Hartmann, E.; Rapoport, T.A. A class of membrane proteins with a C-terminal anchor. *Trends Cell Biol.* **1993**, *3*, 72–75. [CrossRef] [PubMed]
2. Hegde, R.S.; Keenan, R.J. Tail-anchored membrane protein insertion into the endoplasmic reticulum. *Nat. Rev. Mol. Cell Biol.* **2011**, *12*, 787–798. [CrossRef] [PubMed]
3. Chartron, J.W.; Clemons, W.M.; Suloway, C.J.M. The complex process of GETting tail-anchored membrane proteins to the ER. *Curr. Opin. Struct. Biol.* **2012**, *22*, 217–224. [CrossRef] [PubMed]
4. Mateja, A.; Keenan, R.J. A structural perspective on tail-anchored protein biogenesis by the GET pathway. *Curr. Opin. Struct. Biol.* **2018**, *51*, 195–202. [CrossRef]

5. Schuldiner, M.; Metz, J.; Schmid, V.; Denic, V.; Rakwalska, M.; Schmitt, H.D.; Schwappach, B.; Weissman, J.S. The GET complex mediates insertion of tail-anchored proteins into the ER membrane. *Cell* **2008**, *134*, 634–645. [CrossRef]
6. Wang, F.; Whynot, A.; Tung, M.; Denic, V. The mechanism of tail-anchored protein insertion into the ER membrane. *Mol. Cell.* **2011**, *43*, 738–750. [CrossRef]
7. Simpson, P.J.; Schwappach, B.; Dohlman, H.G.; Isaacson, R.L. Structures of get3, get4, and get5 provide new models for TA membrane protein targeting. *Structure* **2010**, *18*, 897–902. [CrossRef]
8. Shen, J.; Hsu, C.M.; Kang, B.K.; Rosen, B.P.; Bhattacharjee, H. The *Saccharomyces cerevisiae Arr4p* is involved in metal and heat tolerance. *Biometals* **2003**, *16*, 369–378. [CrossRef]
9. Voth, W.; Schick, M.; Gates, S.; Li, S.; Vilardi, F.; Gostimskaya, I.; Southworth, D.R.; Schwappach, B.; Jakob, U. The protein targeting factor Get3 functions as ATP-independent chaperone under oxidative stress conditions. *Mol. Cell* **2014**, *56*, 116–127. [CrossRef]
10. Xing, S.; Mehlhorn, D.G.; Wallmeroth, N.; Asseck, L.Y.; Kar, R.; Voss, A.; Denninger, P.; Schmidt, V.A.F.; Schwarzländer, M.; Stierhof, Y.D.; et al. Loss of GET pathway orthologs in *Arabidopsis thaliana* causes root hair growth defects and affects SNARE abundance. *Proc. Natl. Acad. Sci. USA* **2017**, *114*, e1544–e1553. [CrossRef]
11. Srivastava, R.; Zalisko, B.E.; Keenan, R.J.; Howell, S.H. The GET system inserts the tail-anchored protein, SYP72, into endoplasmic reticulum membranes. *Plant Physiol.* **2017**, *173*, 1137–1145. [CrossRef] [PubMed]
12. Mukhopadhyay, R.; Ho, Y.S.; Swiatek, P.J.; Rosen, B.P.; Bhattacharjee, H. Targeted disruption of the mouse *Asna1* gene results in embryonic lethality. *Febs Letters* **2006**, *580*, 3889–3894. [CrossRef] [PubMed]
13. Kumar, T.; Maitra, S.; Rahman, A.; Bhattacharjee, S. A conserved guided entry of tail-anchored pathway is involved in the trafficking of a subset of membrane proteins in *Plasmodium falciparum*. *PLoS Pathog.* **2021**, *17*, e1009595. [CrossRef] [PubMed]
14. Dean, R.; Van Kan, J.A.L.; Pretorius, Z.A.; Hammond-Kosack, K.E.; Di Pietro, A.; Spanu, P.D.; Rudd, J.J.; Dickman, M.; Kahmann, R.; Ellis, J.; et al. The top 10 fungal pathogens in molecular plant pathology. *Mol. Plant Pathol.* **2012**, *13*, 414–430. [CrossRef]
15. Figueroa, M.; Hammond-Kosack, K.E.; Solomon, P.S. A review of wheat diseases-a field perspective. *Mol. Plant Pathol.* **2018**, *19*, 1523–1536. [CrossRef]
16. Powell, A.J.; Vujanovic, V. Evolution of Fusarium head blight management in wheat: Scientific perspectives on biological control agents and crop genotypes protocooperation. *Appl. Sci.* **2021**, *11*, 8960. [CrossRef]
17. Van Egmond, H.P.; Schothorst, R.C.; Jonker, M.A. Regulations relating to mycotoxins in food: Perspectives in a global and European context. *Anal. Bioanal. Chem.* **2007**, *389*, 147–157. [CrossRef]
18. Trail, F. For Blighted Waves of Grain: *Fusarium graminearum* in the postgenomics era. *Plant Physiol.* **2009**, *149*, 103–110. [CrossRef]
19. Kazan, K.; Gardiner, D.M.; Manners, J.M. On the trail of a cereal killer: Recent advances in *Fusarium graminearum* pathogenomics and host resistance. *Mol. Plant Pathol.* **2012**, *13*, 399–413. [CrossRef]
20. Chen, A.H.; Islam, T.; Ma, Z.H. An integrated pest management program for managing Fusarium head blight disease in cereals. *J. Integr. Agric.* **2022**, *21*, 3434–3444. [CrossRef]
21. Niu, G.; Yang, Q.; Liao, Y.; Sun, D.; Tang, Z.; Wang, G.; Xu, M.; Wang, C.W.; Kang, J.G. Advances in understanding *Fusarium graminearum*: Genes involved in the regulation of sexual development, pathogenesis, and deoxynivalenol biosynthesis. *Genes* **2024**, *15*, 475. [CrossRef] [PubMed]
22. Xu, M.; Wang, Q.; Wang, G.; Zhang, X.; Liu, H.; Jiang, C. Combatting Fusarium head blight: Advances in molecular interactions between *Fusarium graminearum* and wheat. *Phytopathol. Res.* **2022**, *4*, 37. [CrossRef]
23. Rampersad, S.N. Pathogenomics and management of Fusarium diseases in plants. *Pathogens* **2020**, *9*, 340. [CrossRef] [PubMed]
24. Leipe, D.D.; Wolf, Y.I.; Koonin, E.V.; Aravind, L. Classification and evolution of P-loop GTPases and related ATPases. *J. Mol. Biol.* **2002**, *317*, 41–72. [CrossRef]
25. Mariappan, M.; Mateja, A.; Dobosz, M.; Bove, E.; Hegde, R.S.; Keenan, R.J. The mechanism of membrane-associated steps in tail-anchored protein insertion. *Nature* **2011**, *477*, 61–69. [CrossRef]
26. Stefer, S.; Reitz, S.; Wang, F.; Wild, K.; Pang, Y.Y.; Schwarz, D.; Bomke, J.; Hein, C.; Löhr, F.; Bernhard, F.; et al. Structural basis for tail-anchored membrane protein biogenesis by the Get3-receptor complex. *Science* **2011**, *333*, 758–762. [CrossRef]
27. Mateja, A.; Szlachcic, A.; Downing, M.E.; Dobosz, M.; Mariappan, M.; Hegde, R.S.; Keenan, R.J. The structural basis of tail-anchored membrane protein recognition by Get3. *Nature* **2009**, *461*, 361–366. [CrossRef]
28. Gristick, H.B.; Rao, M.; Chartron, J.W.; Rome, M.E.; Shan, S.O.; Clemons, W.M. Crystal structure of ATP-bound Get3-Get4-Get5 complex reveals regulation of Get3 by Get4. *Nat. Struct. Mol. Biol.* **2014**, *21*, 437–442. [CrossRef]
29. Boenisch, S.; Boenisch, M.J.; Schäfer, W. *Fusarium graminearum* forms mycotoxin producing infection structures on wheat. *BMC Plant Biol.* **2011**, *11*, 110. [CrossRef]
30. Mateja, A.; Paduch, M.; Chang, H.Y.; Szydlowska, A.; Kossiakoff, A.A.; Hegde, R.S.; Keenan, R.J. Structure of the Get3 targeting factor in complex with its membrane protein cargo. *Science* **2015**, *347*, 1152–1155. [CrossRef]
31. Brown, J.A.; Sherlock, G.; Myers, C.L.; Burrows, N.M.; Deng, C.; Wu, H.I.; McCann, K.E.; Troyanskaya, O.G.; Brown, J.M. Global analysis of gene function in yeast by quantitative phenotypic profiling. *Mol. Syst. Biol.* **2006**, *2*, 2006.0001. [CrossRef] [PubMed]
32. Jiang, C.; Zhang, C.; Wu, C.; Sun, P.; Hou, R.; Liu, H.; Wang, C.; Xu, J.R. *TRI6* and *TRI10* play different roles in the regulation of deoxynivalenol (DON) production by CAMP signaling in *Fusarium graminearum*. *Environ. Microbiol.* **2016**, *18*, 3689–3701. [CrossRef] [PubMed]

33. Chen, Y.; Kistler, H.C.; Ma, Z. Annual review of phytopathology *Fusarium graminearum* trichothecene mycotoxins: Biosynthesis, regulation, and management. *Annu. Rev. Phytopathol.* **2019**, *57*, 15–39. [CrossRef]
34. Qi, P.F.; Zhang, Y.Z.; Liu, C.H.; Zhu, J.; Chen, Q.; Guo, Z.R.; Wang, Y.; Xu, B.J.; Zheng, T.; Jiang, Y.F.; et al. *Fusarium graminearum* ATP-binding cassette transporter gene *FgABCC9* is required for its transportation of salicylic acid, fungicide resistance, mycelial growth and pathogenicity towards wheat. *Int. J. Mol. Sci.* **2018**, *19*, 2351. [CrossRef]
35. Qi, P.F.; Johnston, A.; Balcerzak, M.; Rocheleau, H.; Harris, L.J.; Long, X.Y.; Wei, Y.M.; Zheng, Y.L.; Ouellet, T. Effect of salicylic acid on *Fusarium graminearum*, the major causal agent of fusarium head blight in wheat. *Fungal Biol.* **2012**, *116*, 413–426. [CrossRef]
36. Yun, Y.; Liu, Z.; Yin, Y.; Jiang, J.; Chen, Y.; Xu, J.R.; Ma, Z. Functional analysis of the *Fusarium graminearum* phosphatome. *New Phytol.* **2015**, *207*, 119–134. [CrossRef]
37. Hou, Z.; Xue, C.; Peng, Y.; Katan, T.; Corby Kistler, H.; Xu, J.R. A mitogen-activated protein kinase gene (*MGV1*) in *Fusarium graminearum* is required for female fertility, heterokaryon formation, and plant infection. *Mol. Plant Microbe Interact.* **2002**, *15*, 1119–1127. [CrossRef]
38. Frandsen, R.J.N.; Andersson, J.A.; Kristensen, M.B.; Giese, H. Efficient four fragment cloning for the construction of vectors for targeted gene replacement in filamentous fungi. *BMC Mol. Biol.* **2008**, *9*, 70. [CrossRef]
39. Michielse, C.B.; Hooykaas, P.J.J.; van den Hondel, C.A.M.J.J.; Ram, A.F.J. *Agrobacterium*-mediated transformation as a tool for functional genomics in fungi. *Curr. Genet.* **2005**, *48*, 1–17. [CrossRef]
40. Catlett, N.L.; Lee, B.N.; Yoder, O.C.; Turgeon, B.G. Split-marker recombination for efficient targeted deletion of fungal genes. *Fungal Genet. Rep.* **2003**, *50*, 9–11. [CrossRef]
41. Crespo-Sempere, A.; López-Pérez, M.; Martínez-Culebras, P.V.; González-Candelas, L. Development of a green fluorescent tagged strain of *Aspergillus Carbonarius* to monitor fungal colonization in grapes. *Int. J. Food Microbiol.* **2011**, *148*, 135–140. [CrossRef] [PubMed]
42. Chen, Q.; Lei, L.; Liu, C.H.; Zhang, Y.Z.; Xu, Q.; Zhu, J.; Guo, Z.; Wang, Y.; Li, Q.C.; Li, Y.; et al. Major facilitator superfamily transporter gene *FgMFS1* is essential for *Fusarium graminearum* to deal with salicylic acid stress and for its pathogenicity towards wheat. *Int. J. Mol. Sci.* **2021**, *22*, 8497. [CrossRef] [PubMed]
43. Livak, K.J.; Schmittgen, T.D. Analysis of relative gene expression data using real-time quantitative PCR and the 2(-Delta Delta C(T)) Method. *Methods* **2001**, *25*, 402–408. [CrossRef] [PubMed]
44. Shoji, J.Y.; Arioka, M.; Kitamoto, K. Vacuolar membrane dynamics in the filamentous fungus *Aspergillus oryzae*. *Eukaryot. Cell* **2006**, *5*, 411–442. [CrossRef]

International Journal of
Molecular Sciences

MDPI

Article

The Essentials of *PgPG1*, a Polygalacturonase-Encoding Gene for the Invasion of *Pyrenophora graminea* to *Hordeum vulgare*

Erjing Si [1,2,†], Ming Guo [1,2,†], Haiying Liu [1,2], Chengdao Li [3], Juncheng Wang [1,2], Lirong Yao [1,2], Yaxiong Meng [1,2], Xiaole Ma [1,2], Baochun Li [1,4], Ke Yang [1,2], Xunwu Shang [1] and Huajun Wang [1,2,*]

1 State Key Laboratory of Aridland Crop Science, Gansu Key Laboratory of Crop Improvement and Germplasm Enhancement, Lanzhou 730070, China; sierjing@163.com (E.S.); 18794807552@163.com (M.G.); 19527377701@163.com (H.L.); wangjc@gsau.edu.cn (J.W.); ylr0384@163.com (L.Y.); yxmeng1@163.com (Y.M.); maxl@gsau.edu.cn (X.M.); libc@gsau.edu.cn (B.L.); yk_831116@163.com (K.Y.); 17393294106@163.com (X.S.)
2 College of Agronomy, Gansu Agricultural University, Lanzhou 730070, China
3 Western Barley Genetics Alliance, College of Science, Health, Engineering and Education, Murdoch University, Murdoch, WA 6150, Australia; c.li@murdoch.edu.au
4 College of Life Science and Technology, Gansu Agricultural University, Lanzhou 730070, China
* Correspondence: wanghj@gsau.edu.cn
† These authors contributed equally to this work.

Abstract: Barley leaf stripe, caused by *Pyrenophora graminea*, significantly reduces yield. Polygalacturonase, a key fungal pectinase, facilitates cell wall degradation for nutrition acquisition and colonization. To determine whether *P. graminea* contains polygalacturonase (*PgPG*)-encoding genes and their role in pathogenicity, four *PgPG* genes (*PgPG1–PgPG4*) were identified in the *P. graminea* genome. Quantitative RT-PCR revealed that *PgPG1* had the highest inducible expression during barley infection, suggesting its critical vital role in pathogenesis. *PgPG1* was silenced and overexpressed in *P. graminea* QWC (wild-type) using $CaCl_2$-PEG4000-mediated protoplast transformation. The *PgPG1* RNAi mutants exhibited slower growth, while overexpression mutants grew faster. Relative to the wild-type, the disease incidence of Alexis, a highly susceptible barley variety, decreased by 62.94%, 42.19%, 45.74%, and 40.67% for RNAi mutants, and increased by 12.73%, 12.10%, 12.63%, and 10.31% for overexpression mutants. Pathogenicity analysis showed decreased disease incidence with *PgPG1* RNAi mutants and increased severity with overexpression mutants. Trypan blue staining and polygalacturonase activity assays confirmed that overexpression mutants caused more severe damage compared to wild-type and RNAi mutants. These findings indicate that *PgPG1* plays a vital role in the pathogenicity of *P. graminea* in barley and has great potential as a pathogen target gene to develop a durable resistance variety to *P. graminea*.

Keywords: *Pyrenophora graminea*; polygalacturonase; gene family; pathogenicity

1. Introduction

Barley leaf stripe, caused by *Pyrenophora graminea*, is widespread in many countries and regions, including Europe, North America, North Africa, Russia, India, and China [1]. A 1% increase in disease incidence reduces production by 1% [2]; when the disease incidence is high, it leads to significant yield and economic losses, underscoring the need for effective disease control strategies [3,4]. Despite its importance, the pathogenic mechanisms of *P. graminea* remain poorly understood, with only the mitogen-activated protein kinase kinase gene (*PGPBS*) identified as contributing to *P. graminea* pathogenicity [5]. Research has shown that *P. graminea* mycelia invade the host by spreading within the middle lamella during infection [6,7]. Since the middle lamella primarily comprises pectin, its degradation by *P. graminea*-secreted enzymes is likely critical for colonization. Among the key enzymes involved in pectin degradation are polygalacturonase (PG), pectate lyase, and pectin methyl esterase. Polygalacturonases break down pectin in plant cell walls, facilitating nutrient release to support pathogen growth and development [8–12].

Molecular genetic studies on other plant pathogens have highlighted the role of PG-encoding genes in fungal pathogenicity. For instance, polygalacturonase P2c in *Aspergillus flavus* influences virulence by enhancing host invasion, as demonstrated through gene expression and targeted disruption of the *pecA* gene [13]. Similarly, the *Bcpg1* gene in *Botrytis cinerea* and the *PG1* gene in *Fusarium oxysporum* contributed to host infection [14]. Targeted disruption of PG-encoding genes in other fungi, such as *Alternaria citri* [15], *Claviceps purpurea* [16], and *Xylella fastidiosa* [17], has also resulted in weakened or lost pathogenicity, further underscoring the importance of PGs in fungal virulence.

However, not all *PG* genes are essential for pathogenicity. For instance, disrupting the *enPG1* in *Cryphonectria parasitica* [18] or the *PG5* gene in *F. oxysporum* [19] had no measurable effect on pathogenicity, likely due to functional redundancy among PG isozymes. Similarly, the *AaPG1* gene in *A. alternata* did not impact virulence, as mutants exhibited the same disease symptoms as the wild-type [15]. These findings suggest that while PGs are critical for the virulence of many plant pathogens, their contribution can vary depending on the organism and host–pathogen interaction.

Little is known about PG-encoding genes in *P. graminea*. All known PG genes belong to the glycosylhydrolase gene family (GH28), which hydrolyzes glycosidic bonds. RNA interference (RNAi) technology based on the *pSilent-1* vector has been successfully applied in *P. graminea* [5], offering a valuable tool for functional gene studies in this pathogen. Polygalacturonase production is critical for the success and survival of many fungal pathogens during host infection [20]. However, the specific contribution of PGs to *P. graminea* pathogenicity in barley remains unclear. To address this, we identified members of the GH28 gene family in *P. graminea* and selected *PgPG1* as a key candidate gene. We investigated its subcellular localization, validated its signal peptide secretion function, and evaluated its role in pathogenicity by comparing the virulence of RNAi and overexpression mutants with that of the wild-type strain QWC. This study will provide the pathogen target gene for the creation of disease-resistant germplasm based on host-induced gene silencing.

2. Results

2.1. Identification of PgPG Proteins in P. graminea

Four *Pg* genes, designated *PgPG1–PgPG4*, were identified at the genomic level (Table 1). These genes were predicted to encode polygalacturonase enzymes, suggesting a potential role in pectin degradation in barley. Sequence analysis using the InterPro database confirmed their involvement in carbohydrate metabolism. The proteins encoded by these genes ranged from 370 to 705 aa, with molecular weights between 37.04 and 74.98 kD. Signal peptides were detected in all four proteins (Figure S2), and their subcellular localization was predicted to be extracellular, identifying them as apoplastic proteins (Table 1).

Table 1. Physical and chemical properties of PG genes of *P. graminea*.

Gene	Gene ID	Number of Amino Acids/aa	Molecular Weight/kD	Isoelectric Point	GRAVY Index	Instability Index	Aliphatic Index	Signal Peptide Length	Subcellular Location
PgPG1	P.graminea_ GLEAN_10002343	392	41.62	5.60	−0.174	31.85	81.10	18	extracellular
PgPG2	P.graminea_ GLEAN_10002483	370	37.04	8.31	0.114	21.89	81.73	19	extracellular
PgPG3	P.graminea_ GLEAN_10005830	448	48.31	6.24	−0.276	37.10	79.82	19	extracellular
PgPG4	P.graminea_ GLEAN_10006636	705	74.98	5.98	−0.252	37.08	74.89	19	extracellular

2.2. Gene Structure and Conserved Motif Analysis of PgPG Proteins

The MEME tool (v5.5.7), annotated with InterPro (v87.0), identified ten conserved motifs (Figure 1A), and the conserved amino acid sequences are listed in Table 2. Gene structure analysis showed variability in intron numbers, with *PgPG4* containing five introns, while the other three genes had two introns each (Figure 1B). Conserved amino acid residue analysis revealed that all *PgPG* proteins had at least one NTD, DD, GHG, and RIK: the NTD and DD were located in motif 1, and GHG and RIK were found in motif 2 and motif 8, respectively (Figure 1C, Table 2).

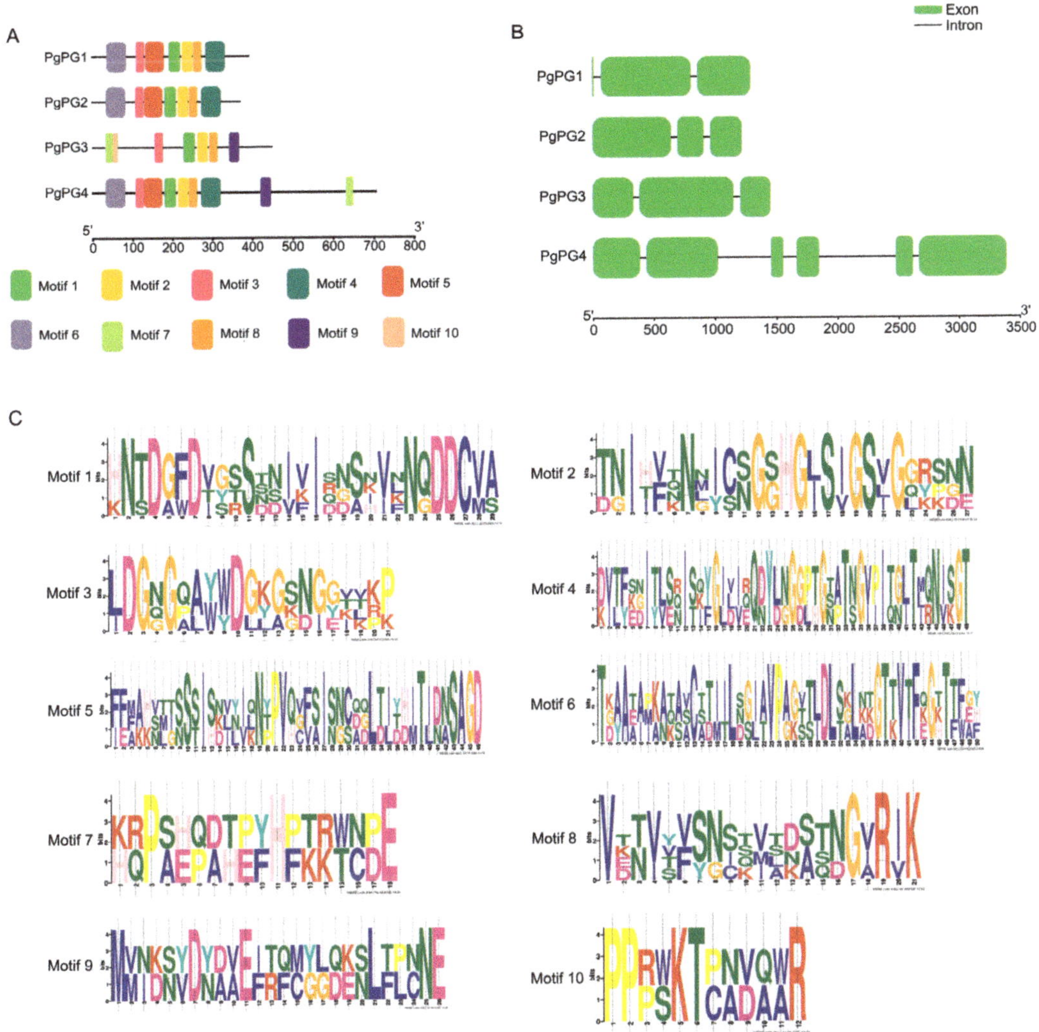

Figure 1. Conserved motif and gene structure analysis of PG family genes in *P. graminea*. (**A**) Distributions of conserved motifs in *PgPG* genes. Each motif is represented by a particular color. (**B**) The exon/intron structure of the *PgPG* genes. Green represents the exon regions, and the black lines represent introns. (**C**) The pattern identification of ten conserved sequences. The letters represent the amino acid sequence, and the size of the letters indicates the degree of sequence conservation; larger letters signify higher conservation.

Table 2. Conserved motifs of amino acids of *PgPG* genes family in *P. graminea*.

Conserved Motifs	E-Value	Number of Amino Acid	Amino Acid Sequence
Motif1	9.8×10^{-13}	29	HNTDGFDVGSSSNIVIQNSNVLNQDDCVA
Motif2	2.2×10^{-7}	27	TNIHFKNLICNGGHGJSIGSVGGRSNN
Motif3	1.5×10^{-4}	21	LDGNGQAWWDGKGSNGGVVKP
Motif4	1.1×10^{-4}	49	DVTFEDITLSRISKYGIVIQQDYLNGGPTGSATNGVPITGLTJQNISGT
Motif5	3.1×10^{-4}	46	FFFAHNLISSSISDIYIQNPPVQVFSISNCDGLTIDHITIDNSAGD
Motif6	1.1×10^{-3}	50	TGAAAAIKAKAACSTIILBGIAVPAGTTLDLSGLKDGTTVTFQGKTTFGY
Motif7	2.8×10^{0}	18	KRPAEPDHPFHFKKWCDE
Motif8	2.1×10^{1}	21	VETVTFSNSSVSDATNGVRIK
Motif9	1.5×10^{1}	26	MMNDNVDNDVEFRFCGGDENLFPCNE
Motif10	5.4×10^{1}	12	PPRWKTCNDQWR

2.3. Phylogenetic Analysis of the PgPG Protein Family

Phylogenetic analysis using the neighbor-joining method classified *PgPG* proteins into eight subfamilies, comprising 20, 2, 15, 2, 2, 1, 6, and 5 members each. *PgPG1* clustered with the PG gene of *Botryotinia fuckeliana*, indicating close evolutionary relatedness, while *PgPG4* shared a branch with *Cochliobolus carbonum* (Figure 2), revealing functional diversification within the *PgPG* protein family.

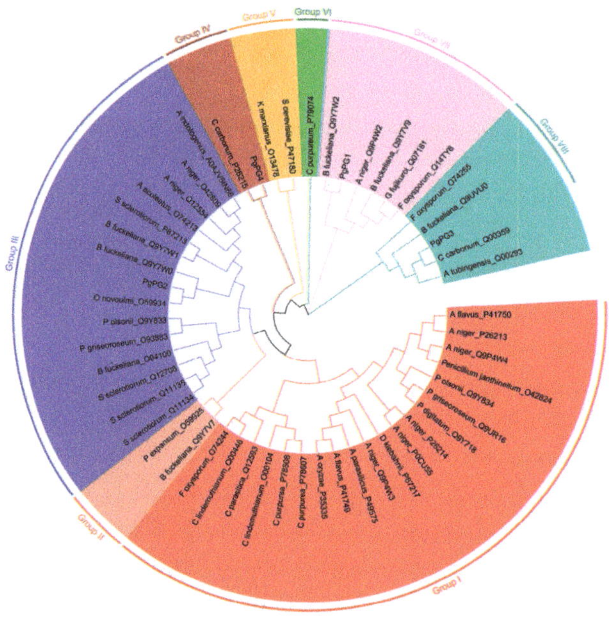

Figure 2. Circular cladogram showing the phylogenetic relationships of the 53 fungal PG proteins. The phylogenetic tree was generated using MEGA with the maximum likelihood method, employing 1000 bootstrap replicates. The same color indicates proteins with the same domain structure.

2.4. PgPG Gene Expression During Infection Stages

All four *PgPG* genes were expressed during *P. graminea* infection of the barley cultivar Alexis. The expression levels of *PgPG1* and *PgPG3* increased consistently over time, while *PgPG2* and *PgPG4* slightly declined by 7 days post-infection before gradually increasing. All four *PgPG* genes reached the highest relative expression at 18 days post-infection

(Figure 3). Among them, *PgPG1* exhibited the highest relative expression across all time points, warranting further investigation into its role in pathogenesis.

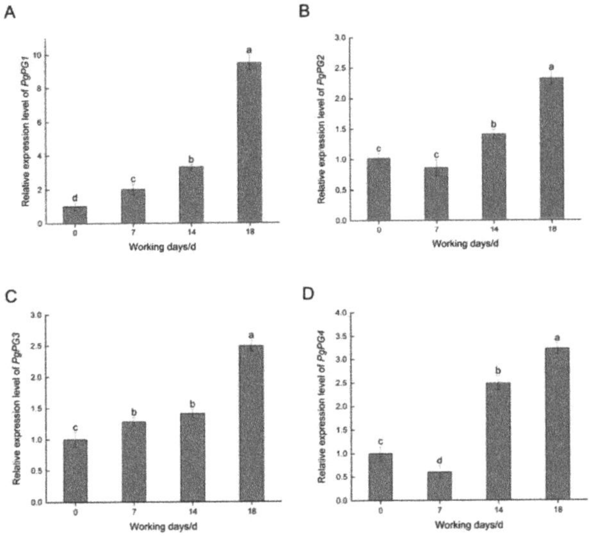

Figure 3. Expression level of *PgPG* genes in the interaction period between *P. graminea* and barley. (**A**) Expression level of *PgPG1* gene at different infection periods. (**B**) Expression level of *PgPG2* gene at different infection periods. (**C**) Expression level of *PgPG3* gene at different infection periods. (**D**) Expression level of *PgPG4* gene at different infection periods. Samples were collected at 0, 7, 14, and 18 d after inoculation. The expression level of each gene is compared to the abundance of *actin*, which was used as a reference gene. Different lowercase letters show standard errors based on three technical replicates.

2.5. Signal Peptide Functionality, Capability of Inducion Cell Death, and Subcellular Location of PgPG1

The TTC staining results showed that Avr1b and the *PgPG1* signal peptide facilitated the secretion of invertase, which degraded sucrose into monosaccharides. In contrast, neither the blank control YTK12 nor pSUC2 exhibited this activity (Figures 4A and S2). These results confirm that the signal peptide from *PgPG1* has a functional secretory role. Transient expression of the *PgPG1* protein in *N. benthamiana* was performed using the potato virus X (PVX) expression system, with the results showing that *PgPG1* could induce cell death in *N. benthamiana* (Figure 4B).

To determine the subcellular localization of *PgPG1*, a fusion protein of *PgPG1* and GFP was transiently expressed in *N. benthamiana*. Confocal microscopy analysis revealed green fluorescence primarily localized to the cell membrane and nucleus, with minor fluorescence detected in the cytoplasm, suggesting that the *PgPG1* gene is expressed predominantly in the cell membrane and nucleus (Figure 4C).

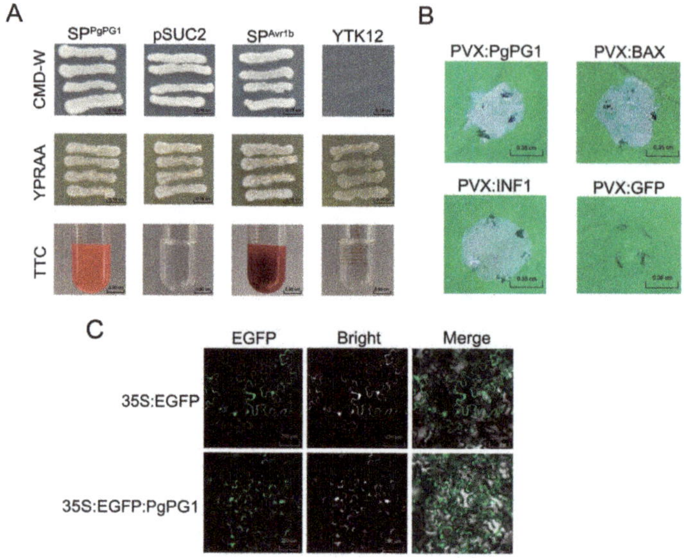

Figure 4. Functional verification of signal peptide, cell death detection and subcellular location of PgPG1. (**A**) The signal peptide (SP) functionality of PgPG1 was validated using a yeast signal trap assay. The yeast strain YTK12 was unable to grow on CMD-W medium (which lacks tryptophan) unless it contained the pSUC2 vector, which allows for the expression of the Trp operon. When the functional SP of PgPG1 is fused in-frame with the mature yeast invertase, it enables the secretion of invertase, allowing the yeast to grow on YPRAA medium. The known functional SP of Avr1b was utilized as a positive control. (**B**) The ability of PgPG1 from *P. graminea* to induce cell death was assessed in *N. benthamiana*. The leaves of 3-week-old *N. benthamiana* were infiltrated with *Agrobacterium tumefaciens* harboring the gene of interest. INF1 (BAX) and green fluorescent protein (GFP) were used as positive and negative controls, respectively. Photographs were taken 72 h after agroinfiltration. (**C**) Subcellular location of PgPG1 in *N. benthamiana*. PgPG1-EGFP fusion proteins or EGFP were expressed in *N. benthamiana*.

2.6. Identification of PgPG1 Transformants

To confirm the presence of positive transformants, including RNAi transformants and expression transformants, we verified the presence of the hygromycin phosphotransferase gene, which acted as a reporter gene for screening transformants containing an exogenous target fragment. PCR amplification produced a band of approximately 1000 bp in transformants containing RNAi and overexpression constructs (Figure S3). Four RNAi transformants (*PgPG1-RNAi-1*, *PgPG1-RNAi-2*, *PgPG1-RNAi-3*, and *PgPG1-RNAi-4*) (Figure S3A) and four overexpression transformants (*PgPG1-OE-1*, *PgPG1-OE-2*, *PgPG1-OE-3*, and *PgPG1-OE-4*) (Figure S3B) were identified.

qRT-PCR analysis was performed on the transformants to assess *PgPG1* gene expression levels. Relative expression levels of *PgPG1-RNAi-1*, *PgPG1-RNAi-2*, *PgPG1-RNAi-3*, and *PgPG1-RNAi-4* were measured at 35.58%, 41.89%, 41.33%, and 54.86%, respectively, indicating decreases of 64.42%, 58.11%, 58.67%, and 45.14% compared to the wild-strain (Figure 5A). In contrast, relative expression levels of the overexpression transformants (*PgPG1-OE-1*, *PgPG1-OE-2*, *PgPG1-OE-3*, and *PgPG1-OE-4*) were significantly elevated, 19.94%, 13.03%, 17.28%, and 16.50% higher than the wild-strain, respectively (Figure 5B).

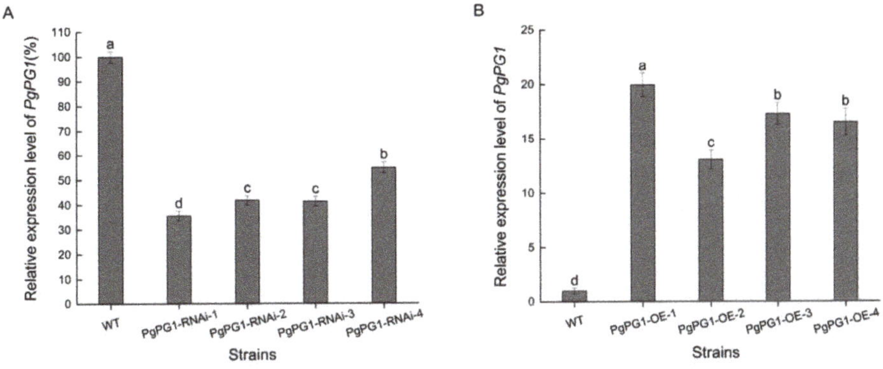

Figure 5. Relative expression of *PgPG1*-gene-transforming strains. (**A**) *PgPG1* relative expression of RNAi mutants. (**B**) *PgPG1* relative expression of overexpression mutants. Different lowercase letters show standard errors based on three technical replicates.

2.7. Role of PgPG1 in Vegetative Growth and Polygalacturonase Activity

The colony morphology of wild-strain and *PgPG1* mutants was observed after 7 days of growth on PDA. All strains exhibited similar morphology, characterized as white, flat, relatively compact, and villose (Figure 6A), indicating that the *PgPG1* gene does not influence colony morphology. Measurements of colony diameter revealed that all *PgPG1* RNAi mutants had smaller diameters than the wild-strain (Figure 6A). Among the overexpression mutants, *PgPG1-OE-2* had a smaller diameter, while *PgPG1-OE-1* and *PgPG1-OE-3* had larger diameters than the wild-strain. The diameter of *PgPG1-OE-4* was comparable to the wild-strain. These findings suggest that the *PgPG1* gene plays a role in *P. graminea* growth.

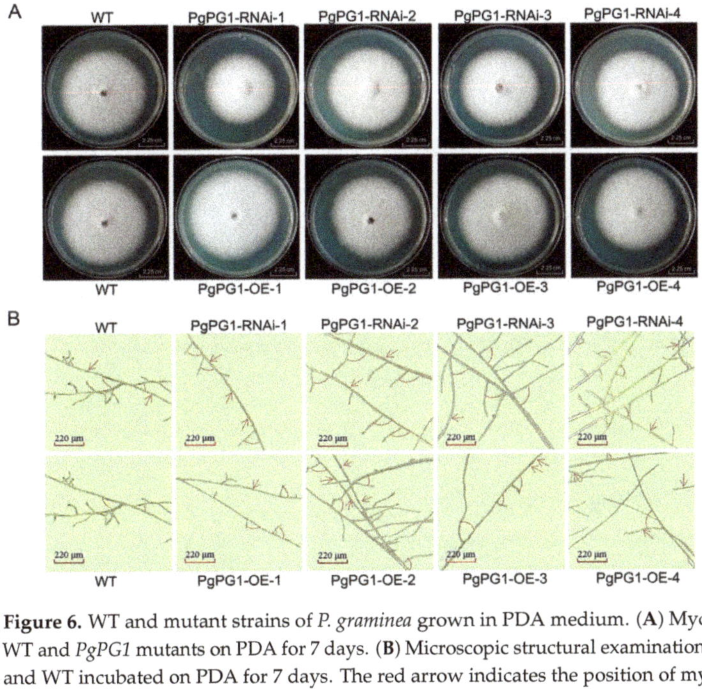

Figure 6. WT and mutant strains of *P. graminea* grown in PDA medium. (**A**) Mycelia growth of the WT and *PgPG1* mutants on PDA for 7 days. (**B**) Microscopic structural examination of *PgPG1* mutants and WT incubated on PDA for 7 days. The red arrow indicates the position of mycelium distortion.

Polygalacturonase activity was measured in the *PgPG1* mutant to assess the role of the *PgPG1* gene in PG activity. The results showed that *PgPG1-OE-1* had 6.11% higher PG activity than the wild-type, while *PgPG1-RNAi-1* had 9.38% lower PG activity (Figure 7D).

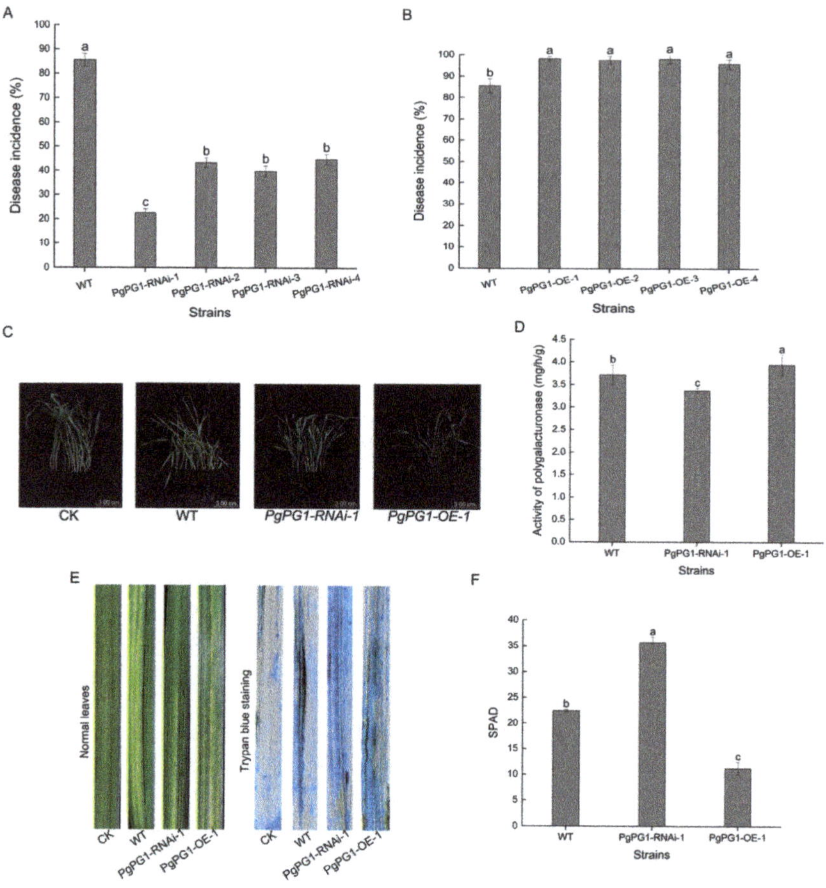

Figure 7. Pathogenicity detection of *PgPG1* mutant strains of *P. graminea*. (**A**,**B**) Disease incidence of barley that had been inoculated with WT and mutant strains for 20 days at 18 °C. (**C**) Growth trends of barley seeds inoculated with fungal mycelium from WT and transformants for 20 days at 4 °C. Barleys were observed after sowing for 14 days. The CK is Alexis plants uninfected by *P. graminea*. (**D**) Activity evaluation of the secreted polygalacturonase of WT, *PgPG1-RNAi-1*, and *PgPG1-OE-1*. (**E**) Trypan blue staining to detect the degree of cell necrosis in WT, *PgPG1-RNAi-1*, and *PgPG1-OE-1* leaves. The CK is Alexis leaves uninfected by *P. graminea*. (**F**) The content of SPAD in Alexis infected by WT, *PgPG1-RNAi-1*, and *PgPG1-OE-1*, respectively. Different letters indicate a significant difference between the WT and transformants ($p < 0.05$). Bars denote standard errors from three repeated experiments.

2.8. Impact of PgPG1 on Pathogenicity

The disease incidence of the barley cultivar Alexis was recorded for the infection of RNAi and overexpression mutants. Infections with RNAi mutants *PgPG1-RNAi-1*, *PgPG1-RNAi-2*, *PgPG1-RNAi-3*, and *PgPG1-RNAi-4* resulted in disease incidences of 22.58%, 43.33%, 39.78%, and 44.85%, respectively (Figure 7A). Infections with overexpression mutants *PgPG1-OE-1*, *PgPG1-OE-2*, *PgPG1-OE-3*, and *PgPG1-OE-4* led to disease incidences of 98.25%, 97.62%, 98.15%, and 95.83%, respectively (Figure 7B). Relative to the wild-type, disease incidence decreased by 62.94%, 42.19%, 45.74%, and 40.67% for RNAi mutants, and increased by 12.73%, 12.10%, 12.63%, and 10.31% for overexpression mutants. These findings indicate that the *PgPG1* gene plays a significant role in the pathogenicity of *P. graminea*.

The growth of plants uninfected by *P. graminea* and those infected by different strains were observed. Because these plants were not infected by *P. graminea*, the growth of the plant in CK (Alexis plants uninfected by *P. graminea*) is the best in Figure 7C; plants infected with *PgPG1-RNAi-1* performed the second best, and plants infected with *PgPG1-OE-1* performed the worst (Figure 7C). To further assess tissue damage, trypan blue dye was used to stain entire barley leaves infected with the wild-type, *PgPG1-RNAi-1*, and *PgPG1-OE-1* strains. Following staining and decolorizing with chloral hydrate, leaves infected with *PgPG1-OE-1* showed greater damage than the control group. In contrast, leaves infected with *PgPG1-RNAi-1* exhibited minimal damage, and no staining was observed in the leaves of the control group (Figure 7E). We examined the chlorophyll content of the leaves infected by the wild-type, *PgPG1-RNAi-1*, and *PgPG1-OE-1* strains, and the SPAD value of leaves infected by *PgPG1-RNAi-1* was the highest; however, that of leaves infected by *PgPG1-OE-1* was the lowest (Figure 7F).

3. Discussion

Fungal pathogens interact with host plants by degrading cell walls, often using PG enzymes as the initial tools for breaking down these structures [21]. We identified and characterized the *PgPG1* gene from *P. graminea*, evaluating its role in growth, infection, and pathogenicity. This work provides insights into the molecular mechanisms underlying fungal pathogenicity, emphasizing the importance of PG genes in host–pathogen interactions.

Polygalacturonase genes play varied roles in fungal pathogenicity. For example, necrotrophic fungi often possess more *GH28* genes than non-pathogenic or biotrophic fungi [22]. *Phytophthora cinnamomi* contains a large *endoPG* gene family with 19 members [23]. In *P. graminea*, we identified four *PgPG* genes at the whole-genome level, demonstrating extensive diversification through differences in intron–exon structures.

Expression analysis during the infection stages revealed that *PgPG1* and *PgPG3* were upregulated over time, with *PgPG1* exhibiting the highest relative increase. This pattern suggests that *PgPG1* is integral to degrading polygalacturonic acid, a primary component of plant cell middle lamellae. These findings identify *PgPG1* as a key player in *P. graminea* pathogenicity.

Fungal *PG* genes exhibit functional differentiation in their roles in pathogenicity. Some studies have demonstrated that certain *PG* genes contribute significantly to fungal virulence, including *pecA* gene in *Aspergillus flavus* [24], *BcPG1* in *Botrytis cinerea* [25], *PG1* in *Fusarium oxysporum* [14], *CpPG1* and *CpPG2* in *Claviceps purpurea* [16], the *PGLA* gene in *Xylella fastidiosa* [17], and many others. Conversely, other studies have shown that some *PG* genes do not contribute to pathogenicity, including *enPG1* in *Cryphonectria parasitica* [18], *PG5* in *F. oxysporum* [19], and *AaPG1* in *A. alternat* [15]. This apparent discrepancy may stem from the functional redundancy of *PG* genes. These findings highlight the controversial role of the *PG* gene in host–pathogen interactions.

Cell-wall-degrading enzymes serve as tools for pathogenic fungi to breach host defenses and as elicitors that trigger plant immune responses, acting as critical warning signals [26]. For instance, PG from *Sclerotinia sclerotiorum* can induce plant cell death [27]. Secreted effectors from diverse organisms have been extensively studied and linked to various physiological functions, including cell proliferation, immune responses, and pathogenic mechanisms [28,29]. To fully understand PG's role in plant–pathogen interactions, it is essential to consider its pathogenic function and capacity to induce immune responses.

In *B. cinerea*, at least five PGs have been confirmed to trigger immune responses in host plants [30]. Similarly, proteins like VmE02 from *Valsa mali* and cerato-platanin SsCP1 from *Sclerotinia sclerotiorum* can induce cell death independently of signal peptide presence [31]. This study identified that *PgPG1* possesses secretory activity and induces cell death. Moreover, subcellular localization studies revealed *PgPG1*'s presence in the cell membrane and nucleus, where maybe the interaction regions between PgPG1 and the host cause an immune response. These findings underscore the importance of the apoplastic space in *PgPG1*'s functional activity, providing new insights into its contribution to *P. graminea* pathogenicity.

The pathogenic process of fungal infection in host plants is complex, with successful invasion and establishment within the host being critical for the pathogen's impact [32,33]. Polygalacturonase, secreted by pathogenic fungi, plays a pivotal role by degrading the middle lamella and the intercellular layers of the plant's primary cell wall, facilitating the pathogen's smooth infiltration [34]. Research has consistently shown the strong association between *PG*s and fungal pathogenicity. For instance, mutating the *PG* gene in *Alternaria citri* significantly reduces the incidence of citrus and horse boll infections [15]. Similarly, the loss of *BcPG1* in *Botrytis cinerea* markedly decreases pathogenicity across various host plants [35]. In *Claviceps purpurea*, deletion of the *endo-PG* gene results in near-complete loss of pathogenicity to rye [16]. In this study, we explored the role of *PgPG1* in *P. graminea* by generating RNAi and overexpression mutants, which revealed its essential function in infection-related morphogenesis. Our results demonstrated that the *PgPG1-RNAi* mutants exhibited reduced pathogenicity, while the *PgPG1-OE* mutants displayed enhanced pathogenicity in barley. These findings were corroborated by trypan blue staining and polygalacturonase activity assays, which confirmed *PgPG1*'s involvement in facilitating pathogenicity.

In conclusion, we identified *PgPG1* as a key pathogenicity factor in *P. graminea*. Our study highlights its critical roles in fungal growth, host infection, and disease development, advancing our understanding of the mechanisms underlying fungal pathogenicity.

4. Materials and Methods

4.1. Plant Materials, Strains, and Culture Conditions

Barley cultivar Alexis was used for inoculation assays. The *P. graminea* strain QWC (wild-type) was cultured at 25 °C on PDA medium, comprising 200 g/L potato decoction, 20 g/L D-glucose anhydrous, and 17 g/L agar.

4.2. Identification of PG Genes from the P. graminea Genome

Genomic data of *P. graminea* were downloaded from NCBI (BioProject: PRJNA373892). The Hidden Markov Model corresponding to the peptidase Glyco_hydro_28 domain (PF00295) was obtained from the Pfam database. Candidate *PG* genes were identified using HMMER 3.3.2 [36] with an E-value threshold $<e^{-10}$, and their Glyco_hydro_28 conserved domain was confirmed using CDD research [37].

4.3. Sequence Feature, Gene Structure, and Motif Analysis

Protein-coding sequences (CDSs) and amino acid sequences of PG were extracted from the *P. graminea* genome. ProtParam (https://web.expasy.org/protparam/, accessed on 14 September 2021) [38] was used to calculate properties such as amino acid count, molecular weight, isoelectric point (pI), grand average of hydropathicity (GRAVY) index, instability index, and aliphatic index of PgPG proteins. SignalP6.0 [39] and Cell-PLoc (v2.0) [40] were used to predict signal peptides and subcellular locations. Exon–intron positions were mapped with TBtools (v2.121) [41], while conserved motifs were analyzed using MEME with default parameters (motif set to 10) and visualized using TBtools.

4.4. Phylogenetic Tree Construction

Protein sequences with locus IDs were retrieved from NCBI to perform a phylogenetic analysis of PgPG and other pathogenic fungi. Phylogenetic trees were constructed using MEGA 7 with the maximum likelihood and neighbor-joining methods, and graphical representations were refined using the Evolview website.

4.5. RNA Extraction and Gene Expression Assay

For gene expression analysis, sterilized barley seeds (cultivar Alexis) were inoculated by sandwiching them between two layers of *P. graminea* mycelia grown for seven days and then incubating at 6 °C. Infected seed embryos were collected at 7, 14, and 18 d post-infection and rapidly frozen in liquid nitrogen. Mycelia pre-infection were used as the control. RNA was extracted using a TRIzol RNA extraction kit, and cDNA was synthesized using a FastKing One Step RT-qPCR kit (Tiangen, Beijing, China, FP313). Relative expression levels were calculated using the $2^{-\Delta\Delta Ct}$ method. All primers used are listed in Table 3, and qPCR analysis was performed with three independent biological replicates.

Table 3. Primers information used in the study.

Code	Primer	Sequence (5′–3′)	Restriction Site	Purpose
1	*PgPG1*^{SP}-*Eco*R I-F	CCGGAATTCATGCGTTCTAA-CGGGATTTT	*Eco*R I	Clone Pgpg1 to pSUC2 for signal peptide functional validation
2	*PgPG1*^{SP}-*Xho* I-R	CCGCTCGAGGGCCTGC-CCTACGGCGGCAA	*Xho* I	
3	*PgPG1*-SL-*Bam*H I-F	CGGGATCCATGCGTTC-TAACGGGATTTT	*Bam*H I	Clone Pgpg1 to EGFP for subcellular localization assay in *N. benthamiana*
4	*PgPG1*-SL-*Sma* I-R	TCCCCCGGGCGCTGCTAGC-TTGGAAACTG	*Sma* I	
5	*PgPG1*-RNAi-*Kpn* I-*Xho* I-F	GGGGTACCCCGCTCGAGG-AGCAGCACCAACATCAAGA	*Kpn* I-*Xho* I	Clone RNAi segment to pSilent-1 for the construction of pSilent-1: *Pgpg1*
6	*PgPG1*-RNAi-*Bgl* II-*Hind* III-R	GAAGATCTCCCAAGCTTCCA-CCGCCAGTAATATCCAC	*Bgl* II-*Hind* III	
7	*PgPG1*-OE-*Eco*R I-F	GGAATTCATGCGTTCTAA-CGGGATTTT	*Eco*R I	Clone *Pgpg1* to pBARGPE1-Hyg-mCherryfor the construction of pBARGPE1-Hyg: *Pgpg1*
8	*PgPG1*-OE-*Xho* I-R	CCGCTCGAGCGCTGCTA-GCTTGGAAACTG	*Xho* I	
9	*Hyg B*-F	ATGAAAAAGCCTGAACTCAC		Generation of *Pgpg1* mutants by the amplificationg segment of hygromycin resistance gene in *Pyrenophora graminea*
10	*Hyg B*-R	CTATTCCTTTGCCCTCGGA		
11	*PgPG1*-RNAi-RT-F	TCAAGATCCGCGACAGCAT		Expression level analysis of *Pgpg1* genes in RNAi mutant of *Pyrenophora graminea*
12	*PgPG1*-RNAi-RT-R	TGGCTGCCGTTGGAGTACA		
13	*PgPG1*-OE-RT-F	CTGGACATTTACCACATCACTCTGA		Expression level analysis of *Pgpg1* genes in OE mutants of *Pyrenophora graminea*
14	*PgPG1*-OE-RT-R	CGTCAAATCCGTCGCTGTTA		
15	*Actin*-F	GCGGTTACACCTCTCTACCAC		Expression level analysis of internal control gene *Actin* in *Pyrenophora graminea*
16	*Actin*-R	AGTCTGGATCTCCTGCTCAAAG		

4.6. Plasmid Construction

Primers for the RNAi segments, including positive and reverse segments, to amplify the *PgPG1* gene were created using the E-RNAi (v3.2) based on the CDS region of the *PgPG1* gene, including responsive restriction sites. The positive segment included *Kpn* I and *Xho* I sites, and the reverse segment included *Bgl* II and *Hind* III sites (Table 3; Figure S1). The amplification process comprised 1 μL cDNA template, 1 μL each of forward and reverse primers, 12.5 μL Premix Taq, and 9.5 μL ddH$_2$O. The RNAi segments were cloned following the protocol for the *PgPG1* gene clone. The RNAi segments and pSilent-1 vector were digested using *Kpn* I/*Xho* I and *Bgl* II/*Hind* III, respectively. Initially, the forward RNAi segment was ligated into pSilent-1 using T4 ligase, and then the reverse RNAi segment was ligated into the pSilent-1 vector containing the forward RNAi segment, resulting in the construction of RNAi pSilent-1: *PgPG1*. For the *PgPG1* gene clone, the fragment was digested with *Xho* I and *Eco*R I (Table 3; Figure S1) and subsequently ligated into the pBARGPE1-Hyg vector, resulting in the construction of the overexpression plasmid pBARGPE1-Hyg: *PgPG1*.

4.7. Transient Expression of PgPG1 in Nicotiana Benthamiana

The PVX: *PgPG1* construct was transformed into *A. tumefaciens* GV3101 and cultured in LB medium at 28 °C at 200 rpm for 48 h. The bacterial pellets were collected after centrifugation at 3500 rpm for 5 min, resuspended in the infiltration buffer (10 mM MgCl$_2$, 10 mM MES, 100 μM AS; pH = 5.7), centrifuged again at 3500 rpm for 5 min, and adjusted to OD$_{600}$ = 0.4. The *Agrobacterium* suspension was infiltrated into the abaxial side of 3-week-old *N. benthamiana* leaves using a 1 mL syringe without a needle. Plants were grown under 16 h light/8 h dark conditions at 24 °C, and symptoms were observed 3–5 days post-infiltration.

4.8. Signal Peptide Functional Validation

Primers for *PgPG1* PCR amplification, including gene clones and subcellular locations, were designed using DNAMAN (v5.2.2). After PCR amplification, the target fragment of the *PgPG1* gene was recovered using a DNA recovery kit (Tiangen, Beijing, China, DP204), ligated to a T-vector, and transformed into *E. coli* DH5α. Following blue–white selection, positive clones were identified, and the *E. coli* cultures containing these clones were sent to a professional sequencing company (Tsing ke, Xi'an, China) for sequencing.

The yeast signal sequence trap system was used to verify the secretion function of the PgPG1 signal peptide. The YTK12 strain, a sucrase-deficient mutant, was used for functional validation. The chemically synthesized *PgPG1* signal peptide sequence was cloned into the pSUC2 vector, resulting in the pSUC2: *PgPG1*^{SP} construct (Table 3; Figure S1). This construction, the negative control vector pSUC2, and the positive control vector pSUC2-Avr1b were transformed into YTK12, respectively. If the PgPG1 signal peptide exhibited secretory functionality, the sucrase from YTK12 would be secreted successfully outside the cell. The hydrolysis of sucrose into monosaccharides by the secreted sucrase would then react with 2, 3, 5-triphenyl tetrazolium chloride (TTC) to produce a red triphenyl tetrazolium chloride precipitate, confirming the peptide's secretion capability.

4.9. Subcellular Localization

The *PgPG1* CDS was inserted into the pCAMBIA1300-35S-EGFP vector (Table 3; Figure S1). This recombinant vector (pCAMBIA1300-35S-EGFP: *PgPG1*) was transformed into *N. benthamiana* leaves using an *Agrobacterium*-mediated transformation method. The leaves of 3–4-week-old tobacco plants were infiltrated and subsequently incubated in a controlled chamber at 22 °C with a 12 h light/dark cycle for 48–72 h. The distribution of green fluorescence in the transformed tobacco leaves was observed and photographed using a laser confocal scanning microscope with an excitation wavelength of 488 nm.

4.10. Protoplast Transformation

Protoplast preparation and genetic transformation were mainly referenced to Aragona and Porta-Puglia [42]. Mycelia cultured for 48 h were harvested by filtration through a sterilized funnel, washed three times, and blotted dry using sterile filter paper. Approximately 0.1 g of mycelia was added to 10 mL enzymatic hydrolysate in a wide-mouth conical flask and incubated at 30 °C with shaking at 80 rpm for 3 h. The enzymatic hydrolysis mixture was filtered using sterilized cellophane, and the filtrate was centrifuged at 1996 rpm for 5 min, discarding the supernatant. Protoplasts were obtained by washing twice with 300 μL of 1.2 mol/L NaCl and resuspended in 100 μL NTC for transformation.

pSilent-1: *PgPG1* DNA (1–10 μg) was added to the 100 μL protoplast suspension, blended, and incubated on ice for 20 min. Sequential additions of 200 μL, 200 μL, and 800 μL PTC were made, mixing well after each addition. Following a 15 min rest at room temperature, 3 mL rPD medium was added and gently mixed. The culture was incubated at 25 °C for 24 h and subsequently mixed with 20 mL rPDA medium before being incubated at 25 °C.

4.11. Selection and Identification of Mutants

Once mycelia appeared on PDA plates, the plates were overlaid with 10 mL PDA medium containing Hygromycin B. Single colonies were picked and subcultured on fresh PDA containing Hygromycin B for three consecutive cycles, resulting in the isolation of transformants. Positive transformants for RNAi and overexpression were identified through Hygromycin phosphotransferase gene segment amplification and qRT-PCR using the primers listed in Table 3.

4.12. Morphology Observation and Transformant Colony Diameter Measurement

Agar disks (5 mm diameter) containing mycelia were placed on blank PDA plates and incubated in the dark at 25 °C. Colony morphology was observed and photographed, and colony diameters were measured. Three biological replicates were used for each strain.

After placing the mycelia in the center of the PDA plate, a coverslip was inserted into the PDA at a 45° angle, 1 cm away from the mycelia. Mycelium morphology was observed under an optical microscope when the mycelia covered two-thirds of the coverslip.

4.13. Pathogenicity Testing of Mutants

The highly susceptible barley cultivar Alexis was used to test the pathogenicity of the transformants. Surface-sterilized Alexis seeds were sandwiched between PDA plates covered with 7-day-old mycelia and incubated in the dark at 25 °C for 20 days. The thirty infected seeds were then planted in pots containing a mixture of vermiculite and nutrient soil. The plants were grown under controlled conditions in a phytotron with alternating 12 h of light at 16 °C and 12 h of dark at 12 °C cycles. When stripe symptoms appeared, diseased plants were counted, and disease incidence was calculated for three biological replicates.

4.14. Trypan Blue Staining

Leaves infected for three weeks were harvested, rinsed with water, and stained with 0.4% (*w*/*v*) trypan blue solution for 3 min in boiling water. The staining solution was then removed before applying a decolorizing solution. Leaves were decolorized for 48 h until they appeared transparent.

4.15. Determination of PG Activity

Wild-type and mutant strains were cultured on PDA medium at 25 °C for seven days. The mycelia were then harvested for PG activity analysis, following the protocol outlined in the polygalacturonase test kit (Aoqing, Nanjing, China, SP7096).

5. Conclusions

In this study, we identified four PgPG genes (PgPG1-PgPG4) in the *P. graminea* genome and obtained the highest inducible expression of the *PgPG1* gene during barley infection. In addition, the PgPG1 was verified to be a secretory protein and had the ability to cause immune response. The *PgPG1* gene was involved in the pathogenicity of *P. graminea* in barley and has great potential as a pathogen target gene to develop a durable resistance variety to *P. graminea*.

Supplementary Materials: The following supporting information can be downloaded at: https://www.mdpi.com/article/10.3390/ijms26062401/s1.

Author Contributions: H.W., Y.M. and C.L. designed the research; E.S., M.G., H.L., J.W., L.Y., B.L. and X.S. performed the research; E.S., M.G., H.L., X.M. and K.Y. analyzed the data; E.S. and M.G. wrote the manuscript. All authors have read and agreed to the published version of the manuscript.

Funding: This work was supported by the Gansu Provincial Key Laboratory of Aridland Crop Science, Gansu Agricultural University (No. GSCS-2018-6); the China Agriculture Research System of MOF and MARA (CARS-05-02A-02); the National Natural Science Foundation of China (No. 32160496); the Industrial Support Project of the Gansu Provincial Education Department (2021 CYZC-12); the Innovation Fund Project of the Education Department of Gansu Province (No. 2021A-055); the Gansu Province Youth Science and Technology Fund (No. 20JR5RA010); and the Gannan Technology Plan Project (22CX8NA047).

Institutional Review Board Statement: Not applicable.

Informed Consent Statement: Not applicable.

Data Availability Statement: Data can be provided by the corresponding author upon reasonable request.

Acknowledgments: The authors thank the editor and the anonymous reviewers for their efforts to improve the manuscript.

Conflicts of Interest: The authors declare no conflicts of interest regarding this article.

References

1. Valè, G.; Tacconi, G.; Francia, E.; Dall'Aglio, E.; Govoni, C.; Pecchioni, N.; Arru, L.; Delogu, G.; Haegi, A.; Porta-Puglia, A.; et al. Advances in understanding barley *Pyrenophora graminea* interactions. In Proceedings of the 2nd International Workshop on Barley Leaf Blights, Aleppo, Syria, 7–11 April 2002.
2. Porta-Puglia, A.; Delogu, G.; Vannacci, G. *Pyrenophora graminea* on Winter Barley Seed: Effect on Disease Incidence and Yield Losses. *J. Phytopathol.* **1986**, *117*, 26–33. [CrossRef]
3. Si, E.J.; Meng, Y.X.; Ma, X.L.; Li, B.C.; Wang, J.C.; Yao, L.R.; Yang, K.; Yu Zhang, Y.; Shang, X.W.; Wang, H.J. Genome resource for barley leaf stripe pathogen *Pyrenophora graminea*. *Plant Dis.* **2020**, *104*, 320–322. [CrossRef]
4. Ackah, M.; Boateng, N.A.S.; Foku, J.M.; Ngea, G.L.N.N.; Godana, E.A.; Zhang, H.Y.; Yang, Q.Y. Genome-wide analysis of the PG gene family in *Penicillium expansum*: Expression during the infection stage in pear fruits (*Pyrus bretschneideri*). *Physiol. Mol. Plant Pathol.* **2024**, *131*, 102270. [CrossRef]
5. Liang, Q.Q.; Li, B.C.; Wang, J.C.; Ren, P.R.; Yao, L.R.; Meng, Y.X.; Shang, X.W.; Wang, H.J. *PGPBS*, a mitogen-activated protein kinase kinase, is required for vegetative differentiation, cell wall integrity, and pathogenicity of the barley leaf stripe fungus *Pyrenophora graminea*. *Gene* **2019**, *696*, 95–104. [CrossRef] [PubMed]
6. Haegi, A.; Bonardi, V.; Dall'Aglio, E.; Glissant, D.; Tumino, G.; Collins, N.C.; Bulgarelli, D.; Infantino, A.; Stanca, M.; Mdelledonne, M.; et al. Histological and molecular analysis of *Rdg2a* barley resistance to leaf stripe. *Mol. Plant Pathol.* **2008**, *9*, 463–478. [CrossRef]
7. Si, E.J.; Meng, Y.X.; Ma, X.L.; Li, B.C.; Wang, J.C.; Ren, P.R.; Yao, L.R.; Yang, K.; Zhang, Y.; Shang, X.W.; et al. Development and characterization of microsatellite markers based on whole-genome sequences and pathogenicity differentiation of *Pyrenophora graminea*, the causative agent of barley leaf stripe. *Eur. J. Plant Pathol.* **2019**, *154*, 227–241. [CrossRef]
8. Xia, K.F.; Pan, X.Q.; Zeng, X.; Zhang, M.Y. *Xoo*-responsive transcriptome reveals the role of the circular RNA133 in disease resistance by regulating expression of *OsARAB* in rice. *Phytopathol. Res.* **2023**, *5*, 33. [CrossRef]

9. Haeger, W.; Pauchet, Y.; Kirsch, R. Evolution of polygalacturonases and polygalacturonase-inhibiting proteins: A view beyond the classical perspective. *Annu. Plant Rev.* **2022**, *5*, 81–122.
10. Xu, M.; Wang, Q.H.; Wang, G.H.; Zhang, X.; Liu, H.Q.; Jiang, C. Combatting Fusarium head blight: Advances in molecular interactions between *Fusarium graminearum* and wheat. *Phytopathol. Res.* **2022**, *4*, 37. [CrossRef]
11. Hamann, T. Plant cell wall integrity maintenance as an essential component of biotic stress response mechanisms. *Front. Plant Sci.* **2012**, *3*, 77. [CrossRef]
12. Duan, W.K.; Huang, Z.N.; Song, X.M.; Liu, T.K.; Liu, H.L.; Hou, X.L.; Li, Y. Comprehensive analysis of the polygalacturonase and pectin methylesterase genes in *Brassica rapa* shed light on their different evolutionary patterns. *Sci. Rep.* **2016**, *6*, 25107. [CrossRef]
13. Shieh, S.Y.; Ikeda, M.; Taya, Y.; Prives, C. DNA damage-induced phosphorylation of p53 alleviates inhibition by MDM2. *Cell* **1997**, *91*, 325–334. [CrossRef] [PubMed]
14. Di Pietro, A.; Roncero, M.I.G. Cloning, expression, and role in pathogenicity of pg1 encoding the major extracellular endopolygalacturonase of the vascular wilt pathogen *Fusarium oxysporum*. *Mol. Plant-Microbe Interact.* **1998**, *11*, 91–98. [CrossRef]
15. Isshiki, A. Endopolygalacturonase is essential for citrus black rot caused by *Alternaria citri* but not brown spot caused by *Alternaria alternata*. *Mol. Plant-Microbe Interact.* **2001**, *14*, 749–757. [CrossRef]
16. Oeser, B.; Heidrich, P.M.; Müller, U.; Tudzynski, P.; Tenberge, K.B. Polygalacturonase is a pathogenicity factor in the *Claviceps purpurearye* interaction. *Fungal Genet. Biol.* **2002**, *36*, 176–186. [CrossRef] [PubMed]
17. Roper, M.C.; Greve, L.C.; Warren, J.G.; Labavitch, J.M.; Kirkpatrick, B.C. *Xylella fastidiosa* requires polygalacturonase for colonization and pathogenicity in *Vitis vinifera* grapevines. *Mol. Plant-Microbe Interact.* **2007**, *20*, 411–419. [CrossRef]
18. Gao, S.; Choi, G.H.; Shain, L.; Nuss, D.L. Cloning and targeted disruption of enpg-1, encoding the major in vitro extracellular endopolygalacturonase of the chestnut blight fungus, *Cryphonectria parasitica*. *Appl. Environ. Microb.* **1996**, *62*, 1984–1990. [CrossRef] [PubMed]
19. García-Maceira, F.I.; Di Pietro, A.; Huertas-González, M.D.; Ruiz-Roldán, M.C.; Roncero, M.I.G. Molecular characterization of an endopolygalacturonase from *Fusarium oxysporum* expressed during early stages of infection. *Appl. Environ. Microb.* **2001**, *67*, 2191–2196. [CrossRef]
20. Mahmood, U.; Fan, Y.H.; Wei, S.Y.; Niu, Y.; Li, Y.H.; Huang, H.L.; Yuling Chen, Y.L.; Tang, Z.L.; Liu, L.Z.; Qu, C.M. Comprehensive analysis of polygalacturonase genes offers new insights into their origin and functional evolution in land plants. *Genomics* **2021**, *113*, 1096–1108. [CrossRef]
21. Chakraborty, A.; Fernando, L.D.; Fang, W.X.; Widanage, M.C.D.; Wei, P.Z.; Jin, C.; Fontaine, T.; Latgé, J.P.; Wang, T. A molecular vision of fungal cell wall organization by functional genomics and solid-state NMR. *Nat. Commun.* **2021**, *12*, 6346. [CrossRef]
22. Sprockett, D.D.; Piontkivska, H.; Blackwood, C.B. Evolutionary analysis of glycosyl hydrolase family 28 (GH28) suggests lineage-specific expansions in necrotrophic fungal pathogens. *Gene* **2011**, *479*, 29–36. [CrossRef]
23. Wu, C.H.; Yan, H.Z.; Liu, L.F.; Liou, R.F. Functional characterization of a gene family encoding polygalacturonases in *Phytophthora parasitica*. *Mol. Plant-Microbe Interact.* **2008**, *21*, 480–489. [CrossRef]
24. Pennerman, K.K.; Yin, G.; Glenn, A.E.; Bennett, J.W. Identifying candidate Aspergillus pathogenicity factors by annotation frequency. *BMC Microbiol.* **2020**, *20*, 1–11. [CrossRef]
25. Have, A.; Mulder, W.; Visser, J.; Kan, J.A.L.V. The endopolygalacturonase gene Bcpg1 is required for full virulence of *Botrytis cinerea*. *Mol. Plant-Microbe Interact.* **1998**, *11*, 1009–1016. [CrossRef]
26. Yang, C.; Liu, R.; Pang, J.H.; Ren, B.; Zhou, H.B.; Wang, G.; Wang, E.T.; Liu, J. Poaceae-specific cell wall-derived oligosaccharides activate plant immunity via *OsCERK1* during *Magnaporthe oryzae* infection in rice. *Nat. Commun.* **2021**, *12*, 2178. [CrossRef]
27. Zuppini, A.; Navazio, L.; Sella, L.; Castiglioni, C.; Favaron, F.; Mariani, P. An endopolygalacturonase from *Sclerotinia sclerotiorum* induces calcium-mediated signaling and programmed cell death in soybean cells. *Mol. Plant-Microbe Interact.* **2005**, *18*, 849–855. [CrossRef]
28. Chen, X.M.; Pan, S.; Bai, H.M.; Fan, J.X.; Batool, W.; Shabbir, A.; Han, Y.J.; Zheng, H.K.; Lu, G.D.; Lin, L.L.; et al. A nonclassically secreted effector of *Magnaporthe oryzae* targets host nuclei and plays important roles in fungal growth and plant infection. *Mol. Plant Pathol.* **2023**, *24*, 1093–1106. [CrossRef]
29. Maricchiolo, E.; Panfili, E.; Pompa, A.; Marchis, F.D.; Bellucci, M.; Pallotta, M.T. Unconventional pathways of protein secretion: Mammals vs. plants. *Front. Cell Dev. Biol.* **2022**, *10*, 895853. [CrossRef]
30. Silva, C.J.; Adaskaveg, J.A.; Mesquida-Pesci, S.D.; Ortega-Salazar, I.B.; Pattathil, S.; Zhang, L.S.; Hahn, M.G.; Kan, J.A.L.V.; Cantu, D.; Powell, A.L.T.; et al. *Botrytis cinerea* infection accelerates ripening and cell wall disassembly to promote disease in tomato fruit. *Plant Physiol.* **2023**, *191*, 575–590. [CrossRef]
31. Nie, J.J.; Yin, Z.Y.; Li, Z.P.; Wu, Y.X.; Huang, L.L. A small cysteine-rich protein from two kingdoms of microbes is recognized as a novel pathogen-associated molecular pattern. *New Phytol.* **2019**, *222*, 995–1011. [CrossRef]
32. Feurtey, A.; Lorrain, C.; McDonald, M.C.; Milgate, A.; Solomon, P.S.; Warren, R.; Puccetti, G.; Scalliet, G.; Torriani, S.F.F.; Gout, L.; et al. A thousand-genome panel retraces the global spread and adaptation of a major fungal crop pathogen. *Nat. Commun.* **2023**, *14*, 1059. [CrossRef]

33. Jaroszuk-Ściseł, J.; Kurek, E. Hydrolysis of fungal and plant cell walls by enzymatic complexes from cultures of *Fusarium* isolates with different aggressiveness to rye (*Secale cereale*). *Arch. Microbiol.* **2012**, *194*, 653–665. [CrossRef]
34. D'Ovidio, R.; Mattei, B.; Roberti, S.; Bellincampi, D. Polygalacturonases, polygalacturonase-inhibiting proteins and pectic oligomers in plant-pathogen interactions. *BBA-Proteins Proteom.* **2004**, *1696*, 237–244. [CrossRef]
35. Kars, I.; Krooshof, G.H.; Wagemakers, L.; Joosten, R.; Benen, J.A.E.; Kan, J.A.L.V. Necrotizing activity of five *Botrytis cinerea* endopolygalacturonases produced in *Pichia pastoris*. *Plant J.* **2010**, *43*, 213–225. [CrossRef]
36. Mistry, J.; Finn, R.D.; Eddy, S.R.; Bateman, A.; Punta, M. Challenges in homology search: HMMER3 and convergent evolution of coiled-coil regions. *Nucleic Acids Res.* **2013**, *41*, e121. [CrossRef]
37. Marchler-Bauer, A.; Bo, Y.; Han, L.; He, J.; Lanczycki, C.J.; Lu, S.; Chitsaz, F.; Derbyshire, M.K.; Geer, R.C.; Gonzales, N.R.; et al. CDD/SPARCLE: Functional classification of proteins via subfamily domain architectures. *Nucleic Acids Res.* **2017**, *45*, 200–203. [CrossRef]
38. Gasteiger, E.; Hoogland, C.; Gattiker, A.; Wilkins, M.R.; Appel, R.D.; Bairoch, A. Protein identification and analysis tools in the ExPASy server. In *The Proteomics Protocols Handbook;* Springer Nature: Berlin, Germany, 2005; pp. 571–607.
39. Teufel, F.; Almagro Armenteros, J.J.; Johansen, A.R.; Gíslason, M.H.; Pihl, S.I.; Tsirigos, K.D.; Winther, O.; Brunak, S.; von Heijne, G.; Nielsen, H. SignalP 6.0 predicts all five types of signal peptides using protein language models. *Nat. Biotechnol.* **2022**, *40*, 1023–1025. [CrossRef]
40. Chou, K.; Shen, H. Cell-PLoc: A package of Web servers for predicting subcellular localization of proteins in various organisms. *Nat. Protoc.* **2008**, *3*, 153–162. [CrossRef]
41. Chen, C.; Wu, Y.; Li, J.; Wang, X.; Zeng, Z.; Xu, J.; Liu, Y.; Feng, F.; Chen, H.; He, Y.; et al. TBtools-II: A "one for all, all for one" bioinformatics platform for biological big-data mining. *Mol. Plant* **2023**, *16*, 1733–1742. [CrossRef]
42. Aragona, M.; Porta-Puglia, A. Genetic transformation of the phytopathogenic fungus *Pyrenophora graminea*. *Mycol. Res.* **1993**, *97*, 1143–1147. [CrossRef]

International Journal of
Molecular Sciences

MDPI

Article

A Novel CFEM Effector in *Fusarium verticillioides* Required for Virulence Involved in Plant Immunity Suppression and Fungal Cell Wall Integrity

Huan Li [1], Shumila Ishfaq [1], Xiaoyan Liang [1], Rui Wang [1], Hailei Wei [2] and Wei Guo [1,3,*]

[1] Institute of Food Science and Technology, Chinese Academy of Agricultural Sciences/Key Laboratory of Agro-Products Quality and Safety Control in Storage and Transport Process, Ministry of Agriculture and Rural Affairs, Beijing 100193, China; lihuancaas@126.com (H.L.); 2020y90100028@caas.cn (S.I.); liangxy_97@163.com (X.L.); wangrui_0121@126.com (R.W.)

[2] Institute of Agricultural Resources and Regional Planning, Chinese Academy of Agricultural Sciences, Beijing 100081, China; weihailei@caas.cn

[3] Ningxia Key Laboratory for the Development and Application of Microbial Resources in Extreme Environments, North Minzu University, Yinchuan 750030, China

* Correspondence: guowei01@caas.cn

Abstract: Common in Fungal Extracellular Membrane (CFEM) effectors, a unique class of fungal-specific proteins, play critical roles in host-pathogen interactions. While CFEM proteins have been extensively characterized in phytopathogens, their presence and functions in *Fusarium verticillioides* remained unexplored. Here, we systematically identified 19 CFEM-containing proteins in *F. verticillioides*, among which FvCFEM12 exhibited secretory activity and plant infection-induced expression. Functional characterization revealed that FvCFEM12 suppressed Bax- and INF1-triggered cell death in *Nicotiana benthamiana* leaves. Furthermore, heterologous expression of FvCFEM12 in maize leaves using *P. syringae* strain D36E can compromise immune responses against bacterial pathogens. Deletion of FvCFEM12 impaired fungal virulence, altered hyphal morphology, and reduced cell wall stress tolerance. Interestingly, FvCFEM12 physically interacted with the maize wall-associated receptor kinase ZmWAK17ET, and targeted silencing of ZmWAK17 in maize enhanced susceptibility to *F. verticillioides*. Our findings revealed that FvCFEM12 is a dual-function effector that suppresses plant immunity and maintains fungal cell wall integrity, thereby orchestrating fungal pathogenicity at the host–pathogen interface.

Keywords: *Fusarium verticillioides*; effectors; CFEM domain; pathogenicity; host–pathogen interaction; maize

1. Introduction

Plants have evolved a multitude of immune mechanisms against pathogen infection and colonization during their long-term interactions with pathogens [1]. The plant immune system primarily involves two distinct layers of immunity: pathogen or microbe-associated molecular pattern (PAMP/MAMP)-triggered immunity (PTI) and effector-triggered immunity (ETI) [1,2]. Evolutionarily conserved PAMPs such as bacterial flagellin, elongation factors, and fungal chitin are evolutionarily conserved and recognized by plant transmembrane pattern recognition receptors (PRRs) and trigger PTI. The PTI response typically triggers a cascade of cellular and physiological responses, including reactive oxygen species (ROS) burst, callose deposition, and regulation of defense-related genes [3]. To counteract host defenses, pathogens often secrete effectors to interfere or disrupt the host PTI response.

However, some effectors may be recognized by intracellular nucleotide binding–leucine rich repeat receptors (NB-LRRs) and induce a more robust and rapid defense response known as ETI. ETI usually results in a hypersensitive response (HR) at the infection sites [4]. Despite these defense mechanisms, adapted pathogens can sometimes evade and overcome both PTI and ETI, ultimately leading to successful infection [5].

Fusarium verticillioides is a globally devastating maize pathogen, causing severe ear rot and stalk rot diseases that result in substantial yield losses worldwide [6]. It can infect all parts of maize plants including seeds, roots, leaves, stalks, and ears at every growth stage. *F. verticillioides* can survive in crop residues and soil, serving as the primary inoculum source in natural infections [7,8]. In addition to its detrimental impact on crop productivity, *F. verticillioides* is notorious for producing toxic secondary metabolites, particularly fumonisins, which pose serious health risks to both animals and humans. Among these, Fumonisin B_1 (FB_1) is the most prevalent and hazardous, recognized globally as a major food and feed contaminant with documented systemic toxicity and potential carcinogenicity [6,9,10]. As a hemibiotrophic fungal pathogen, *F. verticillioides* acquires nutrients from both living and dead tissues [11]. At the initial biotrophic phase, hemibiotrophic pathogens usually synthesize and secrete effector proteins to evade or suppress host immune responses [12,13]. The genome of *F. verticillioides* is predicted to encode numerous effector proteins, including cell wall-degrading enzymes, necrosis-inducing proteins, and small cysteine-rich proteins [14]. However, the specific roles of many of these effectors remain elusive. Notably, the secreted protein FvLcp1, containing a LysM domain, has been shown to suppress BAX-triggered plant cell death in *N. benthamiana* [15]. Additionally, a metalloprotease secreted by *F. verticillioides* can cleave maize class IV chitinases, which are involved in plant defense against fungal pathogens [16].

The CFEM (Common in Fungal Extracellular Membrane) protein has only been identified in fungi to date, characterized by eight cysteine residues that are equally spaced (PxC[A/G]x2Cx8-12Cx1-3[x/T]Dx2-5CxCx9-14Cx3-4Cx15-16C), where x can be any amino acid [17]. CFEM proteins exhibit a significant functional diversity across fungal species [18,19]. For instance, in *Saccharomyces cerevisiae*, the CCW14 protein is essential for maintaining cell wall integrity, while in *Aspergillus fumigatus*, CfmA-C proteins play a crucial role in maintaining the integrity and stability of the cell wall [20]. In the human pathogenic fungus *Candida albicans*, three CFEM proteins, Rbt5, Pga7, and Csa2, are part of a system responsible for iron acquisition from heme and hemoglobin [21]. Recently, the CFEM proteins have gained increased attention for their roles in plant–pathogen interactions. There are two types of CFEM proteins: Pth11-like proteins, which contain multiple transmembrane domains, and non-Pth11-like proteins, which are secreted [17]. For example, the CFEM protein Pth11, with seven transmembrane regions, is critical for appressorium formation in *Magnaporthe oryzae*, a specialized structure essential for host tissue penetration [22]. Moreover, the CFEM protein WISH is involved in mycelial growth, cell wall integrity, appressoria development, and virulence in *M. oryzae* [23]. Secreted CFEM proteins have also been identified as a potential effector in some plant pathogenic fungi [24–26]. For instance, BcCFEM1 in *Botrytis cinerea*, VdSCP76/77 in *Verticillium dahliae*, and PstCFEM1 in *Puccinia striiformis* f. sp. tritici have been reported to contribute to virulence, although their precise mechanisms remain poorly understood [27–29]. Notably, in *Colletotrichum fructicola*, the CFEM containing protein CfEC12 binds NPR1 regulator NIMIN2 to suppress plant immunity [30]. Despite these findings, the roles of CFEM proteins in *F. verticillioides* are largely unexplored, and their mechanisms of action remain largely unknown.

In this study, we identified 19 CFEM proteins in the *F. verticillioides* genome, seven of which were demonstrated to be secreted and capable of suppressing BAX-induced cell death. Among these, FvCFEM12 was found to play a critical role in disrupting plant immune responses and contributing to fungal virulence. Furthermore, FvCFEM12 was found to be involved in regulating fungal colony morphology and cell wall synthesis. Finally, to investigate the molecular mechanisms underlying FvCFEM12's function, we identified an interaction between FvCFEM12 and the maize wall-associated kinase ZmWAK17ET. Collectively, our findings were to identify the CFEM protein in *F. verticillioides* and elucidate the molecular function of FvCFEM12 as a candidate effector in fungal physiology and pathogenicity.

2. Results

2.1. Bioinformatic Identification of CFEM Proteins in F. verticillioides

A total of 19 CFEM proteins (designated FvCFEM1-FvCFEM19) were identified in the *F. verticillioides* genome using BLASTp analysis, using ACI1(AAN64312.1) from *M. oryzae* as the query sequence [31]. Subsequent SMART domain analysis using the SMART tool confirmed the presence of a single CFEM domain in all identified proteins. The protein ID number and associated features retrieved from the NCBI database are shown in Table 1 and Figure 1A, respectively. The predicted length of the CFEM-domain proteins ranged from 95 to 1018 amino acids. Among these, seven proteins were predicted to contain an N-terminal signal peptide, indicating their potential secretory function.

Table 1. Identification of the Common Fungal Extracellular Membrane (CFEM) protein family in *F. verticillioides*.

Name	ID	SP [a]	TM [b]	Amino Acid	Position of CFEM Domain (aa)	GPI-Anchored
FvCFEM1	FVEG_16152	1–17	7	454	24–88	
FvCFEM2	FVEG_07166	1–20	6	460	34–98	
FvCFEM3	FVEG_13472	1–19	7	429	25–88	
FvCFEM4	FVEG_08756	1–19	6	431	27–88	
FvCFEM5	FVEG_17608		7	419	12–75	
FvCFEM6	FVEG_10583	1–22	7	442	23–86	
FvCFEM7	FVEG_04843	1–17	0	314	19–88	A^{291}/A^{314}
FvCFEM8	FVEG_06284		1	313	10–82	
FvCFEM9	FVEG_01052	1–17	0	154	18–85	N^{133}/A^{134}
FvCFEM10	FVEG_02239	1–15	0	196	17–84	S^{166}/S^{167}
FvCFEM11	FVEG_13696	1–18	7	423	20–84	
FvCFEM12	FVEG_07535	1–16	0	193	18–85	G^{168}/S^{173}
FvCFEM13	FVEG_08884	1–18	0	179	33–94	
FvCFEM14	FVEG_13561		7	590	173–231	
FvCFEM15	FVEG_17202	1–21	6	415	38–97	
FvCFEM16	FVEG_02591	1–21	1	306	23–82	
FvCFEM17	FVEG_08871	1–18	0	95	31–91	
FvCFEM18	FVEG_07573	1–19	0	245	21–84	D^{219}
FvCFEM19	FVEG_02431		0	853	633–695	A^{773}/G^{774}

[a] Cleavage site of signal peptide (SP) in the FvCFEM proteins. [b] Number of transmembrane domain (TM).

A multiple sequence alignment of the CFEM domains was performed using the amino acid sequences of the identified proteins (Figure 1B). The alignment revealed a high degree of conservation within the CFEM domain across all sequences, with the domain length consistently ranging from 58 to 71 amino acids. Except FvCFEM 11, 14, 15, 16, 18, and 19, most of the CFEM proteins contained eight cysteine residues, which are probably involved in the formation of disulfide bonds to stabilize the whole protein structure of CFEM domain [21].

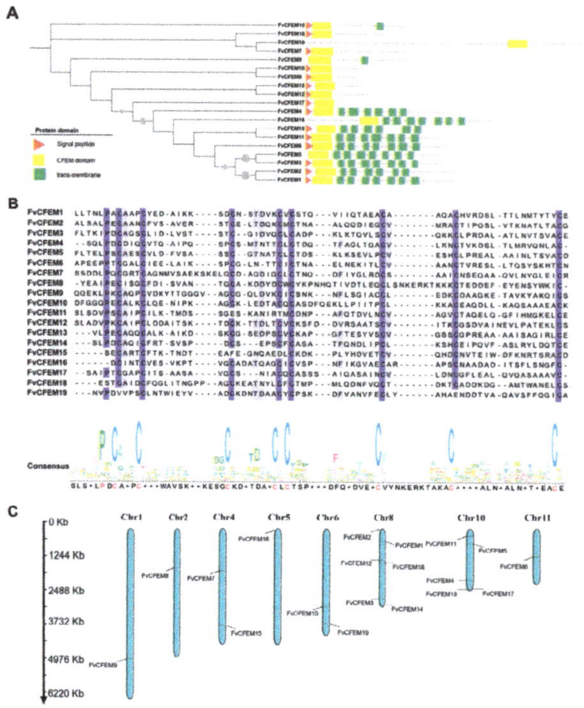

Figure 1. Bioinformatic analysis of Common in Fungal Extracellular Membrane (CFEM) proteins in *F. verticillioides*. (**A**) A neighbor-joining phylogenetic tree was constructed based on the amino acid sequences of FvCFEM proteins. Branches with bootstrap values > 70% are shown, with the size of grey spheres indicating bootstrap values. Protein domain architectures of FvCFEMs are depicted on the right, with different colors representing distinct domains: red triangles for signal peptides (SPs), yellow boxes for CFEM domains and green boxes for transmembrane (TM). (**B**) Multiple sequence alignment of FvCFEM proteins. The alignment was performed using ClustalW 1.1 and visualized using Jalview 2.11.3.2. (**C**) Mapping of the CFEM genes on *F. verticillioides* chromosomes.

To elucidate the chromosomal distribution of putative 19 CFEM genes, we performed genome-wide mapping of *F. verticillioides* using MG2C v2.1 on the website (http://mg2c.iask.in/mg2c_v2.1/index.html, accessed on 20 July 2024). The 19 identified FvCFEM genes were distributed across eight chromosomes, displaying an uneven distribution pattern. Notably, chromosome 8 harbored the highest number of FvCFEM genes (n = 6), while chromosomes 1, 2, 5 and 10 each contained only a single FvCFEM gene (Figure 1C).

2.2. *FvCFEM12 Is an Important Candidate Effector in F. verticillioides CFEM Family*

Fungal pathogen effectors usually contain a signal peptide and lack transmembrane regions, playing crucial roles in pathogenicity [32]. Based on these criteria, seven FvCFEM proteins (FvCFEM7, 9, 10, 12, 13, 17, and 18) were predicted to possess a signal peptide without transmembrane regions. These proteins were considered as candidate effectors and selected for further investigation. To validate the functionality of the predicted signal peptides, we performed a yeast invertase secretion assay [33]. The signal peptides of the seven candidate effectors were successfully cloned into the pSUC2 vector and transformed into yeast strain YTK-12, which is unable to metabolize raffinose as a carbon source. Transformants were streaked onto CMD-W medium and YRPAA medium. Similar to the positive control (Avr1bSP), transformants containing the FvCFEM signal peptide grew on the YRPAA medium (Figure 2A). Furthermore, the enzyme activity of the secreted invertase was confirmed using 2,3,5-triphenyltetrazolium chloride (TTC). Transformants containing signal peptides from Avr1b and FvCFEMs resulted in a red color change in the reaction reagents. These results demonstrated that the predicted signal peptides of FvCFEM proteins are functional and capable of mediating protein secretion.

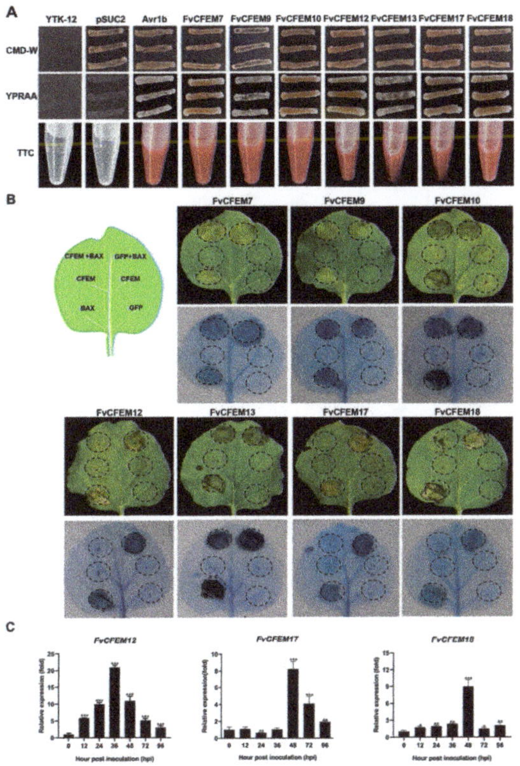

Figure 2. Identification of potential effectors among FvCFEM proteins. (**A**) Functional validation of the putative N-terminal signal peptide of FvCFEM proteins using a yeast signal trap assay. The putative signal peptide sequences of FvCFEMs were fused into the pSUC2 vector, and transformed into *S. cerevisiae* strain YTK12. Growth on CMD-W and YPRAA plates, as well as invertase activity in TTC reagent were assessed. Untransformed YTK12 and the YTK12 carrying the empty pSUC2 vector served as negative controls, while the signal peptide of the oomycete effector Avr1b was used as a positive control. (**B**) Transient expression of seven FvCFEM proteins in *N. benthamiana* leaves 24 h prior to infiltration with *A. tumefaciens* expressing Bax. Photographs were taken at 5 days post-inoculation (dpi), and dead cells were visualized by trypan blue staining. (**C**) Expression profile of FvCFEM12, FvCFEM 17, and FvCFEM 18 during *F. verticillioides* infection of maize. RT-qPCR analysis was performed on samples collected at 0, 12, 24, 36, 48, 72, and 96 h post-inoculation (hpi). *ZmUbiquitin* was used as an internal reference. Error bars represent the standard deviation of three biological replicates. Data are presented as means ± standard error of three biological replicates. *, ** and *** indicate significant differences at $p < 0.05$, $p < 0.01$ and $p < 0.001$, respectively, based on Student's *t*-test.

Some fungal effectors can induce or suppress plant cell death to facilitate infection and survival [34,35]. To investigate whether the candidate FvCFEM proteins modulate programmed cell death (PCD) in plants, we conducted transient expression assays in *N. benthamiana* using an *Agrobacterium*-mediated approach. Negative control (Green fluorescent protein, GFP) and positive control (BCL2-associated X, Bax) were included for comparison, and trypan blue staining was used to visualize cell death. The results revealed that none of the seven candidate FvCFEM proteins induced significant cell death; FvCFEM12, FvCFEM17, and FvCFEM18 were found to suppress Bax-induced PCD (Figure 2B). The expression of these proteins in *N. benthamiana* was confirmed by reverse transcription PCR (RT-PCR) (Figure S1). Based on these results, FvCFEM12, FvCFEM17, and FvCFEM18 were selected for further analysis.

To explore the expression patterns of FvCFEM12, FvCFEM17, and FvCFEM18 during *F. verticillioides* infection, we inoculated maize cultivar B73 with conidia from the wild-type strain FvLNF15-11 and performed quantitative real-time reverse transcription PCR (qRT-PCR) on infected stalk tissues. The results revealed distinct expression profiles for each gene. FvCFEM12 was induced as early as 6 h post-inoculation (hpi) and increased progressively, peaking at 36 hpi. In contrast, FvCFEM17 exhibited high expression levels during the later stages of infection, while FvCFEM18 showed a transient peak at 48 hpi, with low expression at other time points (Figure 2C). These findings suggest that FvCFEM12 may play a critical role during infection.

2.3. *FvCFEM12 Suppresses Cell Death Induced by INF1 and P. syringae DC3000*

To determine whether FvCFEM12 can suppress cell death triggered by INF1, we transiently expressed FvCFEM12 in *N. benthamiana* leaves using *Agrobacterium*-mediated transformation. GFP was used as a negative control. Subsequently, 24 h after infiltration, the leaves were infiltrated with *A. tumefaciens* cells carrying the *INF1* gene. Cell death was significantly suppressed in regions expressing FvCFEM12, whereas no suppression was observed in GFP-expressing control leaves (Figure 3A). These results indicate that FvCFEM12 can inhibit INF1-induced cell death in *N. benthamiana*.

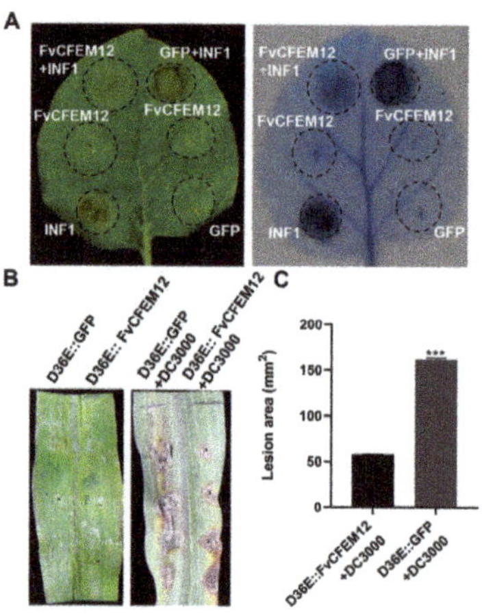

Figure 3. FvCFEM12 suppresses cell death triggered by INF1 and *P. syringae* DC3000. (**A**) FvCFEM12 was transiently expressed in *N. benthamiana* leaves 24 h prior to infiltration with *A. tumefaciens* expressing the *INF1* gene. Photographs were taken at 5 days post-inoculation (dpi), and lesion areas of leaves were visualized by trypan blue staining. (**B**) *P. syringae* strain D36E carrying FvCFEM12 was transiently expressed in maize leaves 24 h prior to infiltration with *P. syringae* strain DC3000. *P. syringae* strain D36E carrying GFP was used as a control. Photographs were taken at 5 dpi after inoculation with *P. syringae* strain DC3000. (**C**) Average lesion area on maize leaves was calculated from three independent biological replicates. Error bars represent the standard deviation (SD). Statistical analysis was determined using Student's test (***, p-value < 0.001).

To further investigate the suppressive activity of FvCFEM12 in the host plant, we utilized a modified bacterial type III secretion system to deliver FvCFEM12 into maize [36]. FvCFEM12 and GFP were expressed in *P. syringae* strain D36E, which lacks all 36 known type III effectors (T3Es) and is non-pathogenic. When FvCFEM12 or GFP was injected alone, neither elicited significant cell death. However, five days after injection of *P. syringae* strain DC3000, regions expressing GFP exhibited extensive necrosis, whereas the region expressing FvCFEM12 showed a decrease in necrotic area (Figure 3B, C). These findings suggest that FvCFEM12 can suppress plant cell death triggered by both INF1 and *P. syringae* DC3000, further supporting its role as a suppressor of plant immune responses.

2.4. Transient Expression of FvCFEM12 Enhances Susceptibility to B.cinerea and Suppresses Plant Immunity

To investigate whether FvCFEM12 regulates plant immunity responses, we transiently expressed FvCFEM12 in *N. benthamiana* leaves, using GFP as a control on the opposite side of the same leaves. Two days after inoculation with *B. cinerea*, the lesion area in leaves expressing FvCFEM12 was significantly larger compared to the GFP-expressing control (Figure 4A,B). These results suggest that FvCFEM12 promotes the infection of *B. cinerea* in *N. benthamiana* plants.

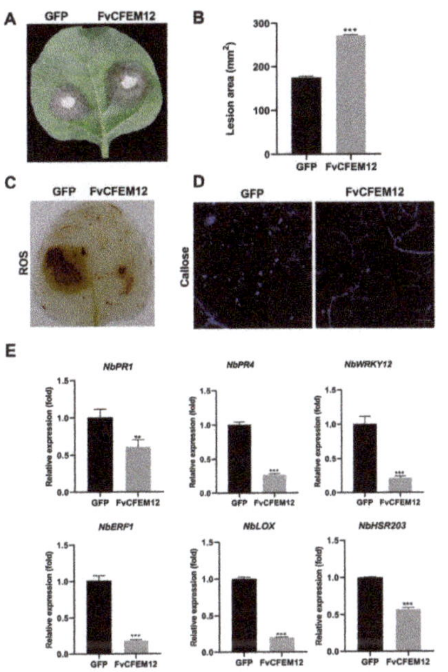

Figure 4. FvCFEM12 suppresses plant resistance responses and enhances susceptibility to *B. cinerea*. (**A**) *B. cinerea* infection assays in 4-week-old *N. benthamiana* leaves. The left side of the leaves was transiently expressed with *A. tumefaciens* carrying a GFP vector as a control, while the right side was transiently expressed with *A. tumefaciens* carrying the FvCFEM12 recombinant vector. Lesion phenotypes (**A**) and diameters (**B**) were examined at 2 days post-inoculation with *B. cinerea*. (**C**) Reactive oxygen species (ROS) accumulation and (**D**) callose deposition in *N. benthamiana* leaves transiently expressing FvCFEM12 were detected using 3,3′-diaminobenzidine (DAB) and aniline blue staining, respectively. Scale bar = 200 μm. (**E**) Relative expression levels of the defense-related genes (*NbRP1*, *NbPR4*, *NbWRKY12*, *NbERF1*, *NbLOX*, and *NbHSR203*) in *N. benthamiana* leaves infiltrated with *A. tumefaciens* carrying the FvCFEM12 recombinant vector and induced by *B. cinerea*. *A. tumefaciens* carrying the GFP vector was used as a control. *NbActin* served as an internal reference. Data are presented as means ± standard error of three biological replicates. ** and *** indicate significant differences at $p < 0.01$ and $p < 0.001$, respectively, based on Student's *t*-test.

Plant defense responses against pathogen infection are typically involved in ROS accumulation, callose deposition, and regulation of defense-related genes [37,38]. To further investigate the impact of FvCFEM12 on these defense responses, we measured ROS accumulation, callose deposition, and the expression of defense-related genes in *N. benthamiana* leaves inoculated with *B. cinerea*. ROS accumulation and callose deposition were significantly reduced in leaves expressing FvCFEM12 compared to the GFP control (Figure 4C,D). Additionally, the expression levels of six defense-related genes, including *NbPR1*, *NbPR4*, *NbWRKY12*, *NbERF1*, *NbLOX* and *NbHSR203*, were markedly downregulated in FvCFEM12-expressing leaves (Figure 4E). Taken together, these findings demonstrate that FvCFEM12 can suppress plant immune responses in *N. benthamiana* plants.

2.5. CFEM Domain Required for FvCFEM12 Cell Death Suppression, Independent of Subcellular Localization

To investigate the functional roles of the signal peptide and the conserved CFEM domain in FvCFEM12-mediated suppression of cell death, we generated a series of deletion mutants (Figure 5A). The first mutant, FvCFEM12$^{\Delta1-17}$, lacked the signal peptide; the second mutant, FvCFEM12$^{\Delta86-193}$, lacked the C-terminal region; and the third mutant, FvCFEM12$^{\Delta18-85}$, lacked the CFEM domain. Transient expression assays in *N. benthamiana* revealed that both FvCFEM12$^{\Delta1-17}$ and FvCFEM12$^{\Delta86-193}$ retained the ability to suppress cell death, similar to the full-length FvCFEM12. In contrast, the FvCFEM12$^{\Delta18-85}$ mutant, which lacks the CFEM domain, completely lost its cell death suppression activity (Figure 5B). These results demonstrate that the CFEM domain is essential for FvCFEM12's function in suppressing cell death, while the signal peptide and C-terminal region are dispensable for this activity.

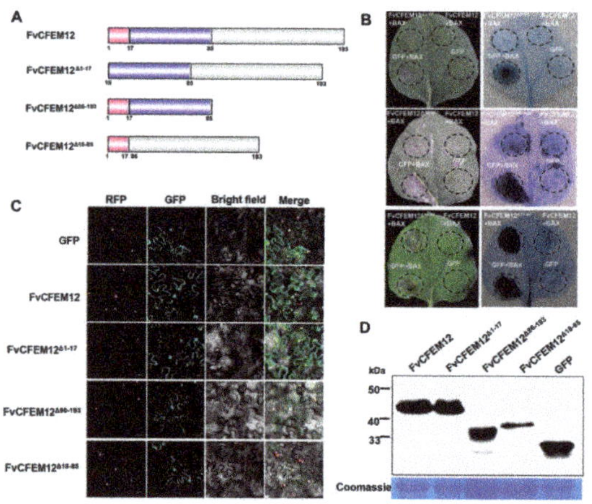

Figure 5. FvCFEM12 suppress Bax-induced cell death dependent on the CFEM domain. (**A**) Schematic representation of each mutant construct of FvCFEM12 generated using the IBS 2.0 website. The first construct, FvCFEM12, contains the full-length sequence of FvCFEM12. The second construct, FvCFEM12$^{\Delta1-17}$, lacks the signal peptide domain. The third construct, FvCFEM12$^{\Delta86-193}$, lacks the C-terminal sequence. The fourth construct, FvCFEM12$^{\Delta17-185}$, lacks the CFEM domain. Pink, blue, and grey blocks represent the signal peptide, CFEM domain, and unidentified sequence, respectively. (**B**) Different FvCFEM12 mutants were transiently expressed in *N. benthamiana* leaves 24 h prior to infiltration with *A. tumefaciens* carrying the *Bax* gene. Photographs were taken at 5 days post-inoculation (dpi), and lesion areas were visualized by trypan blue staining. (**C**) Subcellular localization of FvCFEM12 and its mutants fused with a C-terminal GFP in transgenic *N. benthamiana* leaves expressing H2B-RFP. GFP and RFP fluorescence were observed at 2 days post-agroinfiltration. Scale bar = 20 μm. (**D**) Western blot analysis of proteins extracted from *N. benthamiana* leaves transiently expressing FvCFEM12 and its mutants fused with a GFP tag.

To further explore the subcellular localization of FvCFEM12, we transiently expressed FvCFEM12-GFP fusion protein in *N. benthamiana* leaves. Fluorescence microscopy revealed that the green fluorescence signal from FvCFEM12-GFP was distributed throughout the membrane and co-localized with the red fluorescent signal of the nuclear marker H2B-RFP, indicating that FvCFEM12 localizes to both the membrane and nucleus (Figure 5C). Furthermore, a similar localization pattern was observed for the deletion mutants FvCFEM12$^{\Delta 1-17}$-GFP, FvCFEM12$^{\Delta 86-193}$-GFP, and FvCFEM12$^{\Delta 18-85}$-GFP. Expression of the GFP fusion proteins was verified by Western blot (Figure 5D). These results suggest that FvCFEM12 localized to both the membrane and nucleus in tobacco epidermal cells.

2.6. FvCFEM12 Is Involved in Phenotype Development and Cell Wall Stress Responses

To investigate the biological function of FvCFEM12, we generated two deletion mutants (Δ*FvCFEM12*-1 and Δ*FvCFEM12*-2) using a split-marker approach, in which the FvCFEM12 gene was replaced with a hygromycin resistance (*hph*) cassette (Figure 6A). Complementary strains (Δ*FvCFEM12*-c1 and Δ*FvCFEM12*-c2) were created by reintroducing a functional copy of FvCFEM12 into the respective mutant strains. Southern blot analysis confirmed the successful generation of the mutants and complemented strains (Figure 6B).

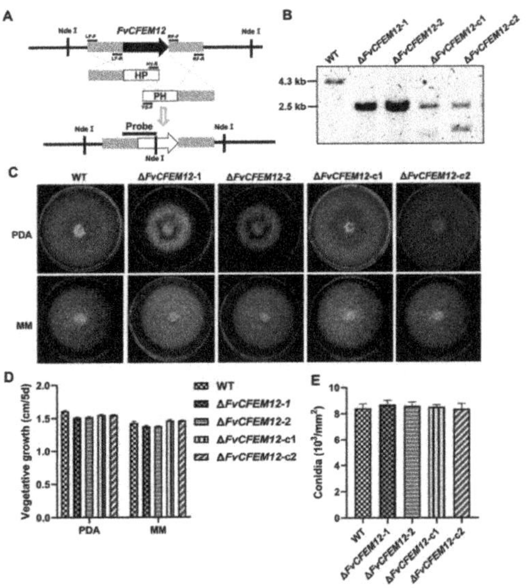

Figure 6. FvCFEM12 deletion affects colony phenotype, but not conidial germination. (**A**) Schematic strategy of the FvCFEM12 deletion based on the split-marker recombination method. NdeI digestion sites and the probe used for Southern hybridization are indicated. HP and PH represent different regions of the hygromycin B phosphotransferase (*hph*) marker gene cassette. Primers LF-F, LF-R, RF-F, RF-R, HY-F, and YG-F were used for gene replacement. (**B**) Southern blot analysis of genomic DNA from the wild-type FvLNF15-11 (lane 1), deletion mutants Δ*FvCFEM12*-1 (lane 2) and Δ*FvCFEM12*-2 (lane 3), and complemented strains Δ*FvCFEM12*-c1 (lane 4) and Δ*FvCFEM12*-c2 (lane 5). (**C**) Mycelial growth phenotypes and (**D**) Colony diameters of the wild-type FvLNF15-11, deletion mutants Δ*FvCFEM12*-1 and Δ*FvCFEM12*-2, and complemented strains Δ*FvCFEM12*-c1 and Δ*FvCFEM12*-c2 on PDA and MM agar plates. Photographs were taken at 5 days post-inoculation (dpi). (**E**) Quantification of conidia produced by the wild-type FvLNF15-11, deletion mutants Δ*FvCFEM12*-1 and Δ*FvCFEM12*-2, and complemented strains Δ*FvCFEM12*-c1 and Δ*FvCFEM12*-c2 on PDA agar plates. Data are presented as means ± standard error of three biological replicates. Data are presented as mean ± standard error from three biological replicates, with no statistically significant differences observed between wild-type and deletion mutants using Student's *t*-test.

To assess the role of FvCFEM12 in fungal growth and development, the wild-type (WT) strain, deletion mutants (Δ*FvCFEM12*-1 and Δ*FvCFEM12*-2), and complemented strains (Δ*FvCFEM12*-c1 and Δ*FvCFEM12*-c2) were cultured on Potato Dextrose Agar (PDA) and minimal medium (MM). Compared to the WT and complemented strains, the deletion mutants exhibited reduced pigment and loss of aerial hyphae at colony margin (Figure 6C,D). However, no significant differences were observed in conidia production, indicating that FvCFEM12 is involved in colony morphology but does not affect conidiation (Figure 6E).

Given the observed phenotypic changes, we hypothesized that FvCFEM12 might play a role in stress responses. To test this, we evaluated the growth of the WT, mutant, and complemented strains in CM medium supplemented with various stressors, including 0.7 M NaCl, 1 M sorbitol, 0.05% H_2O_2, 0.05% SDS, 100 μg/mL calcofluor white (CFW), and 500 μg/mL Congo Red. While no significant differences were observed in sensitivity to NaCl, sorbitol, H_2O_2, or SDS, the Δ*FvCFEM12* mutants exhibited reduced sensitivity to cell wall stressors (CFW and Congo Red) compared to the WT and complemented strains (Figure 7A, B).

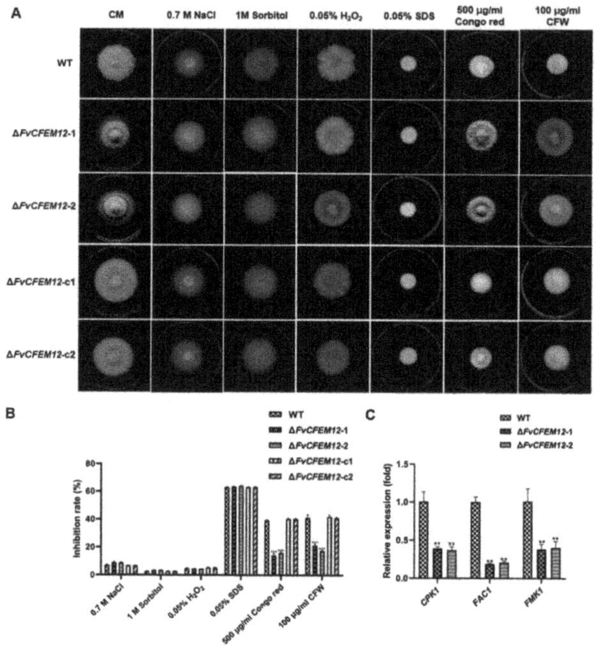

Figure 7. FvCFEM12 deletion alters tolerance to cell wall stress. (**A**) Mycelial growth phenotypes of the wild type FvLNF15-11, the deletion mutant Δ*FvCFEM12*-1 and Δ*FvCFEM12*-2 and complemented strains Δ*FvCFEM12*-c1 and Δ*FvCFEM12*-c12 on CM plates supplemented with 0.7 M NaCl, 1 M sorbitol, 0.05% H_2O_2, 0.05% SDS, 100 μg/mL CFW, and 500 μg/mL Congo red. Photographs were taken at 5 days post-inoculation (dpi). (**B**) Inhibition of mycelial radial growth in response to cell wall stressors. (**C**) Relative expression levels of the genes related to the cAMP-PKA and MAPK pathways in the wild type FvLNF15-11, deletion mutant Δ*FvCFEM12*-1 and Δ*FvCFEM12*-2 and complemented strains Δ*FvCFEM12*-c1 and Δ*FvCFEM12*-c2. *FvTubulin* was used as an internal reference. Data are presented as means ± standard error of three biological replicates. ** and *** indicate significant differences at $p < 0.01$ and $p < 0.001$, respectively, between the wild-type and deletion mutants, as determined by Student's *t*-test.

The PKA and MAPK signaling pathways are known to regulate adaptive responses to environmental stresses [39]. To explore whether FvCFEM12 influences these pathways, we analyzed the expression of key genes (*CPK1*, *FAC1*, and *FMK1*) using qRT-PCR. The Δ*FvCFEM12* mutants showed significantly reduced expression of these genes compared to the WT strain (Figure 7C), indicating that FvCFEM12 may modulate stress responses through the PKA and MAPK pathways. Collectively, these results demonstrate that FvCFEM12 is involved in colony morphology and cell wall stress responses, potentially by regulating the PKA and MAPK signaling pathways.

2.7. *FvCFEM12 Is Required for the Full Virulence of F. verticillioides*

To further investigate the role of FvCFEM12 in pathogenicity, we inoculated the susceptible maize cultivar B73 with the wild-type (WT) strain, FvCFEM12 deletion mutants (Δ*FvCFEM12*-1 and Δ*FvCFEM12*-2), and complemented strains (Δ*FvCFEM12*-c1 and Δ*FvCFEM12*-c2). Lesion areas on infected maize stalks were measured at 7 days post-inoculation. The Δ*FvCFEM12*-1 and Δ*FvCFEM12*-2 mutants exhibited significantly smaller lesion areas compared to the WT strain, while the complemented strains (Δ*FvCFEM12*-c1 and Δ*FvCFEM12*-c2) displayed lesion areas similar to those of the WT (Figure 8A,B). These results demonstrate that FvCFEM12 is essential for the full virulence of *F. verticillioides* during maize infection.

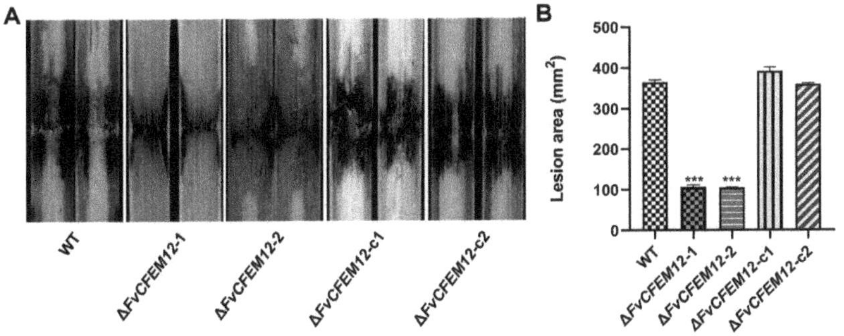

Figure 8. Pathogenicity assays of the wild type FvLNF15-11, deletion mutant Δ*FvCFEM12*-1 and Δ*FvCFEM12*-2 and complemented strains Δ*FvCFEM12*-c1 and Δ*FvCFEM12*-c2 on susceptible maize cultivar B73. (**A**) Phenotypes of maize stalks inoculated with the wild-type (WT), FvCFEM12 deletion mutants (Δ*FvCFEM12*-1 and Δ*FvCFEM12*-2), and complemented strains (Δ*FvCFEM12*-c1 and Δ*FvCFEM12*-c2) at 7 days post-inoculation. (**B**) Lesion areas on maize stalks were quantified using ImageJ1.46. Data are presented as means ± standard error of three biological replicates. *** indicates significant differences at $p < 0.001$ between the wild-type and deletion mutants, as determined by Student's *t*-test.

2.8. *FvCFEM12 Interacts with Wall-Associated Receptor Kinase ZmWAK17*

Previous studies have shown that CFEM proteins in *F. graminearum* can interact with extracellular binding proteins ZmWAK17ET, which is an alternative splicing isoform of ZmWAK17 [40]. Interestingly, FvCFEM12 shares 50.62% amino acid sequence similarity with FgCFEM1, suggesting a potential functional conservation. To investigate whether FvCFEM12 interacts with ZmWAK17ET, we performed co-immunoprecipitation (Co-IP) assays. ZmWAK17ET was fused to GFP and FvCFEM12 was tagged with a Flag epitope, followed by co-expression in *N. benthamiana* leaves. Immunoprecipitation of ZmWAK17ET-GFP using anti-GFP antibodies specifically pulled down FvCFEM12-Flag, as detected by anti-Flag immunoblotting, whereas no signal was observed in the GFP control (Figure 9A), demonstrating a specific interaction between FvCFEM12 and ZmWAK17ET. To further validate this interaction, we conducted a bimolecular fluorescence complementation (BiFC) assay. Recombinant plasmids encoding pUC-SPYCE-FvCFEM12 and pUC-SPYNE-ZmWAK17ET were constructed and transiently co-expressed in *N. benthamiana* leaves via *A. tumefaciens*-mediated transformation. YFP fluorescence signals were observed at the plasma membrane microdomains, indicating a direct interaction between FvCFEM12 and ZmWAK17ET (Figure 9B). These results collectively demonstrate that FvCFEM12 physically interacts with ZmWAK17ET, suggesting a potential role in modulating plant immunity.

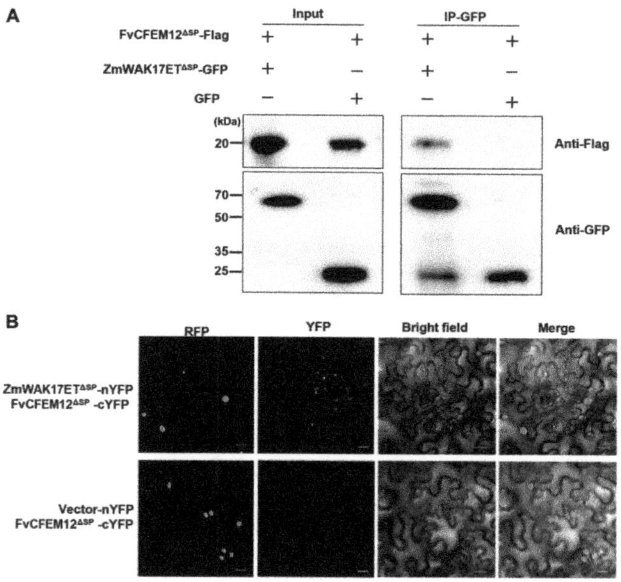

Figure 9. FvCFEM12 interacts with ZmWAK17ET. (**A**) Interaction between FvCFEM12-Flag and ZmWAK17ET-GFP was confirmed by co-immunoprecipitation (Co-IP). Total protein (Input) extracted from *N. benthamiana* leaves expressing FvCFEM12-Flag and ZmWAK17ET-GFP (or FvCFEM12-Flag and GFP) was subjected to SDS-PAGE, and immunoblots were incubated with anti-Flag and anti-GFP antibodies (Input panel). GFP-Trap beads were used to enrich ZmWAK17ET-GFP and GFP. FvCFEM12-Flag was detected in the ZmWAK17ET-GFP immunoprecipitate but not with GFP. (**B**) BiFC visualization of the interaction between FvCFEM12-cYFP and ZmWAK17ET-nYFP in transgenic *N. benthamiana* leaves expressing RFP-H2B. Co-expression of FvCFEM12-cYFP and vector-nYFP served as a control. YFP signals were observed using a confocal microscope at 2 days post-agroinfiltration. Scale bar = 20 μm.

2.9. Silencing ZmWAK17 in Maize Reduce Resistance to F. verticillioides

To investigate the role of ZmWAK17 in maize resistance against *F. verticillioides*, we employed a Cucumber Mosaic Virus (CMV)-mediated virus-induced gene silencing (VIGS) system to downregulate ZmWAK17 expression in maize cultivar B73. Phytoene desaturase (PDS) was used as a positive control for silencing efficiency, while an empty vector (pCMVZ22bN81) served as a negative control. Successful silencing was confirmed by the appearance of photo-bleaching phenotypes in ZmIspH-silenced plants (Figure 10A). Following *F. verticillioides* inoculation, leaves were collected to assess ZmWAK17 expression levels using qRT-PCR. The results showed a significant ~65% reduction in ZmWAK17 expression in silenced plants compared to the control group (Figure 10B). At 5 days post-inoculation, ZmWAK17-silenced plants exhibited increased susceptibility to *F. verticillioides* infection, as evidenced by larger lesion areas compared to control plants (Figure 10 C,D).

These findings demonstrate that ZmWAK17 plays a critical role in maize resistance against *F. verticillioides*.

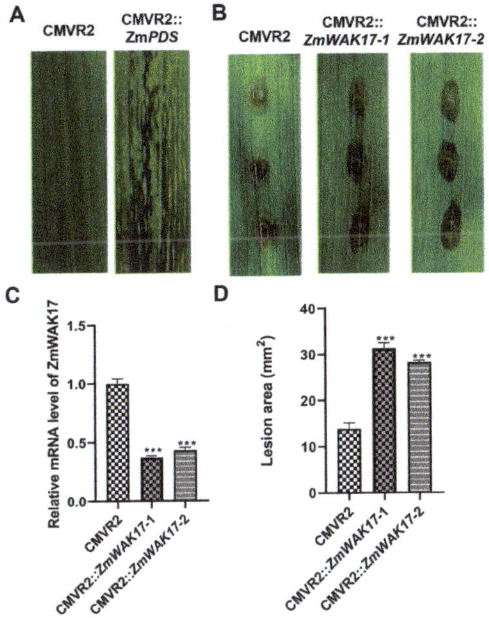

Figure 10. Silencing and functional assays of ZmWAK17 in maize resistance against *F. verticillioides*. (**A**) Leaf phenotypes of maize leaves expressing pCMVZ22bN81 (CMVR2) and CMVR2::*ZmPDS* at 12 days post-agroinfiltration. (**B**) Transcript levels of ZmWAK17 in ZmWAK17-silenced maize leaves. (**C**) Phenotypes and (**D**) lesion area of ZmWAK17 silenced maize leaves infected with *F. verticillioides* at 5 days post-inoculation (dpi). Empty vector CMVR2 silenced maize leaves were used as a control. Data are presented as means ± standard error of three biological replicates. *** indicates significant differences at $p < 0.001$ between CMVR2 and CMVR2::ZmWAK17-1/2, as determined by Student's *t*-test.

3. Discussion

CFEM proteins are uniquely found in fungi, and an increasing number of studies have highlighted their roles as effectors in plant pathogenic fungi. Effectors typically possess signal peptides, lack transmembrane regions, and can modulate plant cell death. These characteristics make it suitable for rapid screening of candidate effectors. Previous studies suggest that some CFEM proteins can suppress plant cell death [25,41], while only a few proteins, such as PstCFEM1 and BcCFEM1, have been reported to induce cell death or chlorosis in *N benthamiana* [27,41]. In this study, we identified FvCFEM12 as a CFEM protein capable of suppressing cell death induced by BAX, INF1 in *N. benthamiana*. Furthermore, we demonstrated that FvCFEM12 also inhibits cell death induced by *P. syringae* strain DC3000 on the host plant (Figures 3B and 5). These findings suggest that FvCFEM12 functions as a potential effector protein.

CFEM effectors are known to play diverse roles in fungal growth and development, with notable functional variations across different fungal species. While certain CFEM effectors, like CfEC12 and FgCFEM1, have no observable impact on fungal morphology, others significantly influence spore formation and development. For instance, PeCFEM5 and PeCFEM8 affect spore morphogenesis without altering colony morphology, whereas CgCFEM1 is crucial for conidial development and invasion structure formation in *Colletotrichum gloeosporioides*. In this study, we demonstrate that FvCFEM12 significantly alters fungal phenotype by reducing pigmentation and inhibiting aerial hyphae formation at colony edges. These observations align with recent studies indicating that certain CFEM effectors affect fungal morphology. In *Neostagonosporella sichuanensis*, *NsCFEM1* deletion mutants exhibited denser mycelium compared to the wild type, with reduced pigmentation and accelerated colony expansion. Similarly, in *C. gloeosporioides*, ΔCgCsa mutants displayed impaired radial growth and defective hyphal tip development, suggesting a crucial role for CFEM proteins in maintaining hyphal polarity. Collectively, these findings highlight the functional diversification of CFEM effectors in regulating fungal morphogenesis.

Glycosylphosphatidylinositol (GPI)-anchored proteins are typically localized to the outer leaflet of the plasma membrane via a GPI anchor and play critical roles in maintaining cell wall integrity and function in fungi [42,43]. For example, the GPI-anchored protein Gas1p in *S. cerevisiae* is essential for cell wall formation [44], while Dfg3 in *A. fumigatus* is involved in the association of galactomannan with the β-(1,3)-glucan–chitin core of the cell wall [45]. These findings are consistent with previous reports on CFEM proteins (e.g., Rbt5, Rbt51, Csa1, and BcCFEM1) that similarly contain GPI anchors and are implicated in the regulation of cell wall biosynthesis [27,46]. In this study, we identified two putative GPI-modification sites within FvCFEM12, and deletion mutants exhibited altered sensitivity to cell wall stressors, suggesting a potential role in cell wall integrity. Notably, the majority of characterized CFEM proteins appear to positively regulate cell wall biosynthesis, while FvCFEM12 seems to function as a negative regulator. This may be due to the fact that different CFEM family proteins may serve distinct roles in various aspects of cell wall biology.

Plant pathogens often secrete effectors to manipulate host immune responses and facilitate infection [47]. Several CFEM proteins including PstCFEM1 from *P. striiformis* f. sp. *tritici* and BcCFEM1 from *B. cinerea* have been reported to be implicated in virulence [27–29,40]. In this study, transient expression of FvCFEM12 in *N. benthamiana* leaves promoted the infection process of *B. cinerea* and suppressed plant defense responses, including ROS accumulation, callose deposition, and the induced expression of defense-related genes (Figure 7). Pathogenicity assays on maize stalks further confirmed that FvCFEM12 contributes to the virulence of *F. verticillioides* (Figure 8). These studies collectively suggest that CFEM proteins play diverse roles in promoting virulence across fungal pathogens, warranting further investigation into the functions of other CFEM family members *F. verticillioides*.

Despite the growing evidence of CFEM proteins' roles in pathogenicity, the molecular mechanisms underlying their interactions with host plants remain poorly understood. In *F. graminearum*, CFEM proteins interact with the maize protein ZmWAK17ET to facilitate infection. Wall-associated kinases (WAKs), as members of the receptor-like kinase (RLK) superfamily, are involved in various biological processes, including growth, development, plant immunity, and abiotic stress responses [48,49]. Previous studies have demonstrated that ZmWAK17 confers resistance to *F. graminearum*; our findings reveal that ZmWAK17 also enhances resistance to *F. verticillioides*. This suggests a potential broad-spectrum resistance mechanism against multiple *Fusarium* species. Furthermore, we show that FvCFEM12 interacts with ZmWAK17ET, although the precise molecular mechanisms of this interaction and the downstream signaling pathways involved in ZmWAK17-mediated resistance remain to be elucidated. Based on our observations, we hypothesize that FvCFEM12 suppresses ZmWAK17-mediated resistance in maize, thereby promoting *F. verticillioides* infection (Figure 11). In summary, FvCFEM12 is identified as a crucial virulence CFEM effector in *F. verticillioides*. We propose that these CFEM effectors, which play significant roles in pathogenicity, could serve as potential targets for the management of plant fungal diseases.

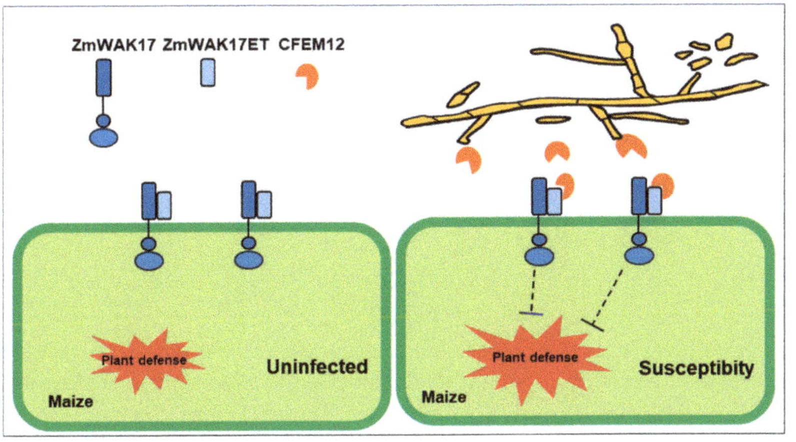

Figure 11. Proposed model of FvCFEM12 suppressing ZmWAK17-mediated plant immunity during *Fusarium verticillioides* infection. Under physiological conditions, ZmWAK17 acts as a pattern recognition receptor that activates basal immune responses through extracellular domain (ET)-mediated perception of pathogen-associated molecular patterns. During *F. verticillioides* infection, the fungal pathogen secretes the effector protein FvCFEM12, which interacts with ZmWAK17ET, thereby disrupting receptor-mediated signaling transduction. The effector–receptor interaction competitively inhibits ZmWAK17's capacity to initiate defense-related pathways, ultimately facilitating fungal infection in maize tissues.

4. Materials and Methods

4.1. Fungal Strains and Plant Materials

The *F. verticillioides* strain FvLNF15-11 was isolated from diseased maize stalks in Liaoning Province, China [50]. For colony morphology analysis, the wild-type strain FvLNF15-11, along with its mutants and complemented strains, were cultured on potato dextrose agar (PDA) and minimal medium (MM) plates. Stress assays were conducted by incubating all strains on complete medium (CM) plates supplemented with specific stressors, including 0.7 M NaCl, 1 M sorbitol, 0.05% hydrogen peroxide (H_2O_2), 0.05% sodium dodecyl sulfate (SDS), 100 µg/mL calcofluor white (CFW), and 500 µg/mL Congo Red (CR), for 5 days. The *B. cinerea* strain was cultured on PDA plates at 25 °C. *N. benthamiana* and maize cultivar B73 plants were grown in a greenhouse at 25 °C with a 16 h light and 8 h dark photoperiod.

4.2. Bioinformatics Identification of CFEM Proteins in F. verticillioides

To identify CFEM proteins in *F. verticillioides*, we employed a search strategy using the CFEM-containing protein ACI1 from M. oryzae as a query in the NCBI database (https://blast.ncbi.nlm.nih.gov/Blast.cgi, accessed on 18 January 2024). The candidate sequences retrieved from this search were subsequently analyzed for the presence of the CFEM domain using the Pfam tool available on the SMART platform (http://smart.embl-heidelberg.de/, accessed on 20 January 2024). For further characterization, the presence of N-terminal signal peptides was predicted using SignalP 5.0 (https://services.healthtech.dtu.dk/service.php?SignalP-5.0, accessed on 2 February 2024). Transmembrane regions within the CFEM proteins were identified using the TMHMM server (https://dtu.biolib.com/DeepTMHMM, accessed on 20 January 2024). Additionally, potential GPI modification sites were predicted using the GPI modification site prediction tool (https://mendel.imp.ac.at/gpi/fungi_server.html, accessed on 22 January 2024).

4.3. Phylogenetic Analysis and Multiple Sequence Alignment

The amino acid sequences of the 19 CFEM-containing proteins from *F. verticillioides* were aligned using ClustalW, and the consensus sequence was visualized using Jalview [51]. A phylogenetic tree was constructed using the neighbor-joining method based on these sequences in MEGA 7.0 [24]. The reliability of the tree nodes was assessed through non-parametric bootstrapping with 1000 pseudo-replicates. The phylogeny and domain architecture of each protein were visualized using iTOL (https://itol.embl.de/, accessed on 12 February 2024). The alignment results of the CFEM domains were analyzed using WebLogo (http://weblogo.berkeley.edu/logo.cgi, accessed on 15 February 2024).

4.4. Analysis of Chromosomal Distribution of CFEM Genes

To systematically investigate the genomic organization characteristics of FvCFEM effector genes in *F. verticillioides*, we performed physical mapping of all 19 identified FvCFEM genes onto their respective chromosomes using MG2C v2.1 on the website (http://mg2c.iask.in/mg2c_v2.1/index.html, accessed on 20 July 2024).

4.5. Yeast Signal Sequence Trap Assays

The functionality of the predicted signal peptides in the seven CFEM proteins was validated using a previously described method [52]. Each predicted signal peptide was fused in-frame to the pSUC2 vector, and transformed into the yeast strain YTK-12 using the Frozen-EZ Yeast Transformation II kit (Zymo Research, Irvine, CA, USA). Transformed yeast cells were cultured on CMD-W medium (0.67% yeast nitrogen base without amino acids, 0.075% tryptophan dropout supplement, 2% sucrose, 0.1% glucose, and 2% agar) and YPRAA medium (1% yeast extract, 2% peptone, 2% raffinose, and 2 μg/mL antimycin A).

Invertase enzymatic activity was further confirmed using 2,3,5-triphenyltetrazolium chloride (TTC) staining assays. Yeast transformants were incubated in sodium acetate buffer (pH 4.7, 100 mM sucrose) for 10 min at 37 °C, followed by centrifugation at 10,000 rpm for 1 min. The supernatant was mixed with an alkaline 1% TTC solution, and a colorimetric change after 5 min at room temperature indicated invertase activity.

4.6. Transient Expression of CFEM Proteins in N. benthamiana and Z. mays

The full-length open reading frames (ORFs) of FvCFEMs genes were amplified and cloned into the expression plasmid pCHF3. All recombinant plasmids were transformed into *A. tumefaciens* strain C58C1 and cultured in Luria–Bertani (LB) medium supplemented with 50 μg/mL spectinomycin. Cultures were grown overnight at 28 °C until reaching an optical density (OD_{600}) of 0.5–1.0. The bacterial cells were then washed three times with the resuspended buffer (10 mM $MgCl_2$, 10 mM MES, and 150 uM acetosyringon) and adjusted to an OD_{600} of 0.4. After incubation for 3 h at room temperature, the bacterial suspensions were infiltrated into leaves of 4-week-old *N. benthamiana* plants. To assess cell death suppression, *A. tumefaciens* carrying BAX or INF1 genes was infiltrated into the same sites 24 h later. For subcellular localization, FvCFEM12-GFP fusion proteins were transiently expressed in transgenic *N. benthamiana* plants expressing RFP-H2B [53]. Leaves were examined at 3 days post-inoculation using a ZEISS LSM980 confocal scanning microscope.

For transient expression in maize, the method was adapted from Tang et al. [54] and Liang et al. [55] with modifications. The FvCFEM12 and GFP were first cloned into the Gateway donor vector pDONR 221 by BP recombination, and then subcloned into the Gateway expression vector pEDV6 through LR recombination (Invitrogen, Carlsbad, CA, USA). The recombinant plasmids were transformed into *P. syringae* strain D36E by electroporation (2.5 kv/6 ms). Transformed acteria were cultured in King's B medium supplemented with 30 μg/mL gentamicin for 24 h at 28 °C. Cells were harvested by centrifugation (4000× g, 5 min), washed three times with 10 mM $MgCl_2$, and resuspended in the same solution. The OD_{600} was adjusted to 0.5 and incubated at room temperature for 3 h. Subsequently, the bacterial suspension was injected into leaves of two-leaf-stage maize plants. After 24 h, *P. syringae* strain DC3000 was infiltrated into the same sites at OD_{600} = 0.8 in 10 mM $MgCl_2$. Plants were maintained under high humidity and darkness for 24 h before being returned to normal conditions. The lesion areas were photographed 5 days post-inoculation and quantified using ImageJ 1.46 software.

4.7. RT-qPCR Analysis

Total RNA was extracted using Trizol reagent (TransGen Biotech, Beijing, China). First-strand cDNA was synthesized from the extracted RNA using the PrimeScript RT reagent Kit (Takara, Dalian, China.). Quantitative real-time PCR (qPCR) was performed using the NovoStart SYBR qPCR SuperMix (Novoprotein, Suzhou, China) on a Real-Time PCR detection system. The relative expression levels of target genes were normalized to internal controls: *FvTubulin* for *F. verticillioides*, *NbActin* for *N. benthamiana*, and *ZmUbiquitin* for *Z. mays*. Gene expression was quantified using the $2^{-\Delta\Delta Ct}$ method.

4.8. Trypan Blue Staining

Infiltrated *N. benthamiana* leaves were collected 3 days post-inoculation. A trypan blue solution was prepared by dissolving 10 mL lactic acid, 10 mL glycerol, 10 g phenol, and 10 mg trypan blue in 10 mL distilled water, with the final volume adjusted to 40 mL by adding distilled water. Leaves were immersed in a 1:1 mixture of trypan blue solution and ethanol (40 mL total volume) and heated at 90 °C for 2 min. After boiling, the stained leaves were incubated overnight at room temperature. Excess stain was removed by washing the leaves several times with distilled water. To reduce background staining, leaves were destained in hydrated trichloroacetaldehyde until the background color was cleared.

4.9. Detection of Oxidative Burst and Callose Deposition

To induce defense responses, a mycelial plug (0.7 cm in diameter) of *B. cinerea* was taken from the edge of the colony and placed on the infiltrated area of *N. benthamiana* leaves. Reactive oxygen species (ROS) accumulation was detected using a 3,3′-diaminobenzidine (DAB) solution kit (Solarbio, Beijing, China). Leaf samples were immersed in DAB solution (1 mg/mL) and shaken at 50 rpm for 12 h. After staining, leaves were bleached in 96% ethanol and photographed.

Callose deposition was analyzed using an aniline blue staining method with minor modifications [56]. Samples were decolorized in absolute ethanol and then incubated in 150 mM phosphate buffer (pH 9.5) containing 0.1% aniline blue for 2 h in the dark. Stained samples were observed under a ZEISS LSM 980 confocal scanning microscope using an excitation filter of 405 nm and an emission filter of 485–515 nm.

4.10. F. verticillioides Gene Knockout and Complementation

Gene deletion constructs were generated using a split-marker recombination approach as previously described [57]. The upstream and downstream flanking regions of the FvCFEM12 gene were amplified using primer pairs LF-F/LF-R and RF-F/RF-R, respectively. Different regions of the hygromycin B phosphotransferase (*hph*) gene were amplified using primers HYG-F/HY-R and YG-F/HYG-R. Overlapping PCR products, amplified with primer pairs LF-F and RF-R, were introduced into protoplasts of the wild-type strain. Protoplast generation and transformation were performed according to the method described by Ridenour et al. [58].

For complementation assays, the ORF of FvCFEM12, including its native promoter, was co-transformed with linearized pFL2 into the yeast strain XK12 to generate the complementation plasmid, as described by Jiang et al. [59].The complementation plasmid was then transformed into protoplasts of the FvCFEM12 deletion mutant to obtain the complemented strain.

4.11. Pathogenicity Assay

The pathogenicity of fungal strains on maize stalks was assessed as previously described [60]. Briefly, fungal strains were cultured on PDA medium at 25 °C for 5 days. Conidia were harvested by washing the fungal culture with sterile water and adjusted to a concentration of 1×10^6 conidia/mL. At the 10-leaf stage, 100 µL of the conidial suspension was injected into maize stalks. Disease symptoms were photographed at 7 days post-inoculation, and lesion areas were quantified using ImageJ software.

4.12. Co-IP Assay

Recombinant plasmids pCHF3-FvCFEM12-Flag and pCHF3-ZmWAK17ET-GFP were transformed into *A. tumefaciens* strain C58C1 and co-infiltrated into *N. benthamiana* leaves. Two days post-infiltration, total protein was extracted from the leaves using lysis buffer (100 mM Tris-Cl, pH 7.5; 150 mM NaCl; 1% Triton X-100; 0.5 mM EDTA, pH 8.0; 0.1% SDS; 10 mM dithiothreitol; and protease inhibitor cocktail). GFP-Trap beads (Chromotek, Martinsried, Germany) were used to enrich GFP-tagged proteins. The beads were washed twice with wash buffer (100 mM Tris-HCl, pH 7.5; 150 mM NaCl; 0.05% Nonidet™ P40 Substitute; and 0.5 mM EDTA). Bound proteins were eluted using 200 mM glycine and analyzed by Western blot.

4.13. BiFC Assay

The FvCFEM12 gene was cloned into the vector pUC-SPYCE, and the ZmWAK17ET gene was cloned into the vector pUC-SPYNE. Recombinant plasmids were transformed into *A. tumefaciens* strain GV3101 and co-infiltrated into transgenic *N. benthamiana* plants expressing RFP-H2B. YFP and RFP fluorescence in infiltrated leaves was observed using a confocal laser microscope 48 to 72 h post-infiltration.

4.14. Virus-Induced Gene Silencing (VIGS) Assay in Maize

The CMV-mediated virus-induced gene silencing (CMV-VIGS) approach was implemented as previously described by Li H [61]. *ZmPDS* (phytoene desaturase gene) was used as the positive control for photobleaching phenotype validation. Gene-specific fragments (300 bp) targeting *ZmWAK17* (amplified with primers WAK17-VIGS-F/R, Supplementary Table S1) were cloned into the pCMVZ22bN81 vector through Kpn I/Xba I restriction sites and electroporated into *A. tumefaciens* strain GV3101. Agrobacterium cells harboring CMV RNAs 1, 2, and 3 constructs or their derivatives were equally mixed and infiltrated into *N. benthamiana* leaves. After four days, *N. benthamiana* leaves were homogenized in a PBS buffer. The homogenized leaf extract was then rub-inoculated onto the second maize leaves pre-dusted with 400-mesh carborundum. Following inoculation, the plants were kept in the dark for two days. After that, the plants were transferred to a growth room with a temperature of 25 °C and 16 h light/8 h dark cycle. The conidia suspension of *F. verticillioides* was adjusted to 1×10^6 condia/mL in 0.02% Tween 20 solution. Silenced maize leaves were punched using a tissue puncher and inoculated with 10 ul conidia suspension. Disease symptoms on the maize leaves were photographed at 5 days post-inoculation and lesions were measured by Image J.

5. Conclusions

In summary, our study identified 19 CFEM proteins in the *F. verticillioides* genome, with FvCFEM12 emerging as a key candidate effector. We demonstrated that FvCFEM12 significantly contributes to fungal virulence and plays a key role in suppressing plant immunity. Additionally, FvCFEM12 was found to influence colony morphology and participate in cell wall synthesis. Importantly, we discovered that FvCFEM12 interacts with the maize protein ZmWAK17ET. Collectively, these findings provide valuable insights into the biological functions of CFEM proteins in *F. verticillioides* and their potential roles in fungal pathogenicity and host–pathogen interactions.

Supplementary Materials: The following Supporting Information can be downloaded at: https://www.mdpi.com/article/10.3390/ijms26094369/s1.

Author Contributions: Conceptualization, W.G.; methodology, H.L.; validation, H.L. and R.W.; formal analysis, H.L.; investigation, H.L. and S.I.; writing original draft preparation, H.L. and S.I.; writing—review and editing, X.L.; methodology, H.W.; funding acquisition, H.W.; supervision, W.G.; project administration, W.G.; funding acquisition, W.G. All authors have read and agreed to the published version of the manuscript.

Funding: This work was financially supported by the National Key Research and Development Program of China (2022YFD1901300), Natural Science Foundation of Ningxia Province, grant numbers 2024AAC02040, National Natural Science Foundation of China, grant numbers 32470213 and 32072377, and the Mission of Konjac Industrial Science and Technology in Malipo County, Yunnan Province, grant numbers 202304BI090027-2.

Institutional Review Board Statement: Not applicable.

Informed Consent Statement: Not applicable.

Data Availability Statement: The original contributions presented in the study are included in the article. Further inquiries can be directed to the corresponding author.

Conflicts of Interest: The authors declare no conflicts of interest.

References

1. Jones, J.D.G.; Dangl, J.L. The plant immune system. *Nature* **2006**, *444*, 323–329. [CrossRef] [PubMed]
2. Tsuda, K.; Katagiri, F. Comparing signaling mechanisms engaged in pattern-triggered and effector-triggered immunity. *Curr. Opin. Plant Biol.* **2010**, *13*, 459–465. [CrossRef] [PubMed]
3. Yu, X.; Feng, B.; He, P.; Shan, L. From chaos to harmony: Responses and signaling upon microbial pattern recognition. *Annu. Rev. Phytopathol.* **2017**, *55*, 109–137. [CrossRef] [PubMed]
4. Toruño, T.Y.; Stergiopoulos, I.; Coaker, G. Plant-pathogen effectors: Cellular probes interfering with plant defenses in spatial and temporal manners. *Annu. Rev. Phytopathol.* **2016**, *54*, 419–441. [CrossRef]
5. Dou, D.; Zhou, J.-M. Phytopathogen effectors subverting host immunity: Different foes, similar battleground. *Cell Host Microbe* **2012**, *12*, 484–495. [CrossRef]
6. Munkvold, G.P.; Desjardins, A.E. Fumonisins in maize: Can we reduce their occurrence? *Plant Dis.* **1997**, *81*, 556–565. [CrossRef]
7. Cotten, T.K.; Munkvold, G.P. Survival of *Fusarium moniliforme*, *F. proliferatum*, and *F. subglutinans* in maize stalk residue. *Phytopathology* **1998**, *88*, 550–555. [CrossRef]
8. Blacutt, A.A.; Gold, S.E.; Voss, K.A.; Gao, M.; Glenn, A.E. *Fusarium verticillioides*: Advancements in understanding the toxicity, virulence, and niche adaptations of a model mycotoxigenic pathogen of maize. *Phytopathology* **2018**, *108*, 312–326. [CrossRef]
9. Ferrigo, D.; Raiola, A.; Causin, R. Fusarium toxins in cereals: Occurrence, legislation, factors promoting the appearance and their management. *Molecules* **2016**, *21*, 627. [CrossRef]
10. Munkvold, G.P. Fusarium species and their associated mycotoxins. In *Mycotoxigenic Fungi: Methods and Protocols*; Moretti, A., Susca, A., Eds.; Springer: New York, NY, USA, 2017; pp. 51–106.
11. Schulz, B.; RÖMmert, A.-K.; Dammann, U.; Aust, H.-J.; Strack, D. The endophyte-host interaction: A balanced antagonism? *Mycol. Res.* **1999**, *103*, 1275–1283. [CrossRef]

12. Dodds, P.N.; Rafiqi, M.; Gan, P.H.P.; Hardham, A.R.; Jones, D.A.; Ellis, J.G. Effectors of biotrophic fungi and oomycetes: Pathogenicity factors and triggers of host resistance. *New Phytol.* **2009**, *183*, 993–1000. [CrossRef] [PubMed]
13. Koeck, M.; Hardham, A.R.; Dodds, P.N. The role of effectors of biotrophic and hemibiotrophic fungi in infection. *Cell. Microbiol.* **2011**, *13*, 1849–1857. [CrossRef] [PubMed]
14. Ma, L.-J.; van der Does, H.C.; Borkovich, K.A.; Coleman, J.J.; Daboussi, M.-J.; Di Pietro, A.; Dufresne, M.; Freitag, M.; Grabherr, M.; Henrissat, B.; et al. Comparative genomics reveals mobile pathogenicity chromosomes in *Fusarium*. *Nature* **2010**, *464*, 367–373. [CrossRef]
15. Zhang, H.; Kim, M.S.; Huang, J.; Yan, H.; Yang, T.; Song, L.; Yu, W.; Shim, W.B. Transcriptome analysis of maize pathogen *Fusarium verticillioides* revealed FvLcp1, a secreted protein with type-D fungal LysM and chitin-binding domains, that plays important roles in pathogenesis and mycotoxin production. *Microbiol. Res.* **2022**, *265*, 127195. [CrossRef]
16. Naumann, T.A.; Wicklow, D.T.; Price, N.P.J. Identification of a chitinase-modifying protein from *Fusarium verticillioides*: Truncation of a host resistance protein by a fungalysin metalloprotease. *J. Biol. Chem.* **2011**, *286*, 35358–35366. [CrossRef]
17. Kulkarni, R.D.; Kelkar, H.S.; Dean, R.A. An eight-cysteine-containing CFEM domain unique to a group of fungal membrane proteins. *Trends Biochem. Sci.* **2003**, *28*, 118–121. [CrossRef]
18. Zhang, Z.-N.; Wu, Q.-Y.; Zhang, G.-Z.; Zhu, Y.-Y.; Murphy, R.W.; Liu, Z.; Zou, C.-G. Systematic analyses reveal uniqueness and origin of the CFEM domain in fungi. *Sci. Rep.* **2015**, *5*, 13032. [CrossRef]
19. Arya, G.C.; Srivastava, D.A.; Pandaranayaka, E.P.J.; Manasherova, E.; Prusky, D.B.; Elad, Y.; Frenkel, O.; Dvir, H.; Harel, A. Characterization of the role of a non-gpcr membrane-bound cfem protein in the pathogenicity and germination of *Botrytis cinerea*. *Microorganisms* **2020**, *8*, 1043. [CrossRef]
20. Moukadiri, I.; Armero, J.; Abad, A.; Sentandreu, R.; Zueco, J. Identification of a mannoprotein present in the inner layer of the cell wall of Saccharomyces cerevisiae. *J. Bacteriol.* **1997**, *179*, 2154–2162. [CrossRef]
21. Nasser, L.; Weissman, Z.; Pinsky, M.; Amartely, H.; Dvir, H.; Kornitzer, D. Structural basis of haem-iron acquisition by fungal pathogens. *Nat. Microbiol.* **2016**, *1*, 16156. [CrossRef]
22. Kou, Y.; Tan, Y.H.; Ramanujam, R.; Naqvi, N.I. Structure–function analyses of the Pth11 receptor reveal an important role for CFEM motif and redox regulation in rice blast. *New Phytol.* **2017**, *214*, 330–342. [CrossRef] [PubMed]
23. Sabnam, N.; Roy Barman, S. WISH, a novel CFEM GPCR is indispensable for surface sensing, asexual and pathogenic differentiation in rice blast fungus. *Fungal Genet. Biol.* **2017**, *105*, 37–51. [CrossRef] [PubMed]
24. Gong, A.-d.; Jing, Z.-y.; Zhang, K.; Tan, Q.-q.; Wang, G.-l.; Liu, W.-d. Bioinformatic analysis and functional characterization of the CFEM proteins in maize anthracnose fungus Colletotrichum graminicola. *J. Integr. Agric.* **2020**, *19*, 541–550. [CrossRef]
25. Wang, J.-x.; Long, F.; Zhu, H.; Zhang, Y.; Wu, J.-y.; Shen, S.; Dong, J.-g.; Hao, Z.-m. Bioinformatic analysis and functional characterization of CFEM proteins in *Setosphaeria turcica*. *J. Integr. Agric.* **2021**, *20*, 2438–2449. [CrossRef]
26. Qian, Y.; Zheng, X.; Wang, X.; Yang, J.; Zheng, X.; Zeng, Q.; Li, J.; Zhuge, Q.; Xiong, Q. Systematic identification and functional characterization of the CFEM proteins in poplar fungus *Marssonina brunnea*. *Front. Cell. Infect. Microbiol.* **2022**, *12*, 1045615. [CrossRef]
27. Zhu, W.; Wei, W.; Wu, Y.; Zhou, Y.; Peng, F.; Zhang, S.; Chen, P.; Xu, X. BcCFEM1, a CFEM domain-containing protein with putative gpi-anchored site, is involved in pathogenicity, conidial production, and stress tolerance in *Botrytis cinerea*. *Front. Microbiol.* **2017**, *8*, 1807. [CrossRef]
28. Wang, D.; Zhang, D.-D.; Song, J.; Li, J.-J.; Wang, J.; Li, R.; Klosterman, S.J.; Kong, Z.-Q.; Lin, F.-Z.; Dai, X.-F.; et al. *Verticillium dahliae* CFEM proteins manipulate host immunity and differentially contribute to virulence. *BMC Biol.* **2022**, *20*, 55. [CrossRef]
29. Bai, X.; Peng, H.; Goher, F.; Islam, M.A.; Xu, S.; Guo, J.; Kang, Z.; Guo, J. A candidate effector protein PstCFEM1 contributes to virulence of stripe rust fungus and impairs wheat immunity. *Stress Biol.* **2022**, *2*, 21. [CrossRef]
30. Shang, S.; Liu, G.; Zhang, S.; Liang, X.; Zhang, R.; Sun, G. A fungal CFEM-containing effector targets NPR1 regulator NIMIN2 to suppress plant immunity. *Plant Biotechnol. J.* **2024**, *22*, 82–97. [CrossRef]
31. Kulkarni, R.D.; Dean, R.A. Identification of proteins that interact with two regulators of appressorium development, adenylate cyclase and cAMP-dependent protein kinase A, in the rice blast fungus *Magnaporthe grisea*. *Mol. Genet. Genom.* **2004**, *270*, 497–508. [CrossRef]
32. Giraldo, M.C.; Valent, B. Filamentous plant pathogen effectors in action. *Nat. Rev. Microbiol.* **2013**, *11*, 800–814. [CrossRef] [PubMed]
33. Jacobs, K.A.; Collins-Racie, L.A.; Colbert, M.; Duckett, M.; Golden-Fleet, M.; Kelleher, K.; Kriz, R.; LaVallie, E.R.; Merberg, D.; Spaulding, V.; et al. A genetic selection for isolating cDNAs encoding secreted proteins. *Gene* **1997**, *198*, 289–296. [CrossRef] [PubMed]
34. Abramovitch, R.B.; Martin, G.B. Strategies used by bacterial pathogens to suppress plant defenses. *Curr. Opin. Plant Biol.* **2004**, *7*, 356–364. [CrossRef] [PubMed]
35. Mukhtar, M.S.; McCormack, M.E.; Argueso, C.T.; Pajerowska-Mukhtar, K.M. Pathogen tactics to manipulate plant cell death. *Curr. Biol.* **2016**, *26*, R608–R619. [CrossRef]

36. Upadhyaya, N.M.; Mago, R.; Staskawicz, B.J.; Ayliffe, M.A.; Ellis, J.G.; Dodds, P.N. A bacterial type iii secretion assay for delivery of fungal effector proteins into wheat. *Mol. Plant-Microbe Interact.* **2014**, *27*, 255–264. [CrossRef]
37. Liang, X.; Bao, Y.; Zhang, M.; Du, D.; Rao, S.; Li, Y.; Wang, X.; Xu, G.; Zhou, Z.; Shen, D.; et al. A *Phytophthora capsici* RXLR effector targets and inhibits the central immune kinases to suppress plant immunity. *New Phytol.* **2021**, *232*, 264–278. [CrossRef]
38. Wu, S.; Shan, L.; He, P. Microbial signature-triggered plant defense responses and early signaling mechanisms. *Plant Sci.* **2014**, *228*, 118–126. [CrossRef]
39. Rajput, N.A.; Zhang, M.; Ru, Y.; Liu, T.; Xu, J.; Liu, L.; Mafurah, J.J.; Dou, D. *Phytophthora sojae* Effector PsCRN70 Suppresses Plant Defenses in Nicotiana benthamiana. *PLoS ONE* **2014**, *9*, e98114. [CrossRef]
40. Klinkert, B.; Narberhaus, F. Microbial thermosensors. *Cell. Mol. Life Sci.* **2009**, *66*, 2661–2676. [CrossRef]
41. Zuo, N.; Bai, W.-Z.; Wei, W.-Q.; Yuan, T.-L.; Zhang, D.; Wang, Y.-Z.; Tang, W.-H. Fungal CFEM effectors negatively regulate a maize wall-associated kinase by interacting with its alternatively spliced variant to dampen resistance. *Cell Rep.* **2022**, *41*, 111877. [CrossRef]
42. Cai, N.; Liu, R.; Yan, D.; Zhang, N.; Zhu, K.; Zhang, D.; Nong, X.; Tu, X.; Zhang, Z.; Wang, G. bioinformatics analysis and functional characterization of the CFEM Proteins of *Metarhizium anisopliae*. *J. Fungi* **2022**, *8*, 661. [CrossRef] [PubMed]
43. Maeda, Y.; Kinoshita, T. Structural remodeling, trafficking and functions of glycosylphosphatidylinositol-anchored proteins. *Prog. Lipid Res.* **2011**, *50*, 411–424. [CrossRef] [PubMed]
44. Plaine, A.; Walker, L.; Da Costa, G.; Mora-Montes, H.M.; McKinnon, A.; Gow, N.A.R.; Gaillardin, C.; Munro, C.A.; Richard, M.L. Functional analysis of *Candida albicans* GPI-anchored proteins: Roles in cell wall integrity and caspofungin sensitivity. *Fungal Genet. Biol.* **2008**, *45*, 1404–1414. [CrossRef]
45. Pittet, M.; Conzelmann, A. Biosynthesis and function of GPI proteins in the yeast *Saccharomyces cerevisiae*. *Biochim. Biophys. Acta (BBA)—Mol. Cell Biol. Lipids* **2007**, *1771*, 405–420. [CrossRef]
46. Muszkieta, L.; Fontaine, T.; Beau, R.; Mouyna, I.; Vogt, M.S.; Trow, J.; Cormack, B.P.; Essen, L.-O.; Jouvion, G.; Latgé, J.-P. The Glycosylphosphatidylinositol-anchored dfg family is essential for the insertion of galactomannan into the β-(1,3)-glucan–chitin core of the cell wall of *Aspergillus fumigatus*. *mSphere* **2019**, *4*, e00397-19. [CrossRef]
47. Vaknin, Y.; Shadkchan, Y.; Levdansky, E.; Morozov, M.; Romano, J.; Osherov, N. The three *Aspergillus fumigatus* CFEM-domain GPI-anchored proteins (CfmA-C) affect cell-wall stability but do not play a role in fungal virulence. *Fungal Genet. Biol.* **2014**, *63*, 55–64. [CrossRef]
48. Lamkanfi, M.; Dixit, V.M. Manipulation of host cell death pathways during microbial infections. *Cell Host Microbe* **2010**, *8*, 44–54. [CrossRef]
49. Wagner, T.A.; Kohorn, B.D. Wall-associated kinases are expressed throughout plant development and are required for cell expansion. *Plant Cell* **2001**, *13*, 303–318. [CrossRef]
50. Zhang, B.; Li, P.; Su, T.; Li, P.; Xin, X.; Wang, W.; Zhao, X.; Yu, Y.; Zhang, D.; Yu, S.; et al. Comprehensive analysis of wall-associated kinase genes and their expression under abiotic and biotic stress in chinese cabbage (*Brassica rapa* ssp. *pekinensis*). *J. Plant Growth Regul.* **2020**, *39*, 72–86. [CrossRef]
51. Abdul Haseeb, H.; Zhang, J.; Guo, Y.-s.; Gao, M.-x.; Guo, W. Proteomic analysis of pathogen-responsive proteins from maize stem apoplast triggered by *Fusarium verticillioides*. *J. Integr. Agric.* **2022**, *21*, 446–459. [CrossRef]
52. Han, Y.; Li, X.; Cheng, L.; Liu, Y.; Wang, H.; Ke, D.; Yuan, H.; Zhang, L.; Wang, L. Genome-wide analysis of soybean jmjc domain-containing proteins suggests evolutionary conservation following whole-genome duplication. *Front. Plant Sci.* **2016**, *7*, 1800. [CrossRef] [PubMed]
53. Fang, A.; Han, Y.; Zhang, N.; Zhang, M.; Liu, L.; Li, S.; Lu, F.; Sun, W. Identification and characterization of plant cell death–inducing secreted proteins from *Ustilaginoidea virens*. *Mol. Plant-Microbe Interact.* **2016**, *29*, 405–416. [CrossRef] [PubMed]
54. Sun, S.; Ren, Y.; Wang, D.; Farooq, T.; He, Z.; Zhang, C.; Li, S.; Yang, X.; Zhou, X. A group I WRKY transcription factor regulates mulberry mosaic dwarf-associated virus-triggered cell death in *Nicotiana benthamiana*. *Mol. Plant Pathol.* **2022**, *23*, 237–253. [CrossRef]
55. Tang, C.; Xu, Q.; Zhao, J.; Yue, M.; Wang, J.; Wang, X.; Kang, Z.; Wang, X. A rust fungus effector directly binds plant pre-mRNA 769 splice site to reprogram alternative splicing and suppress host immunity. *Plant Biotechnol. J.* **2022**, *20*, 1167–1181. [CrossRef]
56. Schenk, S.T.; Hernández-Reyes, C.; Samans, B.; Stein, E.; Neumann, C.; Schikora, M.; Reichelt, M.; Mithöfer, A.; Becker, A.; Kogel, K.-H.; et al. N-acyl-homoserine lactone primes plants for cell wall reinforcement and induces resistance to bacterial pathogens via the salicylic acid/oxylipin pathway. *Plant Cell* **2014**, *26*, 2708–2723. [CrossRef]
57. Choi, Y.-E.; Xu, J.-R. The cAMP Signaling Pathway in *Fusarium verticillioides* is important for conidiation, plant infection, and stress responses but not fumonisin production. *Mol. Plant-Microbe Interact.* **2010**, *23*, 522–533. [CrossRef]
58. Ridenour, J.B.; Hirsch, R.L.; Bluhm, B.H. Identifying genes in *Fusarium verticillioides* through forward and reverse genetics. In *Plant Fungal Pathogens: Methods and Protocols*; Bolton, M.D., Thomma, B.P.H.J., Eds.; Humana Press: Totowa, NJ, USA, 2012; pp. 457–479.

59. Jiang, C.; Hei, R.; Yang, Y.; Zhang, S.; Wang, Q.; Wang, W.; Zhang, Q.; Yan, M.; Zhu, G.; Huang, P.; et al. An orphan protein of *Fusarium graminearum* modulates host immunity by mediating proteasomal degradation of TaSnRK1α. *Nat. Commun.* **2020**, *11*, 4382. [CrossRef]
60. Liang, X.; Zhang, X.; Xi, K.; Liu, Y.; Jijakli, M.H.; Guo, W. Development of an RPA-based CRISPR/Cas12a assay in combination with a lateral flow strip for rapid detection of toxigenic *Fusarium verticillioides* in maize. *Food Control* **2024**, *157*, 110172. [CrossRef]
61. Li, H.; Zhang, D.; Xie, K.; Wang, Y.; Liao, Q.; Hong, Y.; Liu, Y. Efficient and high-throughput pseudorecombinant-chimeric Cucumber mosaic virus-based VIGS in maize. *Plant Physiol.* **2021**, *187*, 2865–2876. [CrossRef]

International Journal of
Molecular Sciences

MDPI

Article

Identification of Host Factors Interacting with Movement Proteins of the 30K Family in *Nicotiana tabacum*

David Villar-Álvarez, Mikhail Oliveira Leastro, Vicente Pallas and Jesús Ángel Sánchez-Navarro *

Instituto de Biología Molecular y Celular de Plantas (IBMCP), Universitat Politècnica de Valencia-CISC, 46022 Valencia, Spain; davilal@posgrado.upv.es (D.V.-Á.); molilea@ibmcp.upv.es (M.O.L.); vpallas@ibmcp.upv.es (V.P.)

* Correspondence: jesanche@ibmcp.upv.es; Tel.: +34-963877989

Abstract: The interaction of viral proteins with host factors represents a crucial aspect of the infection process in plants. In this work, we developed a strategy to identify host factors in *Nicotiana tabacum* that interact with movement proteins (MPs) of the 30K family, a group of viral proteins around 30 kDa related to the MP of tobacco mosaic virus, which enables virus movement between plant cells. Using the alfalfa mosaic virus (AMV) MP as a model, we incorporated tags into its coding sequence, without affecting its functionality, enabling the identification of 121 potential interactors through in vivo immunoprecipitation of the tagged MP. Further analysis of five selected candidates (histone 2B (H2B), actin, 14-3-3A protein, eukaryotic initiation factor 4A (elF4A), and a peroxidase-POX-) were conducted using bimolecular fluorescence complementation (BiFC). The interactions between these factors were also studied, revealing that some form part of protein complexes associated with AMV MP. Moreover, H2B, actin, 14-3-3, and eIF4A interacted with other MPs of the 30K family. This observation suggests that, beyond functional and structural features, 30K family MPs may share common interactors. Our results demonstrate that tagging 30K family MPs is an effective strategy to identify host factors associated with these proteins during viral infection.

Keywords: 30K movement proteins; host factors; AMV; TMV; CPMV; CaMV; CMV

1. Introduction

For a plant virus to establish an infection, it must multiply in the infected cell and move into adjacent cells to eventually invade the vasculature and reach the distal parts of the plant host. Cell-to-cell movement occurs through pores communicating plant cells called plasmodesmata (PD), while long-distance movement takes place in most cases through the phloem [1,2]. Over the course of evolution, viruses have developed different strategies to carry out viral movement. In most cases, these strategies involve the expression of one or more specialized proteins called movement proteins (MPs) [3].

MPs can play different roles during the viral movement. Due to their ability to interact with nucleic acids, these proteins can form ribonucleoprotein complexes (vRNPs) with the viral genome. Along with their association with elements of the cytoskeleton and the endoplasmic reticulum (ER), they facilitate the viral movement towards the cell periphery [3,4]. Once there, MPs assist the passage of the viral progenies into the adjacent cell by different mechanisms that involve the modification of either the size exclusion limit or the whole structure of the PD. Cell-to-cell movement can be executed in different viral forms: as viral particles, vRNPs, or virus replication complexes (VRCs) [2,5].

Plant viruses encode a wide diversity of MPs. Depending on the number of MPs they employ, RNA viruses can be categorized into four distinct groups. The first group consists of viruses that encode multiple MPs of varying size and structure. The second group, known as the triple gene block, is composed of RNA viruses encoding three MPs [6,7]. The third group of viruses encodes two small MPs and is referred to as the double gene block group [8]. Finally, many viruses belonging to numerous viral genera, move using a single MP: this is the case of the 30K family group [3,9,10].

The MPs of the members of the 30K family are characterized by a size of around 30 kDa and share biochemical and structural characteristics with the MPs of tobacco mosaic virus (TMV) [11,12]. Despite their shared characteristics, their primary structure is highly variable. A conserved LXDX50-70G motif is flanked by a variable N-terminus (Nt) and C-terminus (Ct). The secondary structure consists of four α-helices and seven β-elements. The Ct domain is highly unstructured. Its function is related to the interaction with the coat protein (CP), but it is not necessary for intercellular movement [13,14]. The analysis of its three-dimensional structure suggests the presence of a trypsin-resistant "central core" containing two hydrophobic regions [15]. Molecular analysis using the alfalfa mosaic virus (AMV) system has shown that 30K proteins from at least seven different viral genera, including both RNA and DNA viruses, are functionally interchangeable for local and systemic transport [13,16–18]. These results imply that 30K proteins have conserved secondary or tertiary structures that have the same basic functions in all of them [18,19].

MPs can have other functions beyond those strictly related to the viral movement. Through interaction with host factors, MPs are able to modify different cellular processes such as subcellular localization, the regulation of gene expression, signaling, immune response, or metabolism [20–22]. Thus, MP–host factor interactions can influence virus accumulation and pathogenesis in different ways [23]. When these interactions facilitate virus accumulation, the host factors involved are termed pro-viral. Conversely, if the interaction reduces or blocks virus propagation, the host factors are referred to as antiviral [24,25].

Studying the role that these interactions play in the viral cycle is essential to understanding the course of an infection and ultimately developing therapeutic targets. Most of the methods for identifying host factors that interact with MPs are based on the analysis of protein–protein interactions (PPIs) [21]. A wide variety of methods are available to accomplish such approaches. These methods are grouped into three categories: those developed using computational tools or in silico, those performed outside a living organism in vitro, and those performed inside a living organism in vivo. Of these, in vivo methods have the obvious advantage of preserving the native conditions under which the interaction occurs [26,27].

So far, yeast two-hybrid (Y2H) and co-immunoprecipitation (Co-IP) are among the most used methods for host factor identification in vivo [24,25,28]. The Y2H method has enabled the identification of many host factors. Nevertheless, Y2H can also produce a non-negligible number of false positives, and it requires the generation of gene libraries if no previous search has been made and forces proteins to share the same cellular space [29]. On the other hand, Co-IP is helpful for finding new host factors by sequencing the protein or proteins retained following the immunoprecipitation of a previously labeled protein [30]. The main constraint of this approach is the availability of the corresponding antisera to immunoprecipitate the protein of interest. This is particularly relevant in the case of MPs, which are recalcitrant in generating efficient antibodies. However, the use of tags that are translationally fused in-frame to the protein of interest has overcome this limitation and made more universal this technology.

In this work, we developed a method to identify host factors during a viral infection based on Co-IP. By understanding the structure of the 30K family MPs, we inserted a tag into the coding sequence of the AMV MP without altering its functionality. This approach allowed for immunoprecipitation and subsequent sequencing, through which we identified potential host factors that interact with the MP throughout the infection. Since immunoprecipitation was performed from infected tissue, the native conditions under which the interactions occurred were preserved. This strategy allowed us to identify the host factors that interact not only with AMV MP but also with the MPs of other viruses of the 30K superfamily, with DNA and RNA genomes and different intercellular movement mechanisms.

2. Results

2.1. The Introduction of Tags into the Coding Sequence of the AMV MP Alters but Does Not Prevent Intercellular Movement

To identify host factors interacting with the AMV MP during the virus life cycle, two different epitope tags (hemagglutinin (HA) and human c-Myc protein (Myc)) were introduced into the coding sequence of the AMV MP. Tags were inserted between amino acids P256 and S257 (in isolate 425 Leiden, National Center for Biotechnology Information (NCBI); sequence number K03542.1). The insertion site was selected between the N-terminal functional transport domain and the C-terminal CP-interaction region [14]. A series of MP constructs with differing numbers of tags were tested. Three versions of the HA epitope were generated: one containing a single epitope (1HA), another with two epitopes (2HA), and the third with three epitopes (3HA). In the case of the myc epitope, a version of the MP with three myc epitopes repeated consecutively at the aforementioned position (3xmyc) was utilized, since the versions carrying one or two myc epitopes poorly reacted with the corresponding anti-myc antibody. To determine the impact of the tags in the MP-coding sequence on intercellular viral spread and the virus's capability to disseminate the infection, a version of the AMV cDNA3 carrying the GFP gene at its 5′ end was employed [14], thereby enabling the infection to be tracked by detecting fluorescence in the infected cells (Figure 1A).

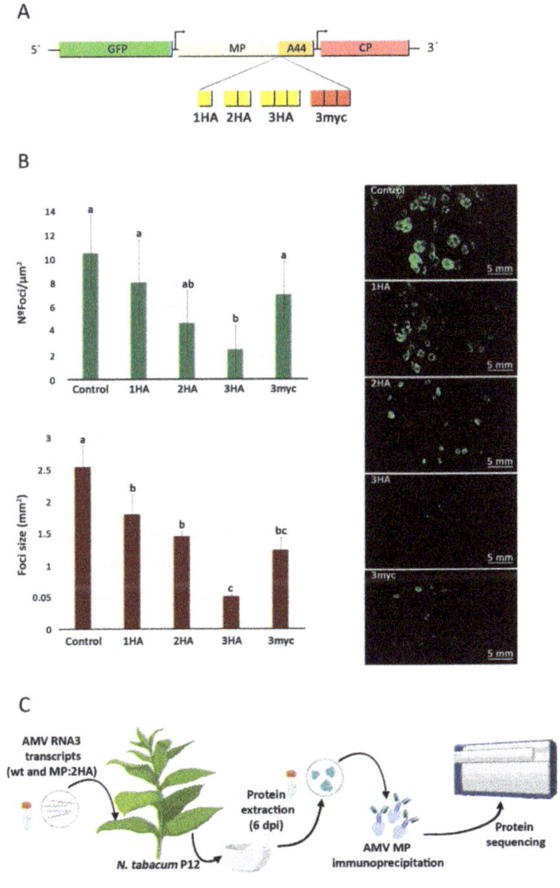

Figure 1. (**A**) Schematic representation of the AMV RNA 3 carrying the green fluorescence protein (GFP). Boxes correspond to the open reading frames of the GFP, the movement protein (MP), and

the coat protein (CP), while arrows indicate the subgenomic promoter. The insertion site for the tags 1HA, 2HA, 3HA, or 3myc, between amino acids P256 and S257, is marked just before the C-terminal 44 amino acids of the MP (A44). (**B**) A quantitative analysis of the number (expressed as N° foci/µm^2) and size (expressed in mm^2) of foci generated upon inoculation of transcripts corresponding to AMV cDNA3 wild type (control) or the constructs whose MPs carried 1HA, 2HA, 3HA, or 3myc epitopes is presented. The images on the right illustrate the infection foci generated by each of the aforementioned constructs. (**C**) Schematic representation of the procedure used to identify AMV MP interactors. *Nicotiana tabacum* p12 plants (plants constitutively overexpressing AMV replicase subunits 1 and 2) were inoculated with AMV cDNA3 transcripts carrying GFP at the 5′ end. Two versions of this cDNA3 were utilized: one with the MP bearing two HA tags and an untagged version, which served as a control. At six days post-inoculation, infected tissue was collected by visualizing the foci of infection. Subsequently, the total protein extraction and successive immunoprecipitation of AMV MP via the HA epitope were performed. The resulting immunoprecipitate was then subjected to mass spectrometry analysis. Experiments were performed in triplicate. Statistical analysis (Student's *t*-test, $p < 0.05$) of the data enabled the identification of differences between groups, which are represented in the graph by the letters a, b, and c.

The different versions of AMV cDNA3, carrying the different epitopes in the MP plus a wild type (wt) construct, were inoculated into *Nicotiana tabacum* P12 plants (plants constitutively expressing the P1 and P2 subunits of the AMV polymerase) [31]. The inoculated leaves were observed under magnification at 3 days post-inoculation (dpi) to analyze both the number and size of infection foci generated with each version of cDNA3. Regarding the number of foci per µm^2 (N° foci/µm^2), the cDNA3 versions carrying 1HA and 3xmyc in the MP did not show a significant decrease in comparison to the wt version. In contrast, the versions with 2HA and 3HA exhibited a more pronounced decrease of 56% and 76% in the number of foci per µm^2, respectively, in comparison to the control (Figure 1B). The size of the foci, expressed in mm^2, decreased in proportion to the number of HA epitopes present, with the versions containing 1HA, 2HA, and 3HA epitopes exhibiting a reduction of 30, 43, and 80%, respectively. The presence of three myc epitopes (3xmyc) resulted in a 51% decrease in mean foci size compared to the version without epitopes (Figure 1B). Based on these results, the AMV MP carrying two HA epitopes (2HA) was selected for the immunoprecipitation experiment and subsequent identification of potential host interactors.

2.2. *Ontological Analysis of the Proteins Immunoprecipitated with the AMV MP*

The Co-IP assay was conducted using the HA epitope and the Pierce Magnetic HA-Tag kit. Both versions of AMV cDNA3-GFP transcripts, namely the 2HA version and the untagged version, were inoculated onto *N. tabacum* P12 leaves by mechanical inoculation. Infection foci were collected at 6 dpi for protein extraction and subsequent MP-2HA immunoprecipitation (Figure 1C). The untagged version was employed as a control. Western blot analysis demonstrated the presence of AMV MP in both the crude protein extract and the final immunoprecipitate obtained from tissue inoculated with the 2HA version but not in those obtained from the tissue inoculated with the untagged version (Figure S1).

The immunoprecipitated proteins were subjected to mass spectrometry sequencing, resulting in the identification of a total of 121 potential interactors (see Table S1). Ontological analysis of these proteins was carried out using Omicsbox software (https://www.biobam.com/omicsbox, accessed on 17 July 2024), generating diagrams corresponding to the ontology divisions: biological process, molecular function, and cellular component. In terms of biological processes, the category to which the majority of proteins were assigned was metabolic processes, with 89.7% of the identified proteins. The second-largest category was that of cellular processes, with 79.4% of the proteins assigned to it. Approximately 7.1% of the proteins were assigned to the category of response to stimulus (Figure 2A). Concerning molecular function, 67.0% of the proteins were assigned to the binding category, 63.3% to catalytic activity, 13.3% to ATP-dependent activity, and 10.8% to structural molecule activity (Figure 2B). With regard to the cellular component, the three categories to which the highest number of proteins were assigned were intracellular anatomical (76.3%), cytoplasm (66.7%), and organelle (60.6%). The remaining categories, to which a smaller number of proteins were assigned, were membrane (31.7%), cytosol (25.9%), membrane protein complex (15.4%), ribonucleoprotein complex (14.4%), extracellular region (14.4%), organelle envelope (12.5%), and catalytic complex (10.6%) (Figure 2C).

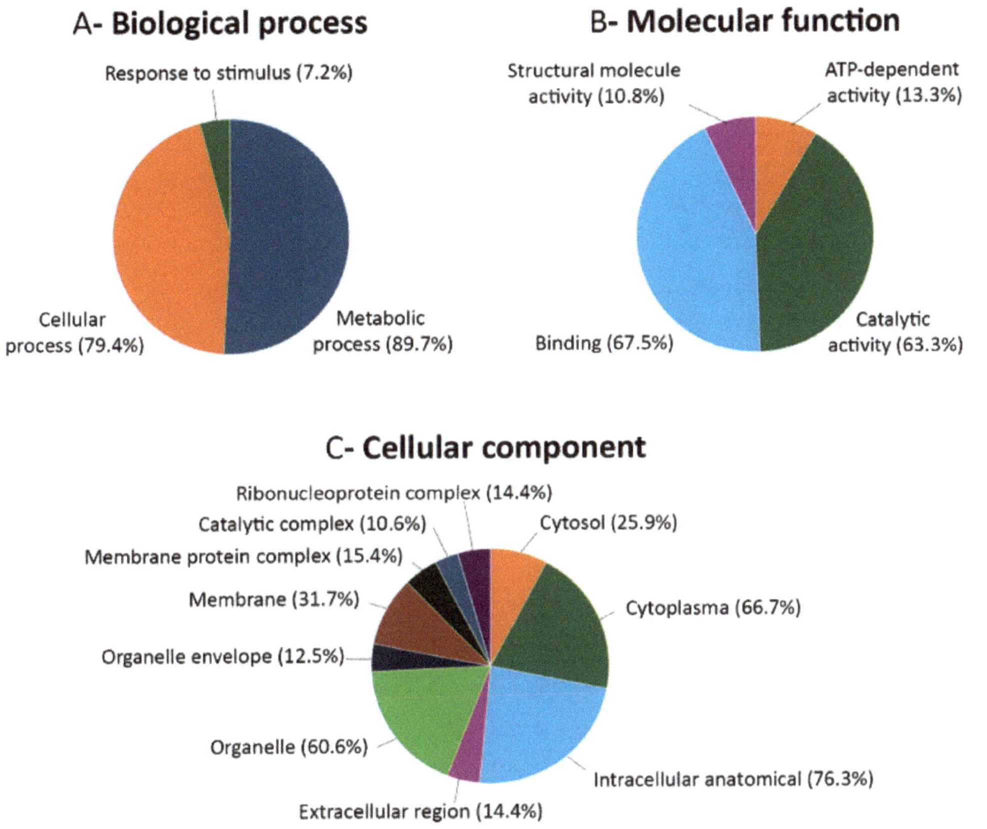

Figure 2. Ontological analysis of the proteins immunoprecipitated with the AMV MP. The graphs corresponding to the three ontological categories, (**A**) biological process, (**B**) molecular function, and (**C**) cellular component, are shown. Each graph illustrates the various subcategories into which the set of proteins under analysis was classified. The proportion of proteins in the sample assigned to each subcategory is indicated in brackets. Each protein can be assigned to more than one subcategory.

2.3. *Some Interactors Identified with the AMV MP Might Be Common in the MP 30K Family*

To validate the interaction of the potential interactors with the AMV MP with a different method, five candidate proteins were selected based on the number of reads obtained with each of them in the sequencing of the eluted peptides and the potential biological significance of the interaction. The selected proteins were histone 2B (H2B, LOC107775690), actin (Act, LOC107831313), 14-3-3A (LOC107765116), eukaryotic initiation factor 4A (eIF4A, LOC107764101), and peroxidase (POX, LOC107808758). Bimolecular fluorescence complementation (BiFC) experiments were conducted between the AMV MP and the five candidate proteins. The reconstitution of the GFP fluorescence signal was visualized in all of them, except for POX (Figure 3A). The nuclear-localized GFP signal was observed in interactions between AMV MP with H2B and eIF4e (Figure 3A,B), as expected. Punctate structures at the cell periphery that resembled the plasmodesmata localization were visualized in all positive interactions: AMV MP-H2B, AMV MP-actin, AMV MP-14-3-3A, and AMV MP-eIF4e (Figure 3A,B).

A

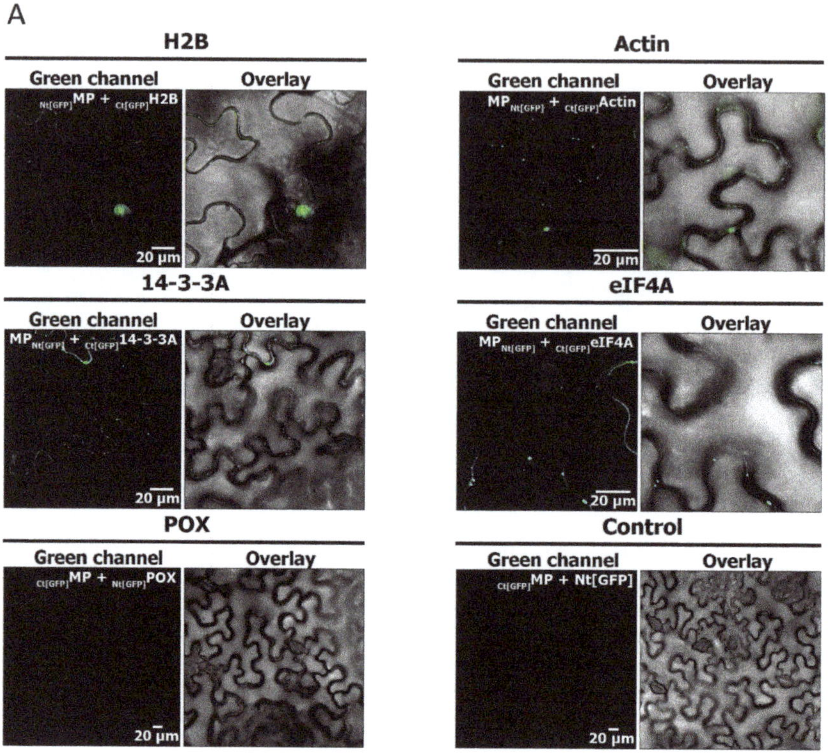

B

		AMV MP				TMV MP				CMV MP				CPMV MP				CaMV MP			
		5´Nt [GFP]	3´Nt [GFP]	5´Ct [GFP]	3´Ct [GFP]	5´Nt [GFP]	3´Nt [GFP]	5´Ct [GFP]	3´Ct [GFP]	5´Nt [GFP]	3´Nt [GFP]	5´Ct [GFP]	3´Ct [GFP]	5´Nt [GFP]	3´Nt [GFP]	5´Ct [GFP]	3´Ct [GFP]	5´Nt [GFP]	3´Nt [GFP]	5´Ct [GFP]	3´Ct [GFP]
H2B	5´Nt [GFP]			+/ N	–			+/ N	+/ N			+++/ C	++/ C,N,P			+++/ C,P,N	+/ C			++/ N	++/ N
	3´Nt [GFP]			++/ N	–			–	–			++/ N	–			+++/ C,N	–			+/ N	++/ C,N
	5´Ct [GFP]	+++/ M,N,P	–			–	++/ C,N,P			++/ C,N	++/ C,N			+++/ C,N	–			–	–		
	3´Ct [GFP]	–	–			–	++/ N			–	++/ C,N			+/ C	–			–	++/ N		
Actin	5´Nt [GFP]			–	–			+/ N	–			+/ M	++/ P			+/ C	+/ C			–	–
	3´Nt [GFP]			+/ M	–			–	–			+/ M	–			–	–			–	–
	5´Ct [GFP]	++/ C	++/ P			–	++/ C			+/ M	+++/ C			++/ C	+++/ C,P			–	–		
	3´Ct [GFP]	–	–			–	–			+/ C	–			++/ C	–			–	–		
eIF4A	5´Nt [GFP]			+++/ M,P	–			–	++/ M,P			+++/ N,M	+++/ P			+++/ M	+/ P			+++/ M,P	+++/ P
	3´Nt [GFP]			++/ M,P	–			–	+/ M			++/ M	++/ M,P			++/ M,P	++/ P			++/ M	++/ M,P
	5´Ct [GFP]	+++/ M,N,P	+++/ M,P			–	+++/ M,N,P			+++/ C	+++/ M,N			+++/ M,N	+++/ P			–	–		
	3´Ct [GFP]	++/ P	++/ N,P			–	++/ M,P			+++/ M,N	+++/ N,P			++/ M,N	+++/ P			–	–		
14-3-3A	5´Nt [GFP]			+++/ M,P	–			–	++/ M,P			++/ M	++/ M			+/ M	++/ M,P			–	–
	3´Nt [GFP]			+++/ M	–			–	++/ M,P			+++/ M	+++/ M			+/ M	+++/ M,N			–	–
	5´Ct [GFP]	+++/ M	+++/ M,P			–	+++/ M,P			+++/ M	+++/ M			+++/ M	+++/ M,P			–	–		
	3´Ct [GFP]	++/ M	+++/ M,P			–	++/ M,P			+++/ M	+++/ M			+/ M	++/ M			–	–		

Figure 3. The bimolecular fluorescence complementation (BiFC) between MPs of AMV, TMV, CMV, CPMV, and CaMV and the selected interactors (histone 2B (H2B), actin (Act), 14-3-3A, eukaryotic initiation

factor 4A (eIF4A), and peroxidase (POX)) is presented. (**A**) Confocal microscopy images of representative combinations of the analysis of each interaction with the AMV MP are shown. The scale bar represents 20 μm. (**B**) Table summarizing the results of the BiFC experiments on the interaction of the MPs of the five viruses with each of the proteins analyzed. The table shows all BiFC combinations in which the Nt or Ct fragments of the GFP are placed at the N-terminus (5′Nt [GFP], 5′Ct [GFP]), or the C-terminus (3′Nt [GFP], 3′Ct [GFP]) of the corresponding protein. The results are indicated as positive (+, green) or negative (−, red), and the intensity of the signal obtained under the same conditions (+, low intensity; ++, medium intensity; +++, high intensity) is noted. The most likely cellular location suggested by the images (C, cytoplasm; M, membrane; N, nucleus; P, plasmodesmata) is also indicated. The free Nt end of GFP was employed in the BiFC negative controls for all proteins analyzed.

To determine whether the interactors identified by the immunoprecipitation of AMV MP were shared by MPs of the 30K family, five more MPs of this family corresponding to tobacco mosaic virus (TMV), cucumber mosaic virus (CMV), cowpea mosaic virus (CPMV), and cauliflower mosaic virus (CaMV), which differ in both their genome and their type of intercellular movement, were also evaluated by BiFC. Similar to the AMV MP, these MPs showed a positive result for interaction with H2B and eIF4A proteins. MP interaction with 14-3-3A and actin proteins was positive for all MPs except the CaMV MP. However, no MP was found to interact with the POX protein (Figure 3 and Figure S2). Taken together, these results suggest that the MPs of the 30K MP family might share common interactors, regardless of how different viral properties are.

2.4. The Strategy Used to Identify Host Factors Interacting with the AMV MP Is Effective in Uncovering Components of a Multi-Protein Complex

BiFC experiments conducted with the AMV MP yielded positive results for the interaction of the AMV MP with four of the five candidate proteins tested. However, neither the AMV MP nor any of the 30K MP family tested yielded positive results for the interaction with the POX protein. One explanation for this result is that POX forms part of a multi-protein complex together with the other candidate proteins. To test whether the identified interactors were part of a protein complex, the interaction of the five selected interactors (H2B, Act, 14-3-3A, eIF4A, and POX) with each other was examined by BiFC. The results confirmed the interaction between all interactors. The fluorescence localization of the interactions was predominantly either nuclear or cytoplasmic. The localization at the plasma membrane or plasmodesmata was less frequent (Figure 4A,B). The interaction of the five interactors with each other suggests that the five protein candidates form part of a multi-protein complex and that POX was carried over during immunoprecipitation as part of this complex. In this hypothetical interactome, all the host proteins, with the exception of POX, interact with the AMV MP (Figure 5). Consequently, the results of this study indicate that the in vivo immunoprecipitation of AMV MP is an effective approach for the identification of interactors of the 30K MP family and for the identification of proteins that may form part of multi-protein complexes associated with them.

A

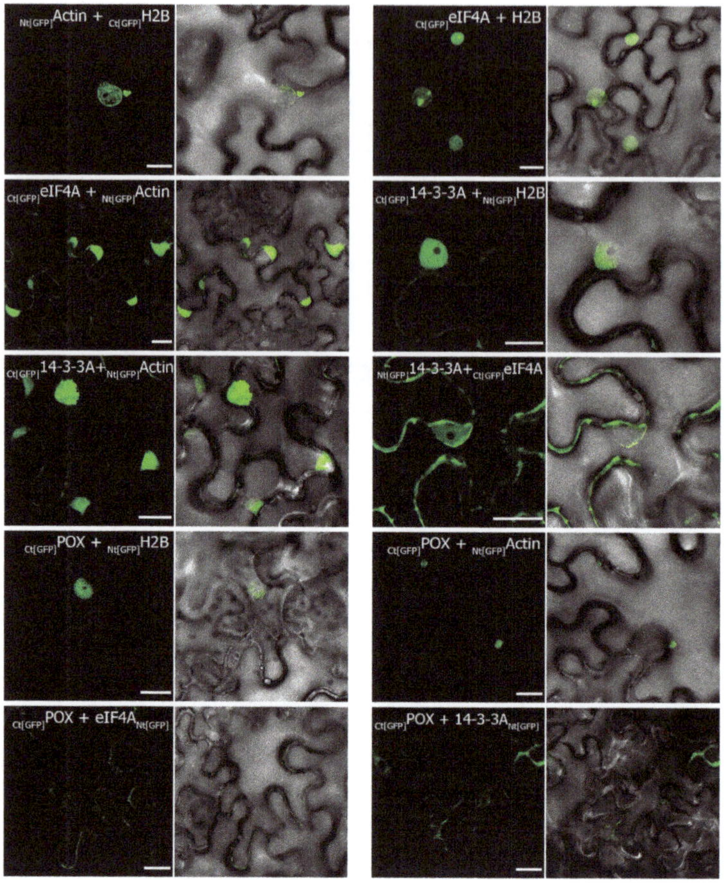

B

		H2B 5´Nt [GFP]	H2B 3´Nt [GFP]	H2B 5´Ct [GFP]	H2B 3´Ct [GFP]	Actin 5´Nt [GFP]	Actin 3´Nt [GFP]	Actin 5´Ct [GFP]	Actin 3´Ct [GFP]	eIF4A 5´Nt [GFP]	eIF4A 3´Nt [GFP]	eIF4A 5´Ct [GFP]	eIF4A 3´Ct [GFP]	14-3-3A 5´Nt [GFP]	14-3-3A 3´Nt [GFP]	14-3-3A 5´Ct [GFP]	14-3-3A 3´Ct [GFP]
Actin	5´Nt [GFP]			+++/C, M,N,P	–												
	3´Nt [GFP]			–	–												
	5´Ct [GFP]	+++/ N	++/ N														
	3´Ct [GFP]	+++/ N	+/ N														
eIF4A	5´Nt [GFP]			+++/ M,P	–			++/ C	+/ C								
	3´Nt [GFP]			++/ M,P	–			+++/ C	++/ C								
	5´Ct [GFP]	+++/ M,N,P	+++/ M,P			+++/ C	+++/ C										
	3´Ct [GFP]	++/ P	++/ N,P			++/ C	++/ C										
14-3-3A	5´Nt [GFP]			+++/ M,P	+/ N			++/ C	++/ C			++/ M	++/ M				
	3´Nt [GFP]			+++/ M	++/ M,N			+/ C	++/ C			+++/ M	+++/ M				
	5´Ct [GFP]	+++/ M,N	++/ M,N			+++/ C	–			+++/ M	+++/ M						
	3´Ct [GFP]	+++/ N	++/ M,N			+++/ C	++/ C			+++/ M	+++/ M						
POX	5´Nt [GFP]			–	–			–	–			–	–			–	–
	3´Nt [GFP]			–	–			–	–			–	–			–	–
	5´Ct [GFP]	++/ N	–			++/ C	–			+/ C	++/ C			+/ C	++/ C		
	3´Ct [GFP]	–	–			–	–			–	–			–	–		

Figure 4. The bimolecular fluorescence complementation (BiFC) experiments conducted among all candidate interactors with each other (histone 2B [H2B], actin [Act], 14-3-3A, eukaryotic initiation factor 4A [eIF4A], and peroxidase [POX]). (**A**) Confocal microscopy images of representative combinations of the analysis of

each interaction are shown. The scale bar represents 20 μm. (**B**) Table summarizing the results of the BiFC experiments conducted among all candidate interactors (H2B, Act, 14-3-3A, eIF4A, and POX). The table shows all BiFC combinations in which the Nt or Ct fragments of the GFP are placed at the N-terminus (5′Nt [GFP], 5′Ct [GFP]) or the C-terminus (3′Nt [GFP], 3′Ct [GFP]) of the corresponding protein. The results are indicated as positive (+, green) or negative (−, red), and the intensity of the signal obtained under the same conditions (+, low intensity; ++, medium intensity +++, high intensity) is noted. The most likely cellular location suggested by the images (C, cytoplasm; M, membrane; N, nucleus; P, plasmodesmata) is also indicated.

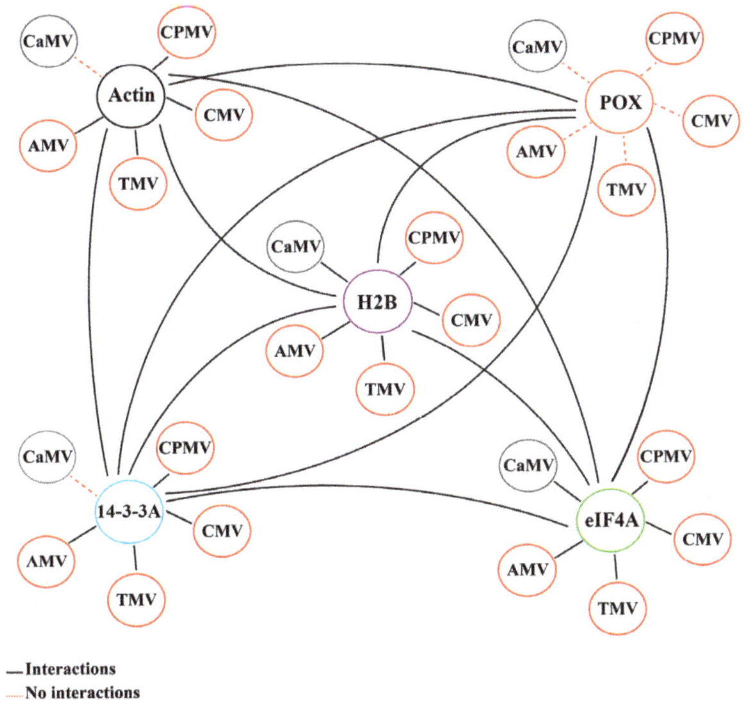

Figure 5. The proposed interactome is presented graphically, comprising the host factors studied with the five MP of the 30K family. The interactors are histone 2B (H2B), actin (Act), 14-3-3A, eukaryotic initiation factor 4A (eIF4A), and peroxidase (POX). MPs are identified by the name of the virus to which they belong: cauliflower mosaic virus (CaMV), cowpea mosaic virus (CPMV), cucumber mosaic virus (CMV), tobacco mosaic virus (TMV), and alfalfa mosaic virus (AMV). Solid and dashed lines between each component indicate positive and negative BiFC interaction, respectively.

3. Discussion

Here, we show that the introduction of tags into the coding sequence of the AMV MP has proven to be an effective strategy for the identification of host factors by immunoprecipitation of the viral protein during the course of infection. This is relevant since most of the studies dealing with the interactome of viral proteins have been carried out by overexpressing the protein of interest but not in the context of a viral infection. This approach has led to the identification of 121 potential candidates. Five of the potential candidates (H2B, Act, 14-3-3A, eIF4A, and POX) were selected based on both statistical and biological considerations. Their corresponding interactions were subsequently analyzed by an alternative method. Of the five candidates, four were confirmed to directly interact with the AMV MP, while POX was proven to interact with the other four candidates but not with the viral protein, suggesting that it forms part of a multi-protein complex related to the AMV MP.

In addition to immunoprecipitation, one of the most employed host factor identification approaches is the yeast-two hybrid (Y2H) system [32]. Zuidmeer-Jongejan (2002) utilized this system to identify host factors related to AMV MP [33]; however, the results were inconclusive. Subsequently, in order to overcome the limitations associated with the hydrophobic domain of the protein within the Y2H system, the same approach was carried out using a version of the MP that lacked the hydrophobic domain. The truncated version enabled the identification of patellins 3 and 6 as AMV MP interactors [15]. In contrast to the Y2H system used by Peiró et al. (2014) [15], the strategy presented in this work has the following advantages: (i) it allows for the use of a complete and functional MP; (ii) it is not necessary to create a protein library; and (iii) it allows for the immunoprecipitation of host factors during the course of infection.

The ontological analysis of the 121 candidates revealed that they exhibited variability within the three ontological categories. Of particular interest are the results obtained in the cellular component category, as they identify host factors associated with different cellular structures, including organelles, membranes, and cellular anatomical structures. The subcellular localization of the five candidates selected reflects this variability. H2B has a nuclear localization [34], whereas the 14-3-3 and eIF4A proteins manifest a variable subcellular localization, which can extend to the nucleus and cytoplasm [35–37]. POX proteins have been identified in several subcellular locations, including the cytosol, stroma, thylakoid membrane, and cell wall [38,39]. Actin, on its part, is primarily associated with the cytoskeleton [40]. The extensive range of locations of these proteins indicates that the system employed is applicable for the identification of interactors associated with any subcellular compartment.

The five selected proteins also display considerable functional diversity. Their potential roles in the viral cycle, which may be influenced by their interactions with MPs, are therefore diverse. The H2B protein has been assigned biological functions that extend beyond its structural role. For example, the post-translational modification of this protein has been linked to the regulation of the genetic response to different types of stress, including that generated by viral infections [41]. An illustrative example of the potential role of H2B in the plant defensive response to RNA viruses is provided by the study of potato virus X (PVX) in Nicotiana benthamiana. The repression of the H2B gene was found to result in a reduction in PVX accumulation, which was attributed to an SA-mediated response [42].

Actin is a crucial protein in the formation of the cytoskeleton in eukaryotic cells. The actin cytoskeleton is involved in numerous cellular processes and is a key player in pathogenesis [43,44]. For instance, Act plays a significant role in the intracellular transport of viral proteins and ribonucleoprotein complexes during infection [45]. Previous studies have already demonstrated that TMV and CMV MPs used actin filaments for intracellular movement [46–48]. The present work has demonstrated that the MPs of these two viruses, as well as those of AMV and CPMV, bind Act. This suggests that viral MPs that use actin microfilaments for intracellular movement may bind Act in addition to other actin-associated proteins.

Notably, 14-3-3 proteins serve as signal transducers in response to a multitude of stimuli. The transduction process is primarily initiated through the binding of these proteins to phosphorylated proteins. Consequently, 14-3-3 proteins are implicated in a plethora of cellular processes, including metabolism, hormone signaling, and stress responses. Furthermore, these proteins play a significant role in the defensive responses to a variety of pathogens, including viruses [49,50]. For instance, the *14-3-3* gene is upregulated in response to TMV infection, while the 14-3-3 protein interacts with the helicase domain of the TMV replicase [51]. Another illustration of the significance of these proteins in cellular responses to viral infections was provided by the beet black scorch virus (BBSV). The CP of this virus targets 14-3-3a to subvert mitogen-activated protein kinase (MAPKKKα)-mediated antiviral immunity in plants [52].

eIF4A is a member of the RNA helicase protein family carrying the amino acid sequence D-E-A-D (asp-glu-ala-asp) or DEAD-box. The primary function of this group of proteins is to facilitate the unfolding of mRNA, which is necessary for ribosome binding during translation initiation. In addition, eIF4 proteins are involved in the non-canonical translation of viral RNAs. The interaction between these factors and viral RNA has been extensively documented, and they are considered attractive targets for the success of antiviral strategies [53].

Finally, POX proteins are involved in cellular responses to different stimuli, catalyzing the oxidation of various organic compounds. In response to biotic stress, they are implicated in the production of H_2O_2, which triggers the activation of pathogen-responsive genes [54]. The expression of POX genes can change as a response to pathogen stress, as it occurs during TMV infection [55] or in apricot seeds infected with prunus necrotic ringspot virus, a virus closely related to AMV [56].

In light of the aforementioned information, it can be concluded that the method presented in this work is effective in identifying host factors relevant to the development of AMV viral infection. It is also important to consider that some of the identified interactors of AMV MP are also interactors of other MPs of the 30K family. The MP 30K family of viruses subjected to this study exhibited a variety of distinct characteristics, including differences in the type of intercellular movement and genome composition. The genomes of AMV, TMV, CMV, and CPMV are composed of RNA, whereas that of CaMV is composed of DNA. All of the aforementioned viruses form tubular structures for intercellular movement, with the exception of TMV. CaMV and CPMV move in the form of viral particles, whereas TMV moves in the form of ribonucleoprotein complexes (vRNPs). AMV and CMV are capable of moving either in the form of vRNPs or viral particles [3,5,12]. The observation that 30K MPs are functionally interchangeable in the AMV system [16] suggests shared functional characteristics despite the great diversity among these viruses. This could explain why they also might share common interactors. Further investigation is required to ascertain the precise role of each of these factors in AMV infection and other viruses within the 30K MPs. Nevertheless, the discovery of common factors interacting with multiple members of the 30K MP family members opens exciting research lines involving the characterization of how these viruses are translocated within the plant.

4. Materials and Methods

4.1. DNA Manipulation

4.1.1. Constructions for Bimolecular Fluorescence Complementation (BiFC) and Competitive Binding Assays

We performed the BiFC assay with the constructions pSK35S-NtGFP:-PoPit, pSK35S-CtGFP:-PoPit, pSK35S-:NtGFP-PoPit, and pSK35S-:CtGFP-PoPit, which permitted N- and C-terminal fusion of the enhanced green fluorescent protein (EGFP) fragments (NtGFP and CtGFP) at the N- or C-terminus of a specific assayed protein by ligation with the NcoI/NheI restriction sites. Detailed handling for obtaining these plasmids was previously described [57]. For the analysis of in vivo interactions, the 30K MPs from AMV (AAA46295.1); TMV (NP_597748); CPMV (CAA25315.1); CMV (BAA01396.1); CaMV (CAA23456.1); *N. tabacum* H2B-like (LOC107775690, hereinafter referred to as H2B only); actin (LOC107831313); peroxidase (LOC107808758); eIF4A (LOC107764101); and 14-3-3A (LOC107765116) were fused with the pSK plasmids described above to obtain the 4 possible combinations. The expression cassettes were subcloned into the pMOG800 binary vector. All genes were amplified using gene-specific oligonucleotides targeting the 5′ and 3′ ends of each gene. All DNA manipulations were confirmed by plasmid DNA sequencing.

4.1.2. Constructions for Immunoprecipitation Assays

For in vivo co-immunoprecipitation assays, HA or Myc tags were introduced into the coding sequence of the AMV MP. For this purpose, the AMV cDNA 3 clone (pGFP/A255/CP) [14], which expresses the GFP, was modified by replacing the wt version of the MP with the HA or Myc tagged version using the NcoI/NheI restriction sites. Epitopes were introduced between amino acids 256 and 257 of the AMV MP by PCR amplification using specific oligonucleotides containing the epitope sequence.

4.2. Inoculation of P12 Plants and Infection Foci Analysis

To analyze cell-to-cell movement, the AMV cDNA3 constructs were amplified from the corresponding plasmid using specific pair primers targeting the 5′ and 3′ termini of the viral sequence, with the sense primer containing the T7 promoter sequence. The generated amplicons were then used directly as templates for in vitro transcription with T7 RNA polymerase (Takara Bio Inc., Shiga, Japan). Transgenic *Nicotiana tabacum* P12 plants, which express the polymerase proteins P1 and P2 of AMV [31] were grown under long-day photoperiods of 16 h light at 25 °C and 8 h dark at 22 °C and inoculated by gently rubbing transcripts onto leaves. The infected areas on P12 plants were photographed using a Lupe microscope (MZZ16F Leica), with measurements taken at 3 dpi using ImageJ software version 2.0r. Graphs depict the average size in mm2 of the infection foci. Briefly, 10 μL of the transcription mixture per leaf was used to inoculate each construct.

4.3. *Agrobacterium tumefaciens*-Mediated Transient Expression and BiFC Assays

In transient expression assays conducted in *N. benthamiana*, the binary vectors acquired previously were introduced into *Agrobacterium tumefaciens* strain C58C1 by electroporation. The transformed bacteria were then cultured overnight in a liquid Luria–Bertani (LB) medium supplemented with kanamycin (50 mg/mL^{-1}) and rifampicin (100 mg/mL^{-1}) antibiotics for agrobacterium and plasmid selection. Subsequently, the bacterial cultures were centrifuged and resuspended to achieve the desired final OD600 value of 0.4 using a solution composed of 10 mM $MgCl_2$ and 10 mM MES adjusted to pH 5.6. These suspensions were gently infiltrated into the abaxial side of two-week-old *N. benthamiana* leaves by applying slight pressure. For experiments involving BiFC, which require the simultaneous expression of two different proteins, bacterial cultures were adjusted to an OD600 value of 0.4 and then mixed in equal proportions before infiltration. The plants were cultivated under long-day photoperiods, consisting of 16 h of light at 25 °C and 8 h of darkness at 22 °C.

4.4. Laser Scanning Confocal Microscopy and Image Analysis

The BiFC and translocation experiments were observed using an inverted Zeiss LSM 780 laser scanning confocal microscope (Carl Zeiss in Oberkochen, Germany). BiFC assays were observed two days after the agroinfiltration of *N. benthamiana* leaves. Leaves were observed under the confocal microscope at 3 dpi. Excitation for eGFP was achieved with lasers emitting light at wavelengths of 488 nm. Emission detection windows were set between 492 and 532 nm. Chlorophyll excitation utilized a 488 nm wavelength, and emission data were collected above 700 nm. Image processing and analysis, including overlays and Z-stack projections, were carried out using the FIJI software (version 2.9.0) [58].

4.5. Co-Immunoprecipitation Assay (Co-IP)

The Co-IP assay was conducted using the HA epitope and the Pierce Magnetic HA-Tag kit (Thermo Fisher Scientific, Waltham, MA USA), following the manufacturer's guidelines. The AMV cDNA3-GFP construct transcripts, comprising the HA epitope within the MP coding sequence, were mechanically inoculated onto *N. tabacum* P12 leaves through gentle rubbing with carborundum. A non-epitope version of cDNA3-GFP was used as the control. Infected tissue weighing 60 mg was obtained at 6 dpi by visualizing GFP under magnification to select infection sites. The tissue was homogenized with liquid nitrogen and 0.5 mL of denaturing buffer (50 mM Tris HCl, pH 8.0; 150 mM NaCl; 1% NP-40; and 0.5% sodium deoxycholate; and 0.1% SDS). The extracts were incubated by constant agitation for 2 h at 4 °C. Briefly, 40 μL of the supernatant was used for the protein expression control, while the remaining supernatant was incubated with 20 μL (0.2 mg) of Pierce Anti-HA Magnetic Beads for 90 min at room temperature with agitation. Following washing steps with non-denaturing buffer (25 mM Tris HCl, pH 7.5; 150 mM NaCl; 1 mM EDTA; 1% NP-

40; and 5% glycerol), the HA-immunoprecipitated proteins were eluted by resuspending the beads with 100 μL of 1× Laemmli buffer and incubated at 100 °C for 10 min. Western blot analysis was conducted using 20 μL of the expression control, 30 μL of IP proteins per line, and a monoclonal anti-HA antibody (Sigma-Aldrich, Steinheim, Germany), in accordance with the manufacturer's instructions. The rest of the IP eluate was sent for mass spectrometry sequencing at the proteomic service of the University of Córdoba. Functional annotation and analysis were conducted using OmicsBox (https://www.blast2go.com/, accessed on 17 July 2024) by using its default settings, followed by Gene Ontology mapping, where each potential GO term was given a GO annotation score.

Supplementary Materials: The following supporting information can be downloaded at: https://www.mdpi.com/article/10.3390/ijms252212251/s1.

Author Contributions: Conceptualization and experiment design, D.V.-Á., M.O.L. and J.Á.S.-N.; experiments, D.V.-Á., M.O.L. and J.Á.S.-N.; data analysis, D.V.-Á., M.O.L., J.Á.S.-N. and V.P.; project supervision, J.Á.S.-N.; writing—original draft preparation, D.V.-Á.; writing—review and editing, M.O.L., J.Á.S.-N. and V.P.; funding acquisition, J.Á.S.-N. and V.P. All authors have read and agreed to the published version of the manuscript.

Funding: D.V.-Á. is a recipient of a predoctoral contract from the Universitat Politècnica of Valencia (PAID-01-19). This work was supported by grants PROMETEO/CIPROM/2022/031 from the Generalitat Valenciana.

Institutional Review Board Statement: This study did not require ethical approval.

Informed Consent Statement: Not applicable.

Data Availability Statement: All the experimental data related to this work are included in the manuscript, figures, or Supporting Information.

Acknowledgments: We are grateful to Lorena Corachán for her excellent technical support.

Conflicts of Interest: The authors declare no conflicts of interest.

References

1. Folimonova, S.Y.; Tilsner, J. Hitchhikers, highway tolls and roadworks: The interactions of plant viruses with the phloem. *Curr. Opin. Plant Biol.* **2018**, *43*, 82–88. [CrossRef] [PubMed]
2. Heinlein, M. Plant virus replication and movement. *Virology* **2015**, *479–480*, 657–671. [CrossRef] [PubMed]
3. Navarro, J.A.; Sanchez-Navarro, J.A.; Pallas, V. Key checkpoints in the movement of plant viruses through the host. *Adv. Virus Res.* **2019**, *104*, 1–64.
4. Hong, J.S.; Ju, H.J. The plant cellular systems for plant virus movement. *Plant Pathol. J.* **2017**, *33*, 213. [CrossRef]
5. Villar-Álvarez, D.; Pallás, V.; Elena, S.F.; Sánchez-Navarro, J.A. An evolved 5′ untranslated region of alfalfa mosaic virus allows the RNA Transport of movement-defective variants. *J. Virol.* **2022**, *96*, e00988-22. [CrossRef]
6. Morozov, S.Y.; Solovyev, A.G. Triple Gene Block: Modular design of a multifunctional machine for plant virus movement. *J. Gen. Virol.* **2003**, *84*, 1351–1366. [CrossRef]
7. Verchot-Lubicz, J.; Torrance, L.; Solovyev, A.G.; Morozov, S.Y.; Jackson, A.O.; Gilmer, D. Varied movement strategies employed by triple gene block–encoding viruses. *Mol. Plant-Microbe Interact.* **2010**, *23*, 1231–1247. [CrossRef] [PubMed]
8. Hull, R. Induction of Disease 1: Virus movement through the plant and effects on plant metabolism. In *Matthews' Plant Virology*; Academic Press: Cambridge, MA, USA, 2002; pp. 373–436.
9. Taliansky, M.; Torrance, L.; Kalinina, N.O. Role of plant virus movement proteins. *Methods Mol. Biol.* **2008**, *451*, 33–54. [PubMed]
10. Tilsner, J.; Taliansky, M.E.; Torrance, L. Plant virus movement. *eLS* **2014**. [CrossRef]
11. Melcher, U. The "30K" Superfamily of viral movement proteins. *J. Gen. Virol.* **2000**, *81*, 257–266. [CrossRef]
12. Mushegian, A.R.; Elena, S.F. Evolution of plant virus movement proteins from the 30k superfamily and of their homologs integrated in plant genomes. *Virology* **2015**, *476*, 304–315. [CrossRef] [PubMed]
13. Aparicio, F.; Pallás, V.; Sánchez-Navarro, J. Implication of the C terminus of the prunus necrotic ringspot virus movement protein in cell-to-cell transport and in its interaction with the coat protein. *J. Gen. Virol.* **2010**, *91*, 1865–1870. [CrossRef] [PubMed]
14. Sanchez-Navarro, J.; Miglino, R.; Ragozzino, A.; Bol, J.F. Engineering of alfalfa mosaic virus RNA 3 into an expression vector. *Arch. Virol.* **2001**, *146*, 923–939. [CrossRef]
15. Peiró, A.; Martínez-Gil, L.; Tamborero, S.; Pallás, V.; Sánchez-Navarro, J.A.; Mingarro, I. The tobacco mosaic virus movement protein associates with but does not integrate into biological membranes. *J. Virol.* **2014**, *88*, 3016–3026. [CrossRef]

16. Fajardo, T.V.M.; Peiró, A.; Pallás, V.; Sánchez-Navarro, J. Systemic transport of alfalfa mosaic virus can be mediated by the movement proteins of several viruses assigned to five genera of the 30K family. *J. Gen. Virol.* **2013**, *94*, 677–681. [CrossRef] [PubMed]
17. Leastro, M.O.; Freitas-Astúa, J.; Kitajima, E.W.; Pallás, V.; Sánchez-Navarro, J.A. Unravelling the involvement of Cilevirus P32 protein in the viral transport. *Sci. Rep.* **2021**, *11*, 2943. [CrossRef]
18. Sánchez-Navarro, J.A.; Herranz, M.C.; Pallás, V. Cell-to-cell movement of alfalfa mosaic virus can be mediated by the movement proteins of ilar-, bromo-, cucumo-, tobamo- and comoviruses and does not require virion formation. *Virology* **2006**, *346*, 66–73. [CrossRef]
19. Pallas, V.; Aparicio, F.; Herranz, M.C.; Sanchez-Navarro, J.A.; Scott, S.W. The molecular biology of ilarviruses. *Adv. Virus. Res.* **2013**, *87*, 139–181.
20. Kumar, G.; Dasgupta, I.; Castillo, N. Variability, functions and interactions of plant virus movement proteins: What do we know so far? *Microorganisms* **2021**, *9*, 365. [CrossRef]
21. Link, K.; Sonnewald, U. Interaction of movement proteins with host factors, mechanism of viral host cell manipulation and influence of MPs on plant growth and development. In *Plant-Virus Interactions: Molecular Biology, Intra- and Intercellular Transport;* Springer: Berlin/Heidelberg, Germany, 2016; pp. 1–37.
22. Atabekova, A.K.; Solovieva, A.D.; Chergintsev, D.A.; Solovyev, A.G.; Morozov, S.Y. Role of plant virus movement proteins in suppression of host RNAi defense. *Int. J. Mol. Sci.* **2023**, *24*, 9049. [CrossRef]
23. Pallas, V.; García, J.A. How do plant viruses induce disease? Interactions and interference with host components. *J. Virol.* **2011**, *92*, 2691–2705. [CrossRef] [PubMed]
24. Garcia-Ruiz, H. Susceptibility genes to plant viruses. *Viruses* **2018**, *10*, 484. [CrossRef] [PubMed]
25. Garcia-Ruiz, H. Host factors against plant viruses. *Mol. Plant. Pathol.* **2019**, *20*, 1588–1601. [CrossRef]
26. Peng, X.; Wang, J.; Peng, W.; Wu, F.-X.; Pan, Y. Protein–protein interactions: Detection, reliability assessment and applications. *Brief. Bioinform.* **2017**, *18*, 798–819. [CrossRef] [PubMed]
27. Xing, S.; Wallmeroth, N.; Berendzen, K.W.; Grefen, C. Techniques for the analysis of protein-protein interactions in vivo. *Plant Physiol.* **2016**, *171*, 727–758. [CrossRef]
28. Rao, V.S.; Srinivas, K.; Sujini, G.N.; Sunand Kumar, G.N. Protein-protein interaction detection: Methods and analysis. *Int. J. Proteom.* **2014**, *2014*, 147648. [CrossRef]
29. Paiano, A.; Margiotta, A.; De Luca, M.; Bucci, C. Yeast two-hybrid assay to identify interacting proteins. *Curr. Protoc. Protein Sci.* **2019**, *95*, e70. [CrossRef]
30. Muñoz, A.; Castellano, M.M. Coimmunoprecipitation of interacting proteins in plants. *Methods Mol. Biol.* **2018**, *1794*, 279–287.
31. Taschner, P.; Van Der Kuyl, A.; Neeleman, L. Replication of an incomplete alfalfa mosaic virus genome in plants transformed with viral replicase genes. *Virology* **1991**, *181*, 445–450. [CrossRef]
32. Farooq, Q.U.A.; Shaukat, Z.; Aiman, S.; Li, C.-H. Protein-protein interactions: Methods, databases, and applications in virus-host study. *World J. Virol.* **2021**, *10*, 288. [CrossRef]
33. Zuidmeer-Jongejan, L. Interactions of Plant Proteins with Alfalfa Mosaic Virus Movement Protein. Ph.D. Thesis, Leiden University, Leiden, The Netherlands, 2002; 137p.
34. Musinova, Y.R.; Lisitsyna, O.M.; Golyshev, S.A.; Tuzhikov, A.I.; Polyakov, V.Y.; Sheval, E.V. Nucleolar localization/retention signal is responsible for transient accumulation of histone H2B in the nucleolus through electrostatic interactions. *Biochim. Biophys. Acta Mol. Cell Res.* **2011**, *1813*, 27–38. [CrossRef] [PubMed]
35. Huang, Y.; Wang, W.; Yu, H.; Peng, J.; Hu, Z.; Chen, L. The role of 14-3-3 proteins in plant growth and response to abiotic stress. *Plant Cell Rep.* **2022**, *41*, 833–852. [CrossRef] [PubMed]
36. Koroleva, O.A.; Brown, J.W.S.; Shaw, P.J. Localization of EIF4A-III in the nucleolus and splicing speckles is an indicator of plant stress. *Plant Signal. Behav.* **2009**, *4*, 1148–1151. [CrossRef]
37. Tyagi, V.; Parihar, V.; Malik, G.; Kalra, V.; Kapoor, S.; Kapoor, M. The DEAD-box RNA helicase EIF4A regulates plant development and interacts with the HnRNP LIF2L1 in Physcomitrella patens. *Mol. Genet. Genom.* **2020**, *295*, 373–389. [CrossRef] [PubMed]
38. Olmos, E.; Piqueras, A.; Ramón Martínez-Solano, J.; Hellín, E. The subcellular localization of peroxidase and the implication of oxidative stress in hyperhydrated leaves of regenerated carnation plants. *Plant Sci.* **1997**, *130*, 97–105. [CrossRef]
39. Teixeira, F.K.; Menezes-Benavente, L.; Galvão, V.C.; Margis, R.; Margis-Pinheiro, M. Rice ascorbate peroxidase gene family encodes functionally diverse isoforms localized in different subcellular compartments. *Planta* **2006**, *224*, 300–314. [CrossRef]
40. Staiger, C.J. Signaling to the actin cytoskeleton in plants. *Annu. Rev. Plant Biol.* **2000**, *51*, 257–288. [CrossRef]
41. Zarreen, F.; Karim, M.J.; Chakraborty, S. The diverse roles of histone 2B monoubiquitination in the life of plants. *J. Exp. Bot.* **2022**, *73*, 3854–3865. [CrossRef]
42. Yang, X.; Lu, Y.; Zhao, X.; Jiang, L.; Xu, S.; Peng, J.; Zheng, H.H.; Lin, L.; Wu, Y.; MacFarlane, S.; et al. Downregulation of nuclear protein H2B induces salicylic acid mediated defense against PVX infection in Nicotiana benthamiana. *Front. Microbiol.* **2019**, *10*, 454484. [CrossRef]
43. Porter, K.; Day, B. From filaments to function: The role of the plant actin cytoskeleton in pathogen perception, signaling and immunity. *J. Integr. Plant Biol.* **2016**, *58*, 299–311. [CrossRef]
44. Yuan, G.; Gao, H.; Yang, T. Exploring the role of the plant actin cytoskeleton: From signaling to cellular functions. *Int. J. Mol. Sci.* **2023**, *24*, 15480. [CrossRef] [PubMed]

45. Takemoto, D.; Hardham, A.R. The cytoskeleton as a regulator and target of biotic interactions in plants. *Plant Physiol.* **2004**, *136*, 3864–3876. [CrossRef] [PubMed]
46. Chen, C.; Zhang, Y.; Zhu, L.; Yuan, M. The actin cytoskeleton is involved in the regulation of the plasmodesmal size exclusion limit. *Plant Signal. Behav.* **2010**, *5*, 1663–1665. [CrossRef] [PubMed]
47. Niehl, A.; Peña, E.J.; Amari, K.; Heinlein, M. Microtubules in viral replication and transport. *Plant J.* **2013**, *75*, 290–308. [CrossRef]
48. Wright, K.M.; Wood, N.T.; Roberts, A.G.; Chapman, S.; Boevink, P.; Mackenzie, K.M.; Oparka, K.J. Targeting of TMV movement protein to plasmodesmata requires the actin/ER network; evidence from FRAP. *Traffic* **2007**, *8*, 21–31. [CrossRef]
49. Lozano-Durán, R.; Robatzek, S. 14-3-3 14-3-3 proteins in plant-pathogen interactions. *Mol. Plant Microbe Interact.* **2015**, *28*, 511–518. [CrossRef]
50. Liu, J.; Cao, S.; Ding, G.; Wang, B.; Li, Y.; Zhao, Y.; Shao, Q.; Feng, J.; Liu, S.; Qin, L.; et al. The Role of 14-3-3 Proteins in cell signalling pathways and virus infection. *J. Cell Mol. Med.* **2021**, *25*, 4173–4182. [CrossRef]
51. Konagaya, K.I.; Matsushita, Y.; Kasahara, M.; Nyunoya, H. Members of 14-3-3 protein isoforms interacting with the resistance gene product N and the elicitor of tobacco mosaic virus. *J. Gen. Plant Pathol.* **2004**, *70*, 221–231. [CrossRef]
52. Gao, Z.; Zhang, D.; Wang, X.; Zhang, X.; Wen, Z.; Zhang, Q.; Li, D.; Dinesh-Kumar, S.P.; Zhang, Y. Coat proteins of necroviruses target 14-3-3a to subvert MAPKKKα-mediated antiviral immunity in plants. *Nat. Commun.* **2022**, *13*, 716. [CrossRef]
53. Shopan, J.; Lv, X.; Hu, Z.; Zhang, M.; Yang, J. Eukaryotic translation initiation factors shape RNA viruses resistance in plants. *Hortic. Plant J.* **2020**, *6*, 81–88. [CrossRef]
54. Kidwai, M.; Ahmad, I.Z.; Chakrabarty, D. Class III Peroxidase: An indispensable enzyme for biotic/abiotic stress tolerance and a potent candidate for crop improvement. *Plant Cell Rep.* **2020**, *39*, 1381–1393. [CrossRef] [PubMed]
55. Hernández, J.A.; Gullner, G.; Clemente-Moreno, M.J.; Künstler, A.; Juhász, C.; Díaz-Vivancos, P.; Király, L. Oxidative stress and antioxidative responses in plant–virus interactions. *Physiol. Mol. Plant Pathol.* **2016**, *94*, 134–148. [CrossRef]
56. Amari, K.; Díaz-Vivancos, P.; Pallás, V.; Sánchez-Pina, M.A.; Hernández, J. A Oxidative stress induction by Prunus necrotic ringspot virus infection in apricot seeds. *Physiol. Plant.* **2007**, *131*, 302–310. [CrossRef] [PubMed]
57. Leastro, M.O.; Pallás, V.; Resende, R.O.; Sánchez-Navarro, J.A. The movement proteins (NSm) of distinct tospoviruses peripherally associate with cellular membranes and interact with homologous and heterologous NSm and nucleocapsid proteins. *Virology* **2015**, *478*, 39–49. [CrossRef]
58. Schindelin, J.; Arganda-Carreras, I.; Frise, E.; Kaynig, V.; Longair, M.; Pietzsch, T.; Preibisch, S.; Rueden, C.; Saalfeld, S.; Schmid, B.; et al. Fiji: An open-source platform for biological-image analysis. *Nat. Methods* **2012**, *9*, 676–682. [CrossRef]

International Journal of
Molecular Sciences

MDPI

Article

Employing Genomic Tools to Explore the Molecular Mechanisms behind the Enhancement of Plant Growth and Stress Resilience Facilitated by a *Burkholderia* Rhizobacterial Strain

Yueh-Long Chang [1], Yu-Cheng Chang [1], Andi Kurniawan [1,2], Po-Chun Chang [1], Ting-Yu Liou [1], Wen-Der Wang [1] and Huey-wen Chuang [1,*]

[1] Department of Agricultural Biotechnology, National Chiayi University, Chiayi 600355, Taiwan
[2] Department of Agronomy, Brawijaya University, Malang 65145, Indonesia
* Correspondence: hwchuang@mail.ncyu.edu.tw

Abstract: The rhizobacterial strain BJ3 showed 16S rDNA sequence similarity to species within the *Burkholderia* genus. Its complete genome sequence revealed a 97% match with *Burkholderia contaminans* and uncovered gene clusters essential for plant-growth-promoting traits (PGPTs). These clusters include genes responsible for producing indole acetic acid (IAA), osmolytes, non-ribosomal peptides (NRPS), volatile organic compounds (VOCs), siderophores, lipopolysaccharides, hydrolytic enzymes, and spermidine. Additionally, the genome contains genes for nitrogen fixation and phosphate solubilization, as well as a gene encoding 1-aminocyclopropane-1-carboxylate (ACC) deaminase. The treatment with BJ3 enhanced root architecture, boosted vegetative growth, and accelerated early flowering in *Arabidopsis*. Treated seedlings also showed increased lignin production and antioxidant capabilities, as well as notably increased tolerance to water deficit and high salinity. An RNA-seq transcriptome analysis indicated that BJ3 treatment significantly activated genes related to immunity induction, hormone signaling, and vegetative growth. It specifically activated genes involved in the production of auxin, ethylene, and salicylic acid (SA), as well as genes involved in the synthesis of defense compounds like glucosinolates, camalexin, and terpenoids. The expression of AP2/ERF transcription factors was markedly increased. These findings highlight BJ3's potential to produce various bioactive metabolites and its ability to activate auxin, ethylene, and SA signaling in *Arabidopsis*, positioning it as a new *Burkholderia* strain that could significantly improve plant growth, stress resilience, and immune function.

Keywords: microbial IAA; nitrogen fixation; microbial osmolytes; auxin signaling; ethylene synthesis; salicylic signaling

1. Introduction

Employing beneficial microorganisms to promote plant growth and reduce stress damage has become a valuable strategy in agricultural practice. Known as plant-growth-promoting rhizobacteria (PGPR), these beneficial microbes in the rhizosphere significantly impact plant health and productivity. PGPR can enhance plant growth by producing phytohormones such as indole acetic acid (IAA), a natural form of auxin, which is an essential regulator of lateral root formation [1]. Rhizobacteria may increase nutrient availability for plant growth through their activities of mineral solubilization, siderophore production, and nitrogen fixation [2]. Alternatively, PGPR enhance plant growth vigor through the production of various antimicrobial metabolites that suppress the viability of pathogens [3]. As a biocontrol agent, PGPR not only directly target pathogens but also produce elicitors that enhance a plant's innate defenses, a response known as induced systemic resistance (ISR). In this defense mechanism, jasmonic acid (JA) and ethylene signals are crucial for reshaping plant metabolism to build resistance against pathogens [4]. ISR, elicited by PGPR, exhibits phenotypic similarities to pathogen-induced systemic acquired resistance (SAR), in which salicylic acid (SA) acts as a key regulator for stimulating cellular pathways to establish disease resistance [5]. In the natural environment, plants interact with rhizobacteria by using their pattern recognition receptors (PRRs) to recognize pathogen-associated molecular patterns (PAMPs), also referred to as microbial-associated molecular patterns (MAMPs) [6]. PRRs recognize the PAMPs/MAMPs and subsequently activate PAMP/MAMP-triggered immunity (PTI/MTI), conferring an effective defense response [7].

PGPR produce various types of bioactive metabolites, which can exert multifaceted effects to improve plant tolerance to unfavorable environments. Throughout their life cycle, plants face a range of environmental stresses at various developmental stages. Root systems are crucial for the uptake of nutrients and water, making them vital for a plant's survival. Consequently, the root system is key to ensuring that the aboveground parts of the plant can adapt to environmental challenges. Lateral roots are important to expand the root system for nutrient acquisition. Auxin is a key regulator of root growth and development [1]. PGPR are known to stimulate root architecture in the host plant by producing significant amounts of IAA [8]. In addition to its main function in controlling plant growth, studies have shown the regulatory role of auxin in plant responses to abiotic stress. For example, auxin biosynthesis and transport are altered in plants under salt stress conditions [9,10]. Additionally, PGPR can enhance the synthesis of IAA in wheat under salt and drought stress conditions [11]. Moreover, treating banana plants with a *Pseudomonas* PGPR strain not only increased their resistance to drought and submergence stress but also activated genes linked to the auxin signaling pathway [12]. As plants encounter an unfavorable environment, ethylene is produced to reduce their growth and proliferation [13]. Studies indicate that PGPR mitigate the adverse effects of stressful conditions by producing ACC deaminase, which in turn reduces ethylene production in the host plants [14]. Ethylene suppresses the development of lateral roots by impacting the polar transport of auxin [15]. However, in collaboration with auxin, ethylene also actively promotes the formation of root hairs [16]. Root hairs, components of the root architecture system, play a role in the adaptive response to abiotic stress [17]. For example, under conditions of phosphate starvation, ethylene signaling increases, resulting in enhanced root hair formation [18]. Moreover, ethylene collaborates with JA to regulate the *AP2/ERF* transcription factor gene family, thereby exerting a positive effect on plant development by enhancing tolerance to abiotic stress and disease resistance [19]. SA and JA signals play crucial roles in triggering plant immunity against pathogens. Additionally, both SA and JA can mitigate the adverse effects of abiotic stress on plant growth by activating antioxidant defenses [20]. For example, SA is a key regulator of the 2,3-butanediol VOC-induced drought stress tolerance [21]. Volatile compounds from a *Bacillus* strain induced salt stress tolerance in *Arabidopsis* due to the activation of the JA signaling pathway [22].

The bacterial strains of the *Burkholderia* genus are widely distributed in the soil. These bacterial strains promote plant growth through their phosphate-solubilizing activities [23], nitrogen-fixation ability [24], and altering phytohormone signaling in the host plants [25]. Genes involved in the synthesis of various antimicrobial metabolites have been identified in the *Burkholderia* genome [26]. Several strains of the *Burkholderia contaminans* species have demonstrated their effectiveness in biocontrol of plant diseases [27,28]. *Burkholderia phytofirmans* PsJN has been reported to activate induced resistance against gray mold and to enhance cold stress tolerance by increasing the levels of stress-related metabolites in a low-temperature environment [29].

In this study, the bacterial strain BJ3 was identified as a newly isolated *Burkholderia contaminans* strain based on its whole-genome sequencing results. Gene clusters associated with plant-growth-promoting traits (PGPTs) were identified; among these, genes involved in the synthesis of IAA, osmolytes, NRPS, and VOCs were particularly significant. This genome also contains genes related to nitrogen fixation and those encoding ACC deaminase. BJ3 not only promoted vegetative growth but also stimulated an early flowering phenotype in the treated plants. Moreover, this treatment improved plant tolerance to drought and salt stress. A transcriptome analysis revealed that BJ3 treatment activated the expression of genes associated with stress tolerance and vegetative growth, as well as genes linked to hormone signaling pathways, including auxin, ethylene, and SA.

2. Results

2.1. Molecular Characterization of BJ3

The sequencing results for the 16S rDNA PCR fragment from BJ3 bacterial genomic DNA showed 100% identity with several strains of the *Burkholderia* genus, including *Burkholderia contaminans*, *Burkholderia* sp., and *Burkholderia cenocepacia* (Table 1). To identify genes associated with plant growth promotion in the BJ3 genome, the whole genome sequence was analyzed using both Illumina and Nanopore sequencing systems. The results revealed 7361 open reading frames (ORFs) in the BJ3 genome, which were assigned to two chromosomes. Of these, 7112 (96.62%) ORFs showed sequence identity to *Burkholderia contaminans*, while 165 (2.24%) ORFs matched sequences from various *Burkholderia* species. Gene clusters responsible for synthesizing metabolites that function in plant growth promotion were identified and are listed in Table 2. These genes are involved in producing IAA, pyrroloquinoline quinone (PQQ), NRPS, siderophores, volatile organic compounds (VOCs), spermidine (SPD), osmolytes, lipopolysaccharides, and hydrolytic enzymes, such as chitinase and cellulase. The identified PGPT-associated genes also play a role in nitrogen fixation and code for ACC deaminase, which has activity for reducing ethylene synthesis (Figure 1). Microbes producing IAA and ACC deaminase may promote plant growth by altering hormone signaling in the host plants [8,14]. The BJ3 genome contains a large number of genes coding for proteins involved in IAA synthesis via three pathways: indolepyruvate ferredoxin oxidoreductase (IOR) in the indole-3-pyruvate (IPyA) pathway [30], iaaH in the indole-3-acetamide (IAM) pathway [31], and nitrile hydratase (nth) in the indole-3-acetaldoxime/indole-3-acetonitrile (IAOx/IAN) pathway [32]. Gene clusters code for proteins functioning in the regulation of vegetative growth in plants, including nitrogen fixation, the synthesis of pyrroloquinoline quinone (PQQ), and siderophore production. Genes identified in the BJ3 genome coding for proteins with functions related to the nitrogen-fixation process included IscU and Nif3, which exhibit nitrogenase activity [33], along with sigma factor 54-interacting protein and glnA, both playing regulatory roles in nitrogen fixation [34,35]. Pyrroloquinoline quinone (PQQ) is known to play a crucial role in stimulating plant growth and facilitating phosphorous solubilization [36] and is a scavenger of ROS [37]. Furthermore, *PvdF* is involved in the synthesis of siderophore and pyoverdine [38]. A large gene cluster for the synthesis of NRPs/polyketides and VOCs was identified; these metabolites possess antimicrobial activities [39–42]. Additionally, two genes, chitinase and cellulase, were identified; the antimicrobial activity of these two hydrolytic enzymes has been characterized [43,44]. Genes involved in the synthesis of lipopolysaccharides were discovered in the BJ3 genome. Bacterial lipopolysaccharides can act as elicitors to trigger plant defense responses [45]. Furthermore, five genes involved in the synthesis of spermidine were identified. Extracellular spermidine produced by *Bacillus amyloliquefaciens* SQR9 has been demonstrated to enhance salt stress tolerance in *Arabidopsis* and maize [46]. The BJ3 genome was found to contain genes involved in the synthesis of metabolites capable of regulating osmotic stress; such metabolites included proline, glycine betaine, trehalose, and ectoine. Bacteria that produce these metabolites have been shown to alleviate osmotic stress in plants [47–50].

Table 1. BLAST search results.

Bacterial Strains	Sequence Length (bp)	Identity (%)
Burkholderia contaminans strain ILQ216	1420	100
Burkholderia contaminans strain XL73	1337	100
Burkholderia sp. strain GB13	1365	100
Burkholderia contaminans strain SK875	1337	100
Burkholderia sp. strain NFB35	1417	100
Burkholderia sp. strain BC	1378	100
Burkholderia sp. strain 337D	1377	100
Burkholderia sp. strain 60H	1368	100
Burkholderia sp. strain 40H	1374	100
Burkholderia sp. strain 1H	1375	100
Burkholderia cenocepacia strain PC184	1338	100

Table 2. Genes related to plant-growth-promoting traits.

Accession No.	Gene Name	Accession No.	Gene Name
IAA synthesis			
WP_039362343	*trpC*	WP_046196360	*iaaH*
WP_039346423	*IOR*	WP_039366865	*IOR*
WP_039358309	*trpB*	WP_039358314	*trpA*
WP_039356730	*nthB*	WP_039356728	*nthA*
WP_027788472	*trpD*		
Ethylene synthesis			
WP_039371479	*ACC deaminase*		
PQQ synthesis			
WP_039369440	*PqqD*	WP_039369437	*PqqC*
WP_039369434	*PqqB*	QFR11588	*PqqE*
Nitrogen fixation			
WP_011352562	*IscU*, *nifU* homolog	WP_039366622	*Nif3-like*
WP_069586144	*sigma 54-interacting*	WP_039364811	*glnA*
WP_039369321	*glnA*	WP_039355335	*glnA*
WP_039342705	*type I methionyl aminopeptidase*		
Siderophore synthesis			
WP_039358638	*PvdF*		
NRPS/PKS			
WP_039359006	*NRPSs*	WP_039358647	*NRPSs*
WP_039366349	*NRPSs*	WP_039366347	*NRPSs*
WP_039364197	*NRPSs*	WP_039361298	*PksD*
WP_039361196	*DsbA_FrnE*		
Hydrolytic enzymes			
MBA9830528	*chitinase*	BBA41828	*cellulase*
VOCs			
-Terpenoid synthesis			
WP_039342858	*ispF*	WP_011352390	*ispD*
WP_039342723	*Dxr*	WP_039360848	*terpene synthase*
-Acetoin biosynthesis			
WP_039351048	*Bdh1*	WP_039344544	*Bdh1*
Lipopolysaccharide synthesis			
WP_039360609	*waaF*	WP_039360073	*LptD*
Spermidine synthesis			
WP_039360784	*speD*	WP_039370372	*SpdS*
WP_039353965	*speB*	WP_039360775	*SPMS*
WP_039339273	*SPMS*		
Osmolytes synthesis			
-Proline			
WP_039361374	*proC*	WP_039353092	*proB*
WP_039361497	*proA*		
-Trehalose			
WP_039348314	*otsB*	WP_039348312	*otsA*
-Glycine betaine			
WP_039369516	*betA*	WP_039354575	*betA*
-Ectoine			
WP_039344448	*ectC*		

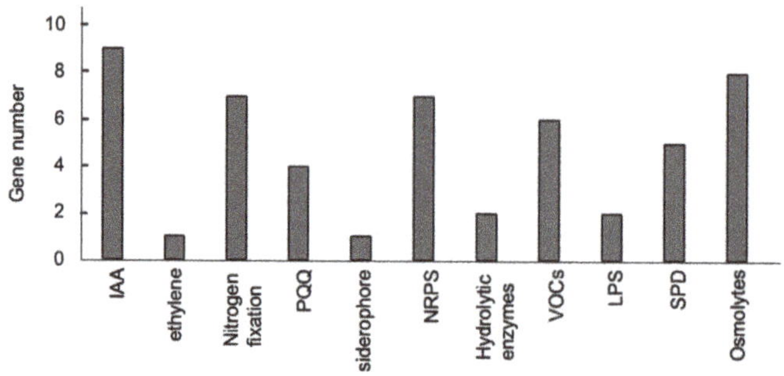

Figure 1. Gene clusters associated with plant-growth-promotion traits. The BJ3 genome contains candidate genes that function in promoting plant growth.

2.2. Biochemical Analysis of Strain BJ3

Biochemical analysis indicated that BJ3 produced a higher amount of IAA on the third day of culture, with approximate average concentrations of 38 ppm (Figure 2A). This bacterial strain exhibited significant phosphate-solubilizing activity, generating a clear zone surrounding the bacterial colony in the NBRIP medium (Figure 2B). Nitrogen-fixation activity was analyzed by inoculating BJ3 in the Jensen's medium, which resulted in the bromothymol blue indicator changing to yellow (Figure 2C). Furthermore, BJ3 also exhibited antifungal activity, suppressing approximately 53% of the mycelial growth of the fungal pathogen *Fusarium oxysporum* f. sp. *cubense* tropical race 4 (*Foc* TR4) (Figure 2D).

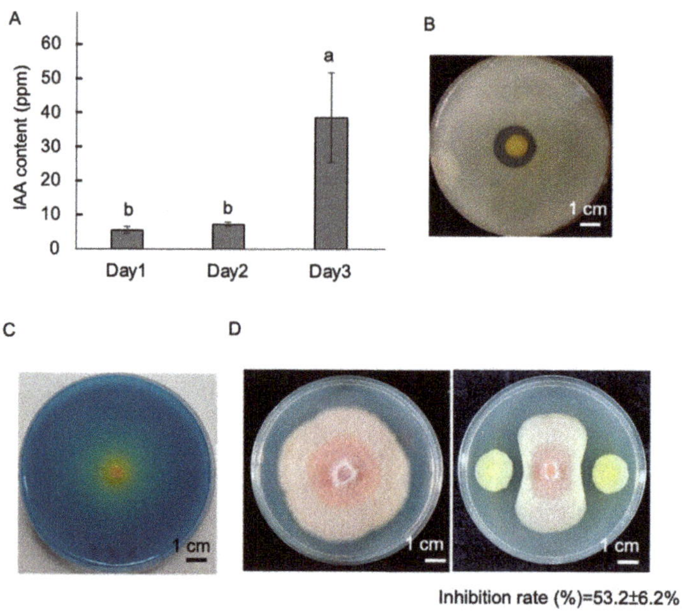

Figure 2. Physiological properties of strain BJ3. The bacterial strain produced IAA; the highest concentration of IAA was detected after a three-day culture period (**A**). Error bars present the SD from the mean of three replicates. Different letters within the histograms represent statistical significance at a *p*-value of 0.05. The phosphate-solubilizing activity of the tested bacterial strain is indicated by the production of an obvious clear zone in the NBRIP medium (**B**). The nitrogen-fixation activity of the tested bacterial strain is indicated by a change in the color of bromothymol blue to yellow (**C**). The tested bacterial strain inhibited the mycelial growth of *Foc* TR4, and the rate of mycelial inhibition is indicated at the bottom of the photo (**D**).

2.3. *Arabidopsis* Growth Promotion Stimulated by Strain BJ3

Metabolites produced by beneficial rhizobacteria can enhance plant growth by affecting root growth [51]. In this study, *Arabidopsis* seedlings cocultured with BJ3 for four days underwent significant changes, as evidenced by an increased number of lateral roots and a reduction in the length of primary roots (Figure 3A). The number of lateral roots in seedlings treated with BJ3 was significantly higher than that in control seedlings (Figure 3A). Furthermore, the soil-grown *Arabidopsis* seedlings treated with BJ3 showed an increase in plant size and fresh weight compared to the control plants (Figure 3B). Three weeks after treatment, the BJ3-treated seedlings displayed visible floral inflorescences earlier than the control seedlings. The days required for the appearance of floral inflorescence in the BJ3-treated plants were significantly fewer than those in the control plants (Figure 3C). Furthermore, seedlings treated with BJ3 exhibited elevated levels of lignin in their leaf tissues (Figure 3D) and increased activity of antioxidant enzymes, such as ascorbate peroxidase (APX) and guaiacol peroxidase (POD) (Figure 3E,F).

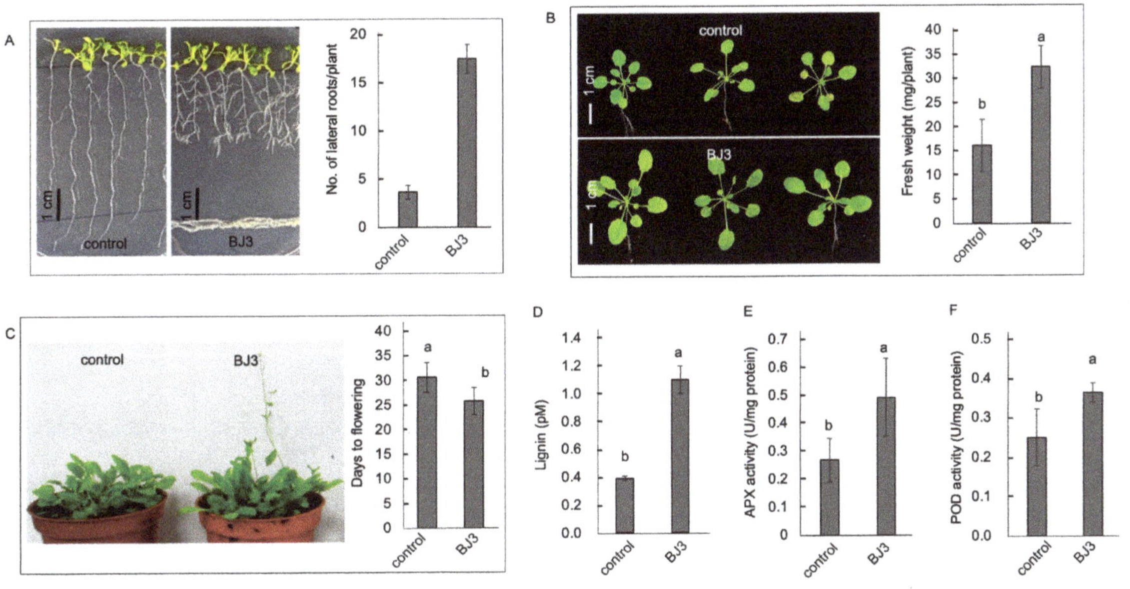

Figure 3. Physiological alterations in *Arabidopsis* seedlings in response to BJ3 treatment. BJ3 inoculation increased the root architecture and lateral root number in seedlings (**A**). The treated seedlings grown in the soil exhibited increased plant size and fresh weight (**B**). An early flowering phenotype was observed, and days to flower development were reduced (**C**). Error bars present the SD from the mean and n = 10. The BJ3 treatment increased lignin content (**D**), as well as the activity of antioxidant enzymes APX (**E**) and POD (**F**). Values in each histogram represent the mean of three replicates ± SD. Different letters within the histograms represent statistical significance at a *p*-value of 0.05.

2.4. BJ3 Treatment Enhanced Seedling Stress Tolerance

Two-week-old *Arabidopsis* seedlings were pre-treated with BJ3 prior to undergoing a five-day period of water deprivation. After the drought stress, the BJ3-treated seedlings exhibited a greater number of unwilted individuals and demonstrated a higher survival rate (Figure 4A,B). Seven days after rehydration in the post-drought period, the BJ3-treated seedlings showed increases in both plant size and fresh weight (Figure 4C,D). Moreover, an analysis of salt stress tolerance was performed in the BJ3-treated seedlings. Two weeks after the salt stress treatment, the seedlings that received the BJ3 treatment demonstrated greater growth in plant size (Figure 5A), elevated levels of chlorophyll content (Figure 5B), and a diminished buildup of H_2O_2 (Figure 5C) in contrast to the untreated control group.

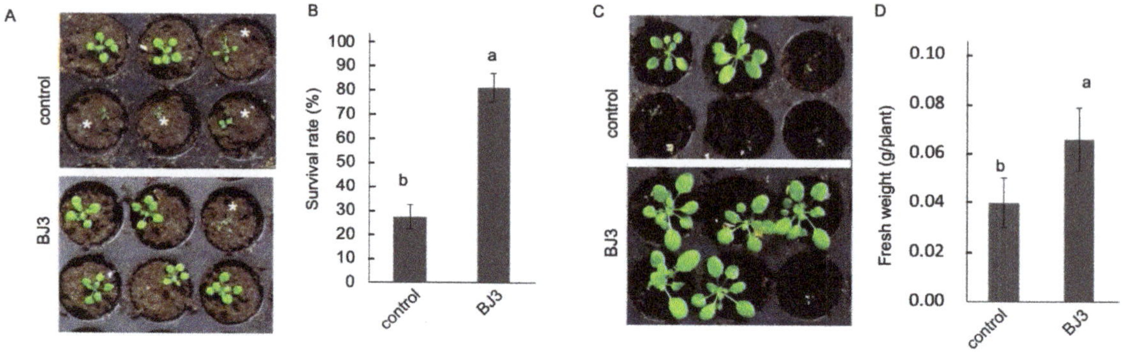

Figure 4. Drought stress tolerance induced by BJ3. After water deprivation, the seedlings treated with BJ3 exhibited a greater number of unwilted individuals (**A**); an asterisk indicates the wilted seedlings. The BJ3-treated seedlings showed a higher survival rate (**B**). Seven days after rewatering following the drought stress treatment, the BJ3-treated seedlings exhibited increases in both plant size (**C**) and fresh weight (**D**). Values in each histogram represent the mean of three replicates ± SD. Different letters within the histograms represent statistical significance at a p-value of 0.05.

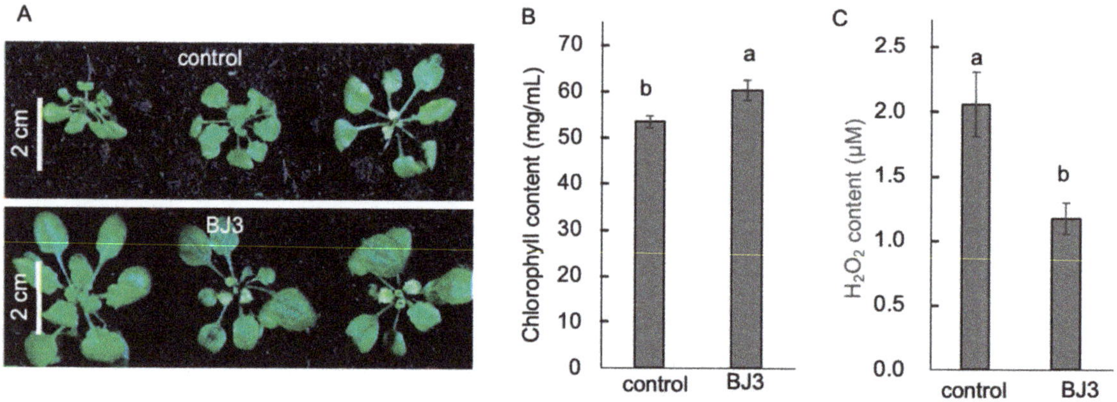

Figure 5. Salt stress tolerance induced by BJ3. Two weeks after the salt stress treatment, the seedlings treated with BJ3 exhibited a larger plant size (**A**), higher chlorophyll content levels (**B**), and reduced H_2O_2 accumulation (**C**) compared to the control. Different letters within the histograms represent statistical significance at a p-value of 0.05.

2.5. BJ3 Altered Arabidopsis Transcription Associated with Plant Development and Stress Response

PGPR exert a strong effect on modulating plant gene expression. To gain insight into the mechanisms involved in plant growth promotion mediated by BJ3, transcriptome alterations of *Arabidopsis* treated with BJ3 were analyzed. In total, 208 *Arabidopsis* genes, which exhibited induction levels greater than $Log_2[FC] = 2.0$, were subjected to further analysis. The detailed gene information is listed in Table S1. As shown in Figure 6A, the upregulated genes are associated with cellular pathways that govern stress tolerance. This includes groups of genes related to disease resistance, stress-associated transcription factors (TFs), and the response to abiotic stress. Upregulated genes also play roles in regulating plant development, such as vegetative and reproductive growth. Furthermore, some of these upregulated genes serve multiple purposes, influencing both stress response and cellular growth by altering hormone signaling, the composition of the cell wall, and chloroplast function. Among these genes, those related to the regulation of disease resistance constituted the largest group (Figure 6B). This group included genes encoding signaling molecules, disease-resistance proteins, pathogenesis-related (PR) proteins, and enzymes involved in the synthesis of defense metabolites, such as glucosinolate, camalexin, and terpenoids (Figure 7A). BJ3 strongly induced the expression of *FLG22-INDUCED RECEPTOR-LIKE KINASE 1* (*FRK1*), a marker for the activation of PTI/MTI [52]. Additionally, it upregulated genes such as *MPK11*, *AtCMPG1*, and *LecRKA4.1*, which play various roles in PTI/MTI [53–55]. Furthermore, *PP2-A5*, encoding a protein belonging to the Phloem Protein2 family, is involved in the induced immunity against insect damage [56]. BJ3 treatment also increased the expression levels of cysteine-rich receptor-like kinase (CRK) gene members. For example, *Arabidopsis CRK13*'s function is involved in the disease resistance against *Pst* DC3000 and SA accumulation [57]. Another upregulated gene, *ACCELERATED CELL DEATH 6* (*ACD6*), is a regulator of the defense response mediated by SA [58]. BJ3 also induced the expression of *AtPME41*, which encodes a pectin methylesterase functioning in the disease resistance against *P. syringae* [59]. Moreover, Arabidopsis *GDSL lipase 1* (*GLIP1*) plays a role in controlling systemic immunity by regulating components of the ethylene signaling pathway [60]. The major *PR* genes induced by BJ3 included members of the defensin-like family (Figure 7A). BJ3 treatment also activated genes involved in the synthesis of secondary metabolites associated with the defense response, including *IGMT2* and *CYP81F2* for glucosinolate synthesis [61,62]; *CYP71A12* and *PHYTOALEXIN DEFICIENT 3* (*PAD3*) for camalexin synthesis [63,64]; *FAD-linked oxidoreductase 1* (*FOX1*) for cyanogenic phytoalexin synthesis [65]; and four genes encoding enzymes for terpenoid synthesis, which include *BARUOL SYNTHASE 1* (*BARS1*), *terpenoid cyclase* (AT1G48820), *CYP705A3*, and *CYP76C2* [66,67].

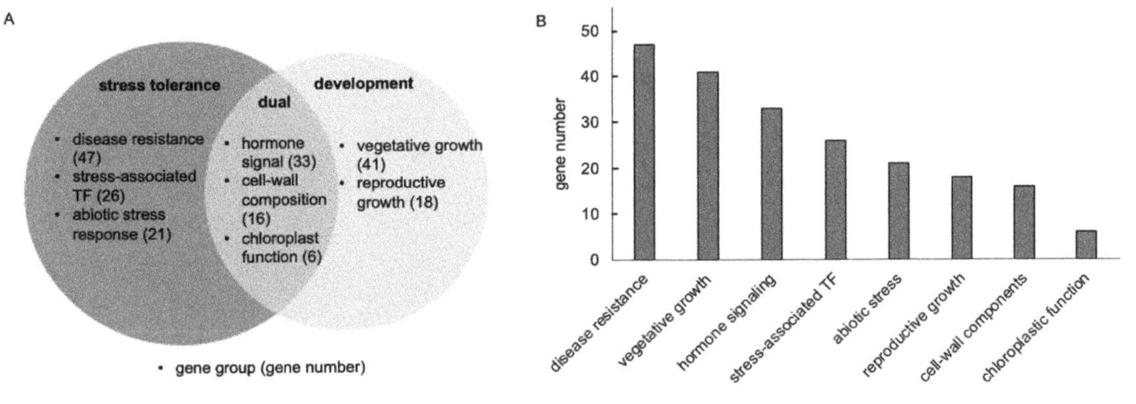

Figure 6. Transcriptomic alterations of Arabidopsis seedlings in response to BJ3 treatment. The BJ3 upregulated genes were classified into eight groups based on their annotated functions; moreover, these groups were assigned to biological functions related to stress tolerance, development, or both (**A**). The number of genes in each group is presented (**B**).

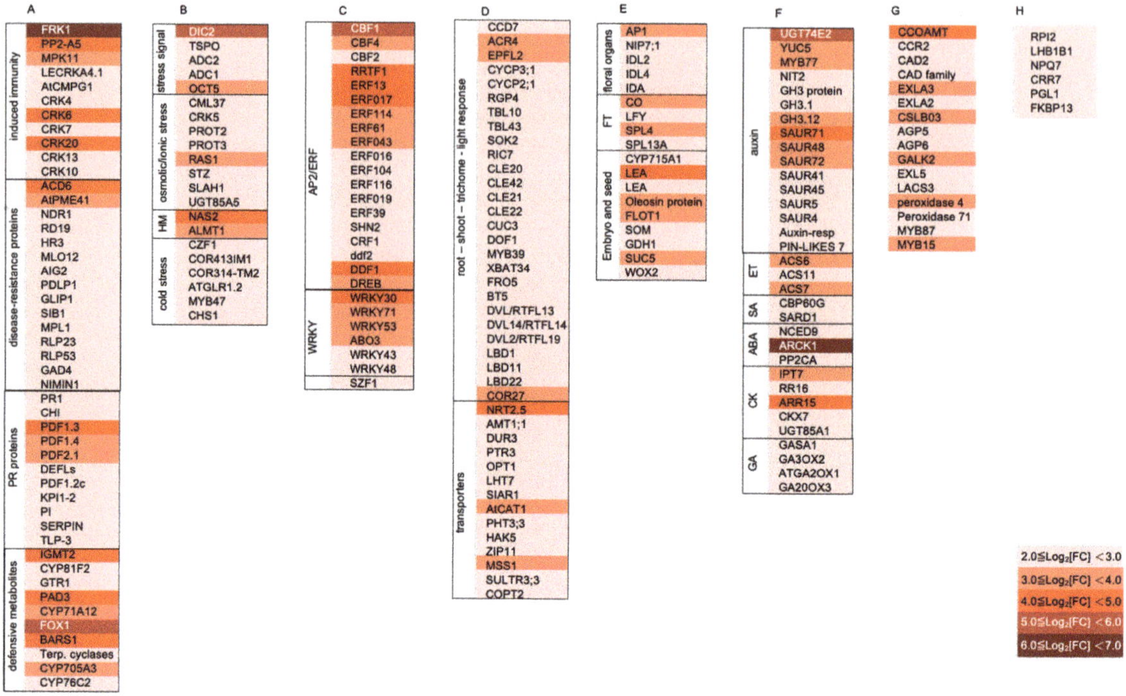

Figure 7. The BJ3 upregulated genes from the Arabidopsis transcriptome analysis. Upregulated genes exhibit annotated functions linked to disease resistance (**A**), abiotic stress (**B**), stress-associated transcription factors (TF) (**C**), vegetative growth (**D**), reproductive growth (**E**), hormone signaling (**F**), cell-wall components (**G**), and chloroplast function (**H**). Different shades of red indicate the fold changes in gene expression.

In Figure 7B, the upregulated genes displaying annotated functions related to abiotic stress responses are highlighted. These include genes regulating the perception and response to stress signals, as well as tolerance to osmotic/ionic stress, heavy metals, and low temperatures. For example, dicarboxylate carrier 2 (DIC2) is important for shuttling malate-oxaloacetate and maintaining redox balance between the mitochondria and the cytosol; genes of DICs are important for plant adaptation to environmental stresses [68,69]. Genes involved in the synthesis of polyamines (PAs) and PA transport include *ADC1*, *ADC2*, and *OCT5*. PAs are polycations and low-molecular-weight metabolites that can exert protective roles on cellular molecules under increased oxidative stress conditions [70]. Furthermore, the expression of *RESPONSE TO ABA AND SALT 1* (*RAS1*) was increased. *Nicotianamine synthase 2* (*NAS2*) is for the synthesis of nicotianamine (NA), acting as a ligand for the detoxification of heavy metals [71]. *Aluminum-activated malate transporter 1* (*ALMT1*) regulates aluminum tolerance in *Arabidopsis* by secreting malate to chelate the Al in acidic soils [72].

BJ3 upregulated a group of genes encoding transcription factors, playing a dual function in the regulation of abiotic and biotic stress-related physiology (Figure 7C). Among these transcription factors, AP2/ERF transcription factors occupied the largest portion of this gene group (Figure 7C). The expression of *CRT/DRE BINDING FACTOR 1* (*CBF1*), *CBF2*, and *CBF3*, members of the AP2/ERF family, is rapidly induced in response to low temperature [73]. *Redox Responsive Transcription Factor 1* (*RRTF1*), also known as *ERF109*, is an important regulator to control redox homeostasis plant cells after exposure to abiotic stress [74]. *ERF114* mediates elicitor-induced disease resistance in *Arabidopsis* [75].

DWARF AND DELAYED FLOWERING 1 (*DDF1*) functions in the adaptation response to multiple abiotic stress [76]. Additionally, BJ3 upregulated the expression of several *WRKY* transcription factors. Among them, *WRKY30* plays a dual function in the regulation of disease resistance [77] and abiotic stress tolerance [78]. The regulatory role of *WRKY53* is implicated in leaf senescence and disease resistance against *P. syringae* in *Arabidopsis* [79,80]. Furthermore, *SALT-INDUCIBLE ZINC FINGER 1* (*SZF1*) encodes a CCCH-type zinc finger protein, displaying functions involved in salt stress and immune responses [81].

2.6. BJ3 Enhanced Arabidopsis Gene Expression Related to Plant Development

The second-largest group of upregulated genes was associated with various aspects of plant growth (Figure 6B). As shown in Figure 7D, genes regulated by BJ3 were involved in regulating root and shoot growth, trichome development, light response, and nutrient transport. Upregulated genes included those encoding the *CLAVATA3/ESR* (*CLE*) peptides, *Devil/Rotundifolia Like* (*RTFL*) peptides, and *LOB domain-containing proteins* (*LBD*). The gene function of the *CLE* family acts as a signaling molecule mediating the *WUSCHEL* (*WUS*) signaling pathway, which is implicated in maintaining the shoot apical meristem (SAM) [82]. The *RTFL* gene family consists of small peptides that regulate development in *Arabidopsis* [83]. *LBD* gene members encode plant-specific transcriptional regulators that play roles in determining organ boundaries during plant development [84]. Another upregulated gene, *EPIDERMAL PATTERNING FACTOR-like protein 2* (*EPFL2*), also plays a role in controlling shoot organ boundaries [85]. The function of *Arabidopsis Crinkly4 Related 4* (*ACR4*) is implicated in regulating root-cell division [86]. *MYB39* is involved in regulating suberin biosynthesis in the root endodermis of *Arabidopsis* [87]. Moreover, *COLD REGULATED GENE 27* (*COR27*) plays a role in controlling the circadian rhythm of plant growth [88]. BJ3 also stimulated the expression of several transporters involved in nitrogen metabolism, including *Nitrate transporter 2.5* (*NRT2.5*), *Ammonium transporter 1 member 1* (*AMT1;1*), *Urea-proton symporter* (*DUR3*), *PEPTIDE TRANSPORTER 3* (*PTR3*), *Oligopeptide transporter 1* (*OPT1*), *Lysine histidine transporter-like 7* (*LHT7*), and *Siliques Are Red 1* (*SIAR1*). Among them, the induction fold of *NRT2.5* in response to BJ3 treatment was the highest in this gene group, with a Log_2[FC] of 4.9.

Additionally, upregulated genes constituted several genes showing annotated functions linked to the regulation of floral organ development, flowering time, and embryo and seed development (Figure 7E). For example, *APETALA 1* (*AP1*) affects floral organ development [89]. The gene function of *INFLORESCENCE DEFICIENT IN ABSCISSION* (*IDA*) and *INFLORESCENCE DEFICIENT IN ABSCISSION* (*IDA*)*-LIKE* (*IDL*) is to regulate floral organ abscission [90]. The transcription factors *CONSTANS* (*CO*), *LEAFY* (*FLY*), and *squamosa-promoter binding protein-like* (*SPL*) are involved in the regulation of flowering time [91,92]. Lea and Oleosin proteins are important during seed maturation [93].

2.7. BJ3 Activated Genes with Multiple Functions in Plant Growth and Stress Tolerance

Rhizobacteria have an impact on plant growth and stress tolerance by influencing the signaling of plant hormones [94]. As shown in Figure 7F, the BJ3 treatment activated the expression of genes involved in hormone signaling transduction pathways, including those for auxin, ethylene, abscisic acid (ABA), SA, gibberellin (GA), and cytokinin (CK). Of these, the genes related to auxin synthesis and response formed the most significant group. The expression of *UGT74E2*, responsible for the production of indole-3-butyric acid (IBA) glucose conjugates [95], was highly induced, with an induction fold of Log_2[FC] greater than 5. Additionally, upregulated genes included *YUC5*, encoding a flavin monooxygenase that catalyzes the conversion of indole-3-pyruvate (IPA) to IAA, and *NIT2*, encoding a nitrilase involved in the last step of the synthesis of *IAA* [96,97]. Furthermore, upregulated genes also included early auxin response genes and *SAUR* and *GH3* gene members [98]. The transcription factor *MYB77* is involved in the regulation of auxin signaling [99]. In the seedlings treated with BJ3, three upregulated genes encoding 1-aminocyclopropane-1-carboxylic acid (ACC) synthase (ACS) proteins were identified, ACS6, ACS7, and ACS11, all of which are involved in the synthesis of ethylene. Additionally, genes associated with SA biosynthesis and response, such as *SARD1* and *CBP60G* [100], were identified. BJ3 treatment activated the expression of genes involved in ABA synthesis, such as *NCED9* [101], and negative regulators of ABA signal transduction, such as *ARCK1* and *PP2CA* [102,103]. The treatment also upregulated genes involved in the cytokinin and GA signaling pathways. However, genes involved in the biosynthesis and degradation of GA and cytokinin had increased expression levels. These results suggested that BJ3 treatment exerted positive effect on auxin, ethylene, and SA signaling in *Arabidopsis*.

During various developmental stages, the structural integrity of the cell wall requires constant adjustments to facilitate plant growth [104]. As shown in Figure 7G, the analysis of transcriptome alterations revealed a group of upregulated genes that show functions associated with the synthesis of metabolites to fortify the cell wall. For example, *Caffeoyl-CoA O-methyltransferase* (*CCoAMT*), *Cinnamoyl-CoA reductase 2* (*CCR*), and *Cinnamyl alcohol dehydrogenase 2* (*CAD2*) encode proteins that are involved in lignin synthesis [105]. The transcription factors MYB87 and MYB15 are regulators of cell-wall strengthening [106,107]. Chloroplasts play a crucial role in energy production via photosynthesis and oxygen release, both of which promote plant growth and enhance crop yields. Additionally, chloroplasts produce biologically active metabolites that enable plants to adapt and respond to stress from harsh environments [108]. As shown in Figure 7H, a gene encoding ribose-5-phosphate isomerase (RPI2) is required for maintaining the chloroplast structure and its photosynthetic capacity [109], and 6-Phosphogluconolactonase 1 (PGL1) is a key enzyme in the oxidative pentose phosphate pathway (OPPP). Moreover, the PGL genes are responsive to abiotic stresses in rice [110].

3. Discussion

The application of *Burkholderia contaminans* strains for biocontrol purposes has been emphasized, which involves the production of diffusible or volatile antimicrobial metabolites [27,28,111]. In this study, the BJ3 genome contains gene clusters involved in the synthesis of diverse metabolites with antimicrobial activity; among these, NRPS/polyketides and VOCs constitute the two largest groups. BJ3's metabolites demonstrated significant antifungal activity in suppressing the banana Fusarium wilt pathogen. According to a previous study, the expression of non-ribosomal peptide synthetases in the *Burkholderia contaminans* strain MS14 is essential for its antifungal activity against *Geotrichum candidum* [112]. Another study has shown that terpenes, produced by the *Burkholderia gladioli* strain, are identified as effective antifungal metabolites against *Fusarium oxysporum* [113]. In addition to exhibiting antifungal activity, BJ3 treatment increased the expression of a large number of *Arabidopsis* genes linked to cellular pathways associated with induced immunity. BJ3 strongly activated the expression of *FRK1*, a marker gene for PTI/MTI [52], and *CRK6* and *CRK20*, regulators of the PTI/MTI response [114,115]. Moreover, the BJ3-upregulated genes encode disease-resistance proteins, PR proteins, and those for the synthesis of antimicrobial metabolites, such as glucosinolates, camalexin, and terpenoids [116]. These results illustrate the ability of BJ3 to serve a biocontrol purpose.

Genes involved in the synthesis of IAA constitute the largest gene group associated with plant-growth-promoting traits in the BJ3 genome. IAA is the most abundant form of auxin. This phytohormone plays a crucial role in the regulation of lateral root development [1]. BJ3 produced a significant amount of IAA (close to 40 ppm) and induced the *Arabidopsis* gene expression involved in auxin synthesis and response. A consistent phenotype has been observed, showing that treatment with BJ3 significantly enhances lateral root development in *Arabidopsis* seedlings. The BJ3 genome contains gene clusters involved in nitrogen fixation and phosphate solubilization; consistently, BJ3 exhibits significant activity in both processes. These microbial activities may contribute to the growth-promoting effects derived from BJ3. Additionally, *Arabidopsis* treated with BJ3 exhibited an early flowering phenotype. This early flowering is consistent with the activated expression of the *Arabidopsis* genes *CO*, *LFY*, and *SPL13A*, which are involved in regulating the transition from vegetative growth to flower initiation [117–119]. Ethylene is a regulator for stimulating the transition of vegetative growth to floral development in *Arabidopsis* [120]. In the BJ3-treated *Arabidopsis* transcriptome, upregulated genes include three members of the ACS family, which is the rate-limiting enzyme in ethylene synthesis [121]. Further evidence supporting an increased ethylene signal in BJ3-treated *Arabidopsis* is that a large number of *AP2/ERF* gene members are stimulated in their expression by BJ3. The expression of these transcription factors is manipulated by the ethylene signal and is crucial for the plant's adaptive response to various abiotic stresses [122]. Our results indicate that BJ3 activates both auxin and ethylene signals to alter plant growth and development. These findings are consistent with a study demonstrating that these two hormonal signals play roles in plant growth induced by *Burkholderia phytofirmans* [123].

The BJ3 genome contains a notable number of genes involved in the synthesis of various osmolytes, which may regulate the osmotic balance in plants under osmotic stress conditions [47–50]. Consistently, the BJ3-treated *Arabidopsis* exhibited increased tolerance to both drought and salt stress. In addition to producing metabolites that act as osmolytes to protect plants from oxidative damage under osmotic/salt stress, microbes can also activate the plant antioxidant defense system to reduce the production of ROS under stressful conditions [124]. The BJ3 genome includes a gene cluster dedicated to producing spermidine. Strains of PGPR that produce spermidine have been shown to enhance salt stress tolerance in host plants [46]. Additionally, applying spermidine externally can stimulate antioxidant activity in plants [125]. Moreover, BJ3 contains genes for the synthesis of terpenoids and activates *Arabidopsis* genes related to terpenoid synthesis. This natural compound, in addition to serving as an antimicrobial agent, can also act as an antioxidant in plants [126]. BJ3 treatment triggered the expression of genes involved in SA biosynthesis and signaling pathways. SA plays a beneficial role in enhancing plant tolerance to various abiotic stresses through the activation of antioxidant activity [127]. In addition to playing a crucial role in the regulation of root development, a regulatory role of auxin in stress adaptation is revealed. For example, transgenic potato overexpressing the auxin biosynthesis gene *YUCCA6* resulted in an increased auxin level and enhanced drought stress tolerance [128]. Moreover, auxin improves plant abiotic stress tolerance by manipulating antioxidant activity in plant cells [129]. In this study, BJ3 treatment activated antioxidant enzymes and enhanced plant tolerance to both drought and salt stress. Therefore, BJ3 may serve a protective function in plants, either by producing various types of osmolytes or by activating signaling pathways that help develop an adaptive response to stress.

4. Materials and Methods

4.1. Bacterial Strain Characterization

The BJ3 strain was isolated from a growing substrate. PCR amplification of the genomic DNA from the bacterial strain was performed using the primers fD1 (5′AGAGTTTGATCCTGGCTCAG3′) and rP1 (5′ACGGTTACCTTGTTACGACTT3′), targeting the 16S rDNA sequence [130]. The PCR fragment was analyzed using a 3730 DNA Analyzer (Applied Biosystems®; Foster City, CA, USA). The obtained 16S rDNA sequences were analyzed using the Basic Local Alignment Search Tool (BLAST) program [131] in the NCBI database (https://blast.ncbi.nlm.nih.gov/Blast.cgi (accessed 1 September 2023)).

4.2. De Novo Whole Genome Assembly

The fragmented genome DNA of BcD1 was prepared using the Celero PCR workflow with an Enzymatic Fragmentation DNA-Seq Kit (Tecan Trading AG, Männedorf, Switzerland). DNA sequencing was performed by the paired-end method on the Illumina MiSeq system (Illumina, Inc., San Diego, CA, USA). The raw sequence data's quality was assessed with NanoPlot v1.28.1 [132]. Following this, adapters, low-quality sequences (Q20), and ambiguous bases were trimmed from the reads. The de novo assembly of the genome sequencing data was carried out using SPAdes v.3.14.1 [133]. Open reading frames (ORFs) were predicted by GlimmerHMM [134]. RNAmmer and tRNAscan SE were used to predict rRNA and tRNA, respectively [135,136]. Gene sequences were annotated through the NCBI database. The functional analysis of the genes was performed using FastAnnotator [137] and the Gene Ontology Consortium (http://geneontology.org/) (accessed on 5 September 2023). The phylogenetic classification of protein families was determined using the Cluster of Orthologous Groups (COG) database (http://www.ncbi.nlm.nih.gov/COG) (accessed 1 September 2023)).

4.3. Analysis of the Bioactivity of Microbial Metabolites

To measure IAA production, BJ3 was cultured for 24, 48, and 72 h in Luria Broth (LB) medium supplemented with 2 mg/mL L-tryptophan. Supernatants from the bacterial cultures were added to Salkowski reagent and incubated in the dark at room temperature for 20 min. The absorbance at 535 nm was determined, and the IAA concentration (in ppm) was calculated based on a standard curve derived from known IAA concentrations. The phosphate-solubilizing activity of BJ3 was determined by following procedures described by Nautiyal (1999) [138]. BJ3 was inoculated on the NBRIP plate containing 1% glucose, 0.5% $Ca_3(PO_4)_2$, 0.5% $MgCl_2{\cdot}6H_2O$, 0.025% $MgSO_4{\cdot}7H_2O$, 0.02% KCl, and 0.01% $(NH_4)_2SO_4$, pH 7.0, and cultured for 14 days. The nitrogen-fixation activity of BJ3 was analyzed on Jensen's media [139] containing × 25 μg/mL of bromothymol blue. After 4 days of culture, the formation of a yellow zone surrounding the bacterial culture was used as an indicator of nitrogen fixation. The detection of antifungal activity in the metabolites produced by the isolated bacterial strain was carried out using methods outlined by Tsai et al. (2023) [140]. Briefly, two filter paper pieces, each soaked with 10 μL of bacterial solutions at a concentration of 1×10^8 CFU/mL, were placed 3 cm away from a mycelial plug of *Fusarium oxysporum* f. sp. *cubense* tropical race 4 (*Foc* TR4) on potato dextrose agar (PDA) plates. Water served as the control. The cultures were then incubated for seven days at 28 °C. The growth inhibition rate was assessed by measuring the diameter of fungal mycelia in the control and treated groups and calculating the inhibition rate (I) = (1 − diameter of mycelia of treatment/diameter of mycelia of control) × 100.

4.4. Analysis of Growth-Promotion Effects

Seeds of *Arabidopsis thaliana* (ecotype Columbia) were germinated on ½ MS medium for 3 days. Seedlings were transferred to a fresh medium without and with bacterial inoculation and cultured for an additional 4 days. Lateral root number per seedling was estimated. A bacterial culture with a density of 1×10^8 CFU/mL was used to treat *Arabidopsis* seedlings that had been germinating in soil for one week. Each pot contained ten seedlings, and four pots in total were subjected to the bacterial treatment, which was performed weekly over a three-week period. After three treatments, the fresh weight and floral inflorescence of the seedlings were recorded. Furthermore, both the lignin content and the activity of antioxidant enzymes, including ascorbate peroxidase (APX) and guaiacol peroxidase (POD), were analyzed, following the procedures described by Chang et al. [141].

4.5. Analysis of Drought and Salt Stress Tolerance

Two-week-old seedlings were treated with BJ3 at a bacterial density of 1×10^8 CFU/mL and then subjected to a water deprivation period of five days. Plants that did not wilt, indicated by stars, were counted. The survival rate was calculated by dividing the number of unwilted plants by the total number of plants. Seven days after the recovery of the water supply, the fresh weight of the seedlings was measured.

For the analysis of salt stress tolerance, after the BJ3 treatment, 125 mM NaCl (30 mL) was applied to both the control and BJ3-treated seedlings. The salt stress treatment was administered three times a week. After two weeks of salt stress treatment, both chlorophyll content and H_2O_2 content were measured by following the procedures described by Kurniawan and Chuang (2022) [142].

4.6. RNA-Seq Transcriptome Analysis

The leaf tissues from 2-week-old Arabidopsis seedlings treated with BJ3 were harvested for total RNA extraction using the method described by Parcy et al. [143]. Five μg of total RNA was used to construct the RNA-seq library following Illumina's protocols. After sequencing the library on the Illumina NextSeq 500 platform, the resulting sequences were analyzed using procedures outlined by Tsai et al. (2023) [140]. Gene expression levels were quantified using FPKM (Fragments Per Kilobase of exons per Million mapped reads) values. The fold change (FC) in gene expression was calculated by dividing the expression levels in the treatment group by those in the control group. Genes with a $\log_2$(FC) greater than 1.0 were considered upregulated; furthermore, genes with a $\log_2$(FC) equal to or greater than 2.0 were selected for further study.

4.7. Statistical Analysis

The differences among the treatments were evaluated using the ANOVA technique within the SAS (3.8) software. Statistical significance was established at a p-value of less than 0.05 from the Tukey test.

5. Conclusions

The BJ3 genome contains four key gene groups crucial for synthesizing IAA, NRPS, terpenoids, and osmolytes, along with a gene cluster associated with nitrogen fixation. BJ3 produces substantial amounts of IAA and antifungal metabolites while demonstrating nitrogen-fixing and phosphate-solubilizing capabilities. It promotes plant growth by strengthening root architecture, encouraging vegetative growth, and inducing earlier flowering. Additionally, BJ3-treated plants exhibit increased antioxidant activity and reduced negative effects from drought and salt stress. The transcriptome analysis indicates that BJ3 is a potent elicitor of gene expression involved in plant hormone signaling pathways, including auxin, ethylene, and SA. These activated pathways may regulate plant growth, disease resistance, and adaptive responses to abiotic stress.

Supplementary Materials: The following supporting information can be downloaded at https://www.mdpi.com/article/10.3390/ijms25116091/s1.

Author Contributions: Y.-L.C.: data analysis and manuscript writing; Y.-C.C.: bacterial characterization; A.K.: plant experiment; P.-C.C.: abiotic stress analysis; T.-Y.L.: bacterial characterization; W.-D.W.: providing comments on manuscript preparation; H.-w.C.: experimental design, data analysis, and manuscript writing. All authors have read and agreed to the published version of the manuscript.

Funding: This research was funded by the Ministry of Agriculture, grant number 113AS-14.2.1-RS-02. The APC was funded by 113AS-14.2.1-RS-02.

Institutional Review Board Statement: Not applicable.

Informed Consent Statement: Not applicable.

Data Availability Statement: Data is contained within the article and Supplementary Materials.

Conflicts of Interest: The authors declare no conflict of interest.

References

1. Lavenus, J.; Goh, T.; Roberts, I.; Guyomarc'h, S.; Lucas, M.; De Smet, I.; Fukaki, H.; Beeckman, T.; Bennett, M.; Laplaze, L. Lateral root development in Arabidopsis: Fifty shades of auxin. *Trends Plant Sci.* **2013**, *18*, 450–458. [CrossRef] [PubMed]
2. Lucke, M.; Correa, M.G.; Levy, A. The role of secretion systems, effectors, and secondary metabolites of beneficial rhizobacteria in interactions with plants and microbes. *Front. Plant Sci.* **2020**, *11*, 589416. [CrossRef] [PubMed]
3. Shameer, S.; Prasad, T. Plant growth promoting rhizobacteria for sustainable agricultural practices with special reference to biotic and abiotic stresses. *Plant Growth Regul.* **2018**, *84*, 603–615. [CrossRef]
4. Wei, G.; Kloepper, J.W.; Tuzun, S. Induction of systemic resistance of cucumber to *Colletotrichum orbiculare* by select strains of plant growth-promoting rhizobacteria. *Phytopathology* **1991**, *81*, 1508–1512. [CrossRef]

5. van Loon, L.C.; Bakker, P.A.; Pieterse, C.M. Systemic resistance induced by rhizosphere bacteria. *Annu. Rev. Phytopath.* **1998**, *36*, 453–483. [CrossRef] [PubMed]
6. Dodds, P.N.; Rathjen, J.P. Plant immunity: Towards an integrated view of plant–pathogen interactions. *Nat. Rev. Genet.* **2010**, *11*, 539–548. [CrossRef] [PubMed]
7. Alhoraibi, H.; Bigeard, J.; Rayapuram, N.; Colcombet, J.; Hirt, H. Plant immunity: The MTI-ETI model and beyond. *Curr. Issues Mol. Biol.* **2019**, *30*, 39–58. [CrossRef]
8. Khalid, A.; Arshad, M.; Zahir, Z.A. Screening plant growth-promoting rhizobacteria for improving growth and yield of wheat. *J. Appl. Microbiol.* **2004**, *96*, 473–480. [CrossRef]
9. Yan, S.; Che, G.; Ding, L.; Chen, Z.; Liu, X.; Wang, H.; Zhao, W.; Ning, K.; Zhao, J.; Tesfamichael, K. Different cucumber CsYUC genes regulate response to abiotic stresses and flower development. *Sci. Rep.* **2016**, *6*, 20760. [CrossRef]
10. Ribba, T.; Garrido-Vargas, F.; O'Brien, J.A. Auxin-mediated responses under salt stress: From developmental regulation to biotechnological applications. *J. Exp. Bot.* **2020**, *71*, 3843–3853. [CrossRef]
11. Barnawal, D.; Bharti, N.; Pandey, S.S.; Pandey, A.; Chanotiya, C.S.; Kalra, A. Plant growth-promoting rhizobacteria enhance wheat salt and drought stress tolerance by altering endogenous phytohormone levels and TaCTR1/TaDREB2 expression. *Physiol. Plant.* **2017**, *161*, 502–514. [CrossRef] [PubMed]
12. Maulidah, N.I.; Tseng, T.-S.; Chen, G.-H.; Hsieh, H.-Y.; Chang, S.-F.; Chuang, H.-W. Transcriptome analysis revealed cellular pathways associated with abiotic stress tolerance and disease resistance induced by *Pseudomonas aeruginosa* in banana plants. *Plant Gene* **2021**, *27*, 100321. [CrossRef]
13. Pierik, R.; Sasidharan, R.; Voesenek, L.A.C.J. Growth control by ethylene: Adjusting phenotypes to the environment. *J. Plant Growth Regul.* **2007**, *26*, 188–200. [CrossRef]
14. Singh, R.P.; Pandey, D.M.; Jha, P.N.; Ma, Y. ACC deaminase producing rhizobacterium *Enterobacter cloacae* ZNP-4 enhance abiotic stress tolerance in wheat plant. *PLoS ONE* **2022**, *17*, e0267127. [CrossRef] [PubMed]
15. Lewis, D.R.; Negi, S.; Sukumar, P.; Muday, G.K. Ethylene inhibits lateral root development, increases IAA transport and expression of PIN3 and PIN7 auxin efflux carriers. *Development* **2011**, *138*, 3485–3495. [CrossRef]
16. Strader, L.C.; Chen, G.L.; Bartel, B. Ethylene directs auxin to control root cell expansion. *Plant J.* **2010**, *64*, 874–884. [CrossRef]
17. Karlova, R.; Boer, D.; Hayes, S.; Testerink, C. Root plasticity under abiotic stress. *Plant Physiol.* **2021**, *187*, 1057–1070. [CrossRef]
18. Song, L.; Yu, H.; Dong, J.; Che, X.; Jiao, Y.; Liu, D. The molecular mechanism of ethylene-mediated root hair development induced by phosphate starvation. *PLoS Genet.* **2016**, *12*, e1006194. [CrossRef] [PubMed]
19. Kazan, K. Diverse roles of jasmonates and ethylene in abiotic stress tolerance. *Trends Plant Sci.* **2015**, *20*, 219–229. [CrossRef]
20. Karimi, N.; Ghasempour, H.-R. Salicylic acid and jasmonic acid restrains nickel toxicity by ameliorating antioxidant defense system in shoots of metallicolous and non-metallicolous *Alyssum inflatum* Náyr. Populations. *Plant Physiol. Biochem.* **2019**, *135*, 450–459.
21. Cho, S.M.; Kang, B.R.; Han, S.H.; Anderson, A.J.; Park, J.-Y.; Lee, Y.-H.; Cho, B.H.; Yang, K.-Y.; Ryu, C.-M.; Kim, Y.C. 2R, 3R-butanediol, a bacterial volatile produced by *Pseudomonas chlororaphis* O6, is involved in induction of systemic tolerance to drought in *Arabidopsis thaliana*. *Mol. Plant Microbe Interact.* **2008**, *21*, 1067–1075. [CrossRef] [PubMed]
22. Liu, S.; Tian, Y.; Jia, M.; Lu, X.; Yue, L.; Zhao, X.; Jin, W.; Wang, Y.; Zhang, Y.; Xie, Z. Induction of salt tolerance in *Arabidopsis thaliana* by volatiles from *Bacillus amyloliquefaciens* FZB42 via the jasmonic acid signaling pathway. *Front. Microbiol.* **2020**, *11*, 562934. [CrossRef] [PubMed]
23. Ghosh, R.; Barman, S.; Mukherjee, R.; Mandal, N.C. Role of phosphate solubilizing *Burkholderia* spp. for successful colonization and growth promotion of *Lycopodium cernuum* L. (Lycopodiaceae) in lateritic belt of Birbhum district of West Bengal, India. *Microbiol. Res.* **2016**, *183*, 80–91. [CrossRef] [PubMed]
24. Elliott, G.N.; Chen, W.M.; Chou, J.H.; Wang, H.C.; Sheu, S.Y.; Perin, L.; Reis, V.M.; Moulin, L.; Simon, M.F.; Bontemps, C. *Burkholderia phymatum* is a highly effective nitrogen-fixing symbiont of *Mimosa* spp. and fixes nitrogen ex planta. *New Phytol.* **2007**, *173*, 168–180. [CrossRef] [PubMed]
25. Kurepin, L.V.; Park, J.M.; Lazarovits, G.; Bernards, M.A. *Burkholderia phytofirmans*-induced shoot and root growth promotion is associated with endogenous changes in plant growth hormone levels. *Plant Growth Regul.* **2015**, *75*, 199–207. [CrossRef]
26. Bach, E.; Passaglia, L.M.P.; Jiao, J.; Gross, H. Burkholderia in the genomic era: From taxonomy to the discovery of new antimicrobial secondary metabolites. *Crit. Rev. Microbiol.* **2022**, *48*, 121–160. [CrossRef]
27. Zaman, N.R.; Chowdhury, U.F.; Reza, R.N.; Chowdhury, F.T.; Sarker, M.; Hossain, M.M.; Akbor, M.A.; Amin, A.; Islam, M.R.; Khan, H. Plant growth promoting endophyte *Burkholderia contaminans* NZ antagonizes phytopathogen *Macrophomina phaseolina* through melanin synthesis and pyrrolnitrin inhibition. *PLoS ONE* **2021**, *16*, e0257863. [CrossRef]
28. Heo, A.Y.; Koo, Y.M.; Choi, H.W. Biological control activity of plant growth promoting rhizobacteria *Burkholderia contaminans* AY001 against tomato fusarium wilt and bacterial speck diseases. *Biology* **2022**, *11*, 619. [CrossRef]
29. Theocharis, A.; Bordiec, S.; Fernandez, O.; Paquis, S.; Dhondt-Cordelier, S.; Baillieul, F.; Clément, C.; Barka, E.A. *Burkholderia phytofirmans* PsJN primes *Vitis vinifera* L. and confers a better tolerance to low nonfreezing temperatures. *Mol. Plant Microbe In.* **2012**, *25*, 241–249. [CrossRef]
30. Imada, E.L.; de Oliveira, A.L.M.; Hungria, M.; Rodrigues, E.P. Indole-3-acetic acid production via the indole-3-pyruvate pathway by plant growth promoter *Rhizobium tropici* CIAT 899 is strongly inhibited by ammonium. *Res. Microbiol.* **2017**, *168*, 283–292. [CrossRef]

31. Lehmann, T.; Hoffmann, M.; Hentrich, M.; Pollmann, S. Indole-3-acetamide-dependent auxin biosynthesis: A widely distributed way of indole-3-acetic acid production? *Eur. J. Cell Biol.* **2010**, *89*, 895–905. [CrossRef] [PubMed]
32. Duca, D.; Rose, D.R.; Glick, B.R. Characterization of a nitrilase and a nitrile hydratase from *Pseudomonas* sp. strain UW4 that converts indole-3-acetonitrile to indole-3-acetic acid. *Appl. Environ. Microbiol.* **2014**, *80*, 4640–4649. [CrossRef] [PubMed]
33. Dos Santos, P.C.; Johnson, D.C.; Ragle, B.E.; Unciuleac, M.-C.; Dean, D.R. Controlled expression of nif and isc iron-sulfur protein maturation components reveals target specificity and limited functional replacement between the two systems. *J. Bacteriol.* **2007**, *189*, 2854–2862. [CrossRef] [PubMed]
34. do Carmo Dias, B.; da Mota, F.F.; Jurelevicius, D.; Seldin, L. Genetics and regulation of nitrogen fixation in *Paenibacillus brasilensis* PB24. *Microbiol. Res.* **2021**, *243*, 126647. [CrossRef] [PubMed]
35. Kullik, I.; Fritsche, S.; Knobel, H.; Sanjuan, J.; Hennecke, H.; Fischer, H. *Bradyrhizobium japonicum* has two differentially regulated, functional homologs of the sigma 54 gene (rpoN). *J. Bacteriol.* **1991**, *173*, 1125–1138. [CrossRef] [PubMed]
36. Ludueña, L.M.; Anzuay, M.S.; Angelini, J.G.; Barros, G.; Luna, M.F.; Monge, M.D.P.; Fabra, A.; Taurian, T. Role of bacterial pyrroloquinoline quinone in phosphate solubilizing ability and in plant growth promotion on strain *Serratia* sp. S119. *Symbiosis* **2017**, *72*, 31–43. [CrossRef]
37. Ludueña, L.M.; Bianucci, E.; Anzuay, M.S.; Angelini, J.G.; Fabra, A.; Taurian, T. First insights into the role of PQQ cofactor in the modulation of bacterial redox state and in the early interaction with peanut (*Arachis hypogaea* L.). *Appl Soil Ecol* **2020**, *152*, 103560. [CrossRef]
38. Kenjić, N.; Hoag, M.; Moraski, G.C.; Caperelli, C.A.; Moran, G.R.; Lamb, A.L. PvdF of pyoverdin biosynthesis is a structurally unique N10-formyltetrahydrofolate-dependent formyltransferase. *Arch. Biochem. Biophys.* **2019**, *664*, 40–50. [CrossRef] [PubMed]
39. Gu, G.; Smith, L.; Wang, N.; Wang, H.; Lu, S.-E. Biosynthesis of an antifungal oligopeptide in *Burkholderia contaminans* strain MS14. *Biochem. Biophys. Res. Commun.* **2009**, *380*, 328–332. [CrossRef]
40. Singh, D.; Devappa, V.; Yadav, D.K. Suppression of tomato bacterial wilt incited by *Ralstonia pseudosolanacearum* using polyketide antibiotic-producing *Bacillus* spp. isolated from rhizospheric soil. *Agriculture* **2022**, *12*, 2009. [CrossRef]
41. Garcia-Gonzalez, E.; Müller, S.; Ensle, P.; Süssmuth, R.D.; Genersch, E. Elucidation of sevadicin, a novel non-ribosomal peptide secondary metabolite produced by the honey bee pathogenic bacterium *P aenibacillus* larvae. *Environ. Microbiol.* **2014**, *16*, 1297–1309. [CrossRef] [PubMed]
42. Farag, M.A.; Zhang, H.; Ryu, C.-M. Dynamic chemical communication between plants and bacteria through airborne signals: Induced resistance by bacterial volatiles. *J. Chem. Ecol.* **2013**, *39*, 1007–1018. [CrossRef] [PubMed]
43. Singh, R.P.; Saati, A.A.; Faidah, H.; Bantun, F.; Jalal, N.A.; Haque, S.; Rai, A.K.; Srivastava, M. Prospects of microbial cellulase production using banana peels wastes for antimicrobial applications. *Intern. J. Food Microbiol.* **2023**, *388*, 110069. [CrossRef] [PubMed]
44. Loc, N.H.; Huy, N.D.; Quang, H.T.; Lan, T.T.; Thu, H.T.T. Characterisation and antifungal activity of extracellular chitinase from a biocontrol fungus, *Trichoderma asperellum* PQ34. *Mycology* **2020**, *11*, 38–48. [CrossRef] [PubMed]
45. Sun, A.; Li, Z. Regulatory role of nitric oxide in lipopolysaccharides-triggered plant innate immunity. *Plant Signal Behav.* **2013**, *8*, 1081–1096. [CrossRef] [PubMed]
46. Chen, L.; Liu, Y.; Wu, G.; Zhang, N.; Shen, Q.; Zhang, R. Beneficial rhizobacterium *Bacillus amyloliquefaciens* SQR9 induces plant salt tolerance through spermidine production. *Mol. Plant-Microbe Interact.* **2017**, *30*, 423–432. [CrossRef]
47. Rodríguez-Salazar, J.; Suarez, R.; Caballero-Mellado, J.; Iturriaga, G. Trehalose accumulation in *Azospirillum brasilense* improves drought tolerance and biomass in maize plants. *FEMS Microbiol. Lett.* **2009**, *296*, 52–59. [CrossRef]
48. Elsakhawy, T.A.; Fetyan, N.; Ghazi, A.A. The potential use of ectoine produced by a moderately halophilic bacteria *Chromohalobacter salexigens* KT989776 for enhancing germination and primary seedling of flax "*Linum usitatissimum* L." under salinity conditions. *Biotechnol. J. Int.* **2019**, *23*, 1–12. [CrossRef]
49. Yasin, N.A.; Akram, W.; Khan, W.U.; Ahmad, S.R.; Ahmad, A.; Ali, A. Halotolerant plant-growth promoting rhizobacteria modulate gene expression and osmolyte production to improve salinity tolerance and growth in *Capsicum annum* L. *Environ. Sci. Pollut. Res.* **2018**, *25*, 23236–23250. [CrossRef]
50. Mukhtar, S.; Zareen, M.; Khaliq, Z.; Mehnaz, S.; Malik, K. Phylogenetic analysis of halophyte-associated rhizobacteria and effect of halotolerant and halophilic phosphate-solubilizing biofertilizers on maize growth under salinity stress conditions. *J. Appl. Microbiol.* **2020**, *128*, 556–573. [CrossRef]
51. Vacheron, J.; Desbrosses, G.; Bouffaud, M.-L.; Touraine, B.; Moënne-Loccoz, Y.; Muller, D.; Legendre, L.; Wisniewski-Dyé, F.; Prigent-Combaret, C. Plant growth-promoting rhizobacteria and root system functioning. *Front. Plant Sci.* **2013**, *4*, 57135. [CrossRef] [PubMed]
52. Asai, T.; Tena, G.; Plotnikova, J.; Willmann, M.R.; Chiu, W.-L.; Gomez-Gomez, L.; Boller, T.; Ausubel, F.M.; Sheen, J. MAP kinase signalling cascade in Arabidopsis innate immunity. *Nature* **2002**, *415*, 977–983. [CrossRef] [PubMed]
53. Heise, A.; Lippok, B.; Kirsch, C.; Hahlbrock, K. Two immediate-early pathogen-responsive members of the AtCMPG gene family in *Arabidopsis thaliana* and the W-box-containing elicitor-response element of AtCMPG1. *Proc. Natl. Acad. Sci. USA* **2002**, *99*, 9049–9054. [CrossRef] [PubMed]
54. Bethke, G.; Pecher, P.; Eschen-Lippold, L.; Tsuda, K.; Katagiri, F.; Glazebrook, J.; Scheel, D.; Lee, J. Activation of the *Arabidopsis thaliana* mitogen-activated protein kinase MPK11 by the flagellin-derived elicitor peptide, flg22. *Mol. Plant Microbe Interact.* **2012**, *25*, 471–480. [CrossRef] [PubMed]

55. Singh, P.; Kuo, Y.-C.; Mishra, S.; Tsai, C.-H.; Chien, C.-C.; Chen, C.-W.; Desclos-Theveniau, M.; Chu, P.-W.; Schulze, B.; Chinchilla, D. The lectin receptor kinase-VI. 2 is required for priming and positively regulates Arabidopsis pattern-triggered immunity. *Plant Cell* **2012**, *24*, 1256–1270. [CrossRef] [PubMed]
56. Santamaría, M.E.; Martínez, M.; Arnaiz, A.; Rioja, C.; Burow, M.; Grbic, V.; Díaz, I. An Arabidopsis TIR-lectin two-domain protein confers defense properties against *Tetranychus urticae*. *Plant Physiol.* **2019**, *179*, 1298–1314. [CrossRef] [PubMed]
57. Acharya, B.R.; Raina, S.; Maqbool, S.B.; Jagadeeswaran, G.; Mosher, S.L.; Appel, H.M.; Schultz, J.C.; Klessig, D.F.; Raina, R. Overexpression of CRK13, an Arabidopsis cysteine-rich receptor-like kinase, results in enhanced resistance to *Pseudomonas syringae*. *Plant J.* **2007**, *50*, 488–499. [CrossRef] [PubMed]
58. Lu, H.; Rate, D.N.; Song, J.T.; Greenberg, J.T. ACD6, a novel ankyrin protein, is a regulator and an effector of salicylic acid signaling in the Arabidopsis defense response. *Plant Cell* **2003**, *15*, 2408–2420. [CrossRef]
59. Bethke, G.; Grundman, R.E.; Sreekanta, S.; Truman, W.; Katagiri, F.; Glazebrook, J. Arabidopsis PECTIN METHYLESTERASEs contribute to immunity against *Pseudomonas syringae*. *Plant Physiol.* **2014**, *164*, 1093–1107. [CrossRef]
60. Kim, H.G.; Kwon, S.J.; Jang, Y.J.; Nam, M.H.; Chung, J.H.; Na, Y.-C.; Guo, H.; Park, O.K. GDSL LIPASE1 modulates plant immunity through feedback regulation of ethylene signaling. *Plant Physiol.* **2013**, *163*, 1776–1791. [CrossRef]
61. Pfalz, M.; Mikkelsen, M.D.; Bednarek, P.; Olsen, C.E.; Halkier, B.A.; Kroymann, J. Metabolic engineering in *Nicotiana benthamiana* reveals key enzyme functions in Arabidopsis indole glucosinolate modification. *Plant Cell* **2011**, *23*, 716–729. [CrossRef] [PubMed]
62. Pfalz, M.; Vogel, H.; Kroymann, J. The gene controlling the indole glucosinolate modifier1 quantitative trait locus alters indole glucosinolate structures and aphid resistance in Arabidopsis. *Plant Cell* **2009**, *21*, 985–999. [CrossRef] [PubMed]
63. Schuhegger, R.; Nafisi, M.; Mansourova, M.; Petersen, B.L.; Olsen, C.E.; Svatoš, A.; Halkier, B.A.; Glawischnig, E. CYP71B15 (PAD3) catalyzes the final step in camalexin biosynthesis. *Plant Physiol.* **2006**, *141*, 1248–1254. [CrossRef] [PubMed]
64. Millet, Y.A.; Danna, C.H.; Clay, N.K.; Songnuan, W.; Simon, M.D.; Werck-Reichhart, D.; Ausubel, F.M. Innate immune responses activated in Arabidopsis roots by microbe-associated molecular patterns. *Plant Cell* **2010**, *22*, 973–990. [CrossRef] [PubMed]
65. Rajniak, J.; Barco, B.; Clay, N.K.; Sattely, E.S. A new cyanogenic metabolite in Arabidopsis required for inducible pathogen defence. *Nature* **2015**, *525*, 376–379. [CrossRef] [PubMed]
66. Mizutani, M.; Ohta, D. Diversification of P450 genes during land plant evolution. *Annu. Rev. Plant Biol.* **2010**, *61*, 291–315. [CrossRef] [PubMed]
67. Lodeiro, S.; Xiong, Q.; Wilson, W.K.; Kolesnikova, M.D.; Onak, C.S.; Matsuda, S.P. An oxidosqualene cyclase makes numerous products by diverse mechanisms: A challenge to prevailing concepts of triterpene biosynthesis. *J. Am. Chem. Soc.* **2007**, *129*, 11213–11222. [CrossRef] [PubMed]
68. Barreto, P.; Arcuri, M.L.C.; Lima, R.P.M.; Marino, C.L.; Maia, I.G. Comprehensive in silico analysis and transcriptional profiles highlight the importance of mitochondrial dicarboxylate carriers (DICs) on hypoxia response in both *Arabidopsis thaliana* and *Eucalyptus grandis*. *Plants* **2022**, *11*, 181. [CrossRef]
69. Palmieri, L.; Picault, N.; Arrigoni, R.; Besin, E.; Palmieri, F.; Hodges, M. Molecular identification of three *Arabidopsis thaliana* mitochondrial dicarboxylate carrier isoforms: Organ distribution, bacterial expression, reconstitution into liposomes and functional characterization. *Biochem. J.* **2008**, *410*, 621–629. [CrossRef]
70. Alcázar, R.; Bueno, M.; Tiburcio, A.F. Polyamines: Small amines with large effects on plant abiotic stress tolerance. *Cells* **2020**, *9*, 2373. [CrossRef]
71. Schuler, M.; Bauer, P. Heavy metals need assistance: The contribution of nicotianamine to metal circulation throughout the plant and the Arabidopsis NAS gene family. *Front Plant Sci.* **2011**, *2*, 69. [CrossRef] [PubMed]
72. Wang, J.; Yu, X.; Ding, Z.J.; Zhang, X.; Luo, Y.; Xu, X.; Xie, Y.; Li, X.; Yuan, T.; Zheng, S.J. Structural basis of ALMT1-mediated aluminum resistance in Arabidopsis. *Cell Res.* **2022**, *32*, 89–98. [CrossRef] [PubMed]
73. Gilmour, S.J.; Zarka, D.G.; Stockinger, E.J.; Salazar, M.P.; Houghton, J.M.; Thomashow, M.F. Low temperature regulation of the Arabidopsis CBF family of AP2 transcriptional activators as an early step in cold-induced COR gene expression. *Plant J.* **1998**, *16*, 433–442. [CrossRef] [PubMed]
74. Bahieldin, A.; Atef, A.; Edris, S.; Gadalla, N.O.; Ramadan, A.M.; Hassan, S.M.; Al Attas, S.G.; Al-Kordy, M.A.; Al-Hajar, A.S.M.; Sabir, J.S.M. Multifunctional activities of ERF109 as affected by salt stress in Arabidopsis. *Sci. Rep.* **2018**, *8*, 6403. [CrossRef] [PubMed]
75. Li, Z.; Zhang, Y.; Ren, J.; Jia, F.; Zeng, H.; Li, G.; Yang, X. Ethylene-responsive factor ERF114 mediates fungal pathogen effector PevD1-induced disease resistance in *Arabidopsis thaliana*. *Mol. Plant Pathol.* **2022**, *23*, 819–831. [CrossRef] [PubMed]
76. Kang, H.-G.; Kim, J.; Kim, B.; Jeong, H.; Choi, S.H.; Kim, E.K.; Lee, H.-Y.; Lim, P.O. Overexpression of FTL1/DDF1, an AP2 transcription factor, enhances tolerance to cold, drought, and heat stresses in *Arabidopsis thaliana*. *Plant Sci.* **2011**, *180*, 634–641. [CrossRef] [PubMed]
77. Zou, L.; Yang, F.; Ma, Y.; Wu, Q.; Yi, K.; Zhang, D. Transcription factor WRKY30 mediates resistance to Cucumber mosaic virus in Arabidopsis. *Biochem. Biophys. Res. Commun.* **2019**, *517*, 118–124. [CrossRef] [PubMed]
78. Scarpeci, T.E.; Zanor, M.I.; Mueller-Roeber, B.; Valle, E.M. Overexpression of AtWRKY30 enhances abiotic stress tolerance during early growth stages in *Arabidopsis thaliana*. *Plant Mol. Biol.* **2013**, *83*, 265–277. [CrossRef] [PubMed]
79. Zentgraf, U.; Doll, J. Arabidopsis WRKY53, a node of multi-layer regulation in the network of senescence. *Plants* **2019**, *8*, 578. [CrossRef]

80. Hu, Y.; Dong, Q.; Yu, D. Arabidopsis WRKY46 coordinates with WRKY70 and WRKY53 in basal resistance against pathogen *Pseudomonas syringae*. *Plant Sci.* **2012**, *185*, 288–297. [CrossRef]
81. Kong, L.; Feng, B.; Yan, Y.; Zhang, C.; Kim, J.H.; Xu, L.; Rack, J.G.M.; Wang, Y.; Jang, J.-C.; Ahel, I. Noncanonical mono (ADP-ribosyl) ation of zinc finger SZF proteins counteracts ubiquitination for protein homeostasis in plant immunity. *Mol. Cell* **2021**, *81*, 4591–4604. [CrossRef] [PubMed]
82. Miwa, H.; Kinoshita, A.; Fukuda, H.; Sawa, S. Plant meristems: CLAVATA3/ESR-related signaling in the shoot apical meristem and the root apical meristem. *J. Plant Res.* **2009**, *122*, 31–39. [CrossRef] [PubMed]
83. Wen, J.; Lease, K.A.; Walker, J.C. DVL, a novel class of small polypeptides: Overexpression alters Arabidopsis development. *Plant J.* **2004**, *37*, 668–677. [CrossRef] [PubMed]
84. Yruela, I. Plant development regulation: Overview and perspectives. *J. Plant Physiol.* **2015**, *182*, 62–78. [CrossRef]
85. Fujihara, R.; Uchida, N.; Tameshige, T.; Kawamoto, N.; Hotokezaka, Y.; Higaki, T.; Simon, R.; Torii, K.U.; Tasaka, M.; Aida, M. The boundary-expressed EPIDERMAL PATTERNING FACTOR-LIKE2 gene encoding a signaling peptide promotes cotyledon growth during *Arabidopsis thaliana* embryogenesis. *Plant Biotechnol.* **2021**, *38*, 317–322. [CrossRef] [PubMed]
86. De Smet, I.; Vassileva, V.; De Rybel, B.; Levesque, M.P.; Grunewald, W.; Van Damme, D.; Van Noorden, G.; Naudts, M.; Van Isterdael, G.; De Clercq, R. Receptor-like kinase ACR4 restricts formative cell divisions in the Arabidopsis root. *Science* **2008**, *322*, 594–597. [CrossRef] [PubMed]
87. Cohen, H.; Fedyuk, V.; Wang, C.; Wu, S.; Aharoni, A. SUBERMAN regulates developmental suberization of the Arabidopsis root endodermis. *Plant J.* **2020**, *102*, 431–447. [CrossRef] [PubMed]
88. Li, X.; Ma, D.; Lu, S.X.; Hu, X.; Huang, R.; Liang, T.; Xu, T.; Tobin, E.M.; Liu, H. Blue light-and low temperature-regulated COR27 and COR28 play roles in the Arabidopsis circadian clock. *Plant Cell* **2016**, *28*, 2755–2769. [CrossRef] [PubMed]
89. Wellmer, F.; Graciet, E.; Riechmann, J.L. Specification of floral organs in Arabidopsis. *J. Exp. Bot.* **2014**, *65*, 1–9. [CrossRef]
90. Stenvik, G.-E.; Tandstad, N.M.; Guo, Y.; Shi, C.-L.; Kristiansen, W.; Holmgren, A.; Clark, S.E.; Aalen, R.B.; Butenko, M.A. The EPIP peptide of INFLORESCENCE DEFICIENT IN ABSCISSION is sufficient to induce abscission in Arabidopsis through the receptor-like kinases HAESA and HAESA-LIKE2. *Plant Cell* **2008**, *20*, 1805–1817. [CrossRef]
91. Kaufmann, K.; Pajoro, A.; Angenent, G.C. Regulation of transcription in plants: Mechanisms controlling developmental switches. *Nat. Rev. Genet.* **2010**, *11*, 830–842. [CrossRef] [PubMed]
92. Preston, J.C.; Jorgensen, S.A.; Orozco, R.; Hileman, L.C. Paralogous SQUAMOSA PROMOTER BINDING PROTEIN-LIKE (SPL) genes differentially regulate leaf initiation and reproductive phase change in petunia. *Planta* **2016**, *243*, 429–440. [CrossRef] [PubMed]
93. Santos-Mendoza, M.; Dubreucq, B.; Baud, S.; Parcy, F.; Caboche, M.; Lepiniec, L. Deciphering gene regulatory networks that control seed development and maturation in Arabidopsis. *Plant J.* **2008**, *54*, 608–620. [CrossRef] [PubMed]
94. Khan, N.; Bano, A.; Ali, S.; Babar, M.A. Crosstalk amongst phytohormones from planta and PGPR under biotic and abiotic stresses. *Plant Growth Regul.* **2020**, *90*, 189–203. [CrossRef]
95. Tognetti, V.B.; Van Aken, O.; Morreel, K.; Vandenbroucke, K.; Van De Cotte, B.; De Clercq, I.; Chiwocha, S.; Fenske, R.; Prinsen, E.; Boerjan, W. Perturbation of indole-3-butyric acid homeostasis by the UDP-glucosyltransferase UGT74E2 modulates Arabidopsis architecture and water stress tolerance. *Plant Cell* **2010**, *22*, 2660–2679. [CrossRef] [PubMed]
96. Won, C.; Shen, X.; Mashiguchi, K.; Zheng, Z.; Dai, X.; Cheng, Y.; Kasahara, H.; Kamiya, Y.; Chory, J.; Zhao, Y. Conversion of tryptophan to indole-3-acetic acid by TRYPTOPHAN AMINOTRANSFERASES OF ARABIDOPSIS and YUCCAs in Arabidopsis. *Proc. Natl. Acad. Sci. USA* **2011**, *108*, 18518–18523. [CrossRef]
97. Hillebrand, H.; Bartling, D.; Weiler, E.W. Structural analysis of the nit2/nit1/nit3 gene cluster encoding nitrilases, enzymes catalyzing the terminal activation step in indole-acetic acid biosynthesis in *Arabidopsis thaliana*. *Plant Mol. Biol.* **1998**, *36*, 89–99. [CrossRef] [PubMed]
98. Woodward, A.W.; Bartel, B. Auxin: Regulation, action, and interaction. *Ann. Bot.* **2005**, *95*, 707–735. [CrossRef] [PubMed]
99. Shin, R.; Burch, A.Y.; Huppert, K.A.; Tiwari, S.B.; Murphy, A.S.; Guilfoyle, T.J.; Schachtman, D.P. The Arabidopsis transcription factor MYB77 modulates auxin signal transduction. *Plant Cell* **2007**, *19*, 440–2453. [CrossRef]
100. Wang, L.; Tsuda, K.; Truman, W.; Sato, M.; Nguyen, L.V.; Katagiri, F.; Glazebrook, J. CBP60g and SARD1 play partially redundant critical roles in salicylic acid signaling. *Plant J.* **2011**, *67*, 1029–1041. [CrossRef]
101. Lefebvre, V.; North, H.; Frey, A.; Sotta, B.; Seo, M.; Okamoto, M.; Nambara, E.; Marion-Poll, A. Functional analysis of Arabidopsis NCED6 and NCED9 genes indicates that ABA synthesized in the endosperm is involved in the induction of seed dormancy. *Plant J.* **2006**, *45*, 309–319. [CrossRef]
102. Tanaka, H.; Osakabe, Y.; Katsura, S.; Mizuno, S.; Maruyama, K.; Kusakabe, K.; Mizoi, J.; Shinozaki, K.; Yamaguchi-Shinozaki, K. Abiotic stress-inducible receptor-like kinases negatively control ABA signaling in Arabidopsis. *Plant J.* **2012**, *70*, 599–613. [CrossRef]
103. Rodrigues, A.; Adamo, M.; Crozet, P.; Margalha, L.; Confraria, A.; Martinho, C.; Elias, A.; Rabissi, A.; Lumbreras, V.; González-Guzmán, M. ABI1 and PP2CA phosphatases are negative regulators of Snf1-related protein kinase1 signaling in Arabidopsis. *Plant Cell* **2013**, *25*, 3871–3884. [CrossRef]
104. Rui, Y.; Dinneny, J.R. A wall with integrity: Surveillance and maintenance of the plant cell wall under stress. *New Phytol.* **2020**, *225*, 1428–1439. [CrossRef]

105. Goujon, T.; Sibout, R.; Eudes, A.; MacKay, J.; Jouanin, L. Genes involved in the biosynthesis of lignin precursors in *Arabidopsis thaliana*. *Plant Physiol. Biochem.* **2003**, *41*, 677–687. [CrossRef]
106. Fujiwara, S.; Mitsuda, N.; Nakai, Y.; Kigoshi, K.; Suzuki, K.; Ohme-Takagi, M. Chimeric repressor analysis identifies MYB87 as a possible regulator of morphogenesis via cell wall organization and remodeling in Arabidopsis. *Biotechnol. Lett.* **2014**, *36*, 1049–1057. [CrossRef] [PubMed]
107. Chezem, W.R.; Memon, A.; Li, F.-S.; Weng, J.-K.; Clay, N.K. SG2-type R2R3-MYB transcription factor MYB15 controls defense-induced lignification and basal immunity in Arabidopsis. *Plant Cell* **2017**, *29*, 1907–1926. [CrossRef]
108. Mamaeva, A.; Taliansky, M.; Filippova, A.; Love, A.J.; Golub, N.; Fesenko, I. The role of chloroplast protein remodeling in stress responses and shaping of the plant peptidome. *New Phytol.* **2020**, *227*, 1326–1334. [CrossRef] [PubMed]
109. Xiong, Y.; DeFraia, C.; Williams, D.; Zhang, X.; Mou, Z. Deficiency in a cytosolic ribose-5-phosphate isomerase causes chloroplast dysfunction, late flowering and premature cell death in Arabidopsis. *Physiol. Plant.* **2009**, *137*, 249–263. [CrossRef] [PubMed]
110. Hou, F.Y.; Huang, J.; Yu, S.L.; Zhang, H.S. The 6-phosphogluconate dehydrogenase genes are responsive to abiotic stresses in rice. *J. Integr. Plant Biol.* **2007**, *49*, 655–663. [CrossRef]
111. Tagele, S.B.; Kim, S.W.; Lee, H.G.; Kim, H.S.; Lee, Y.S. Effectiveness of multi-trait *Burkholderia contaminans* KNU17BI1 in growth promotion and management of banded leaf and sheath blight in maize seedling. *Microbiol. Res.* **2018**, *214*, 8–18. [CrossRef] [PubMed]
112. Gu, G.; Wang, N.; Chaney, N.; Smith, L.; Lu, S.-E. AmbR1 is a key transcriptional regulator for production of antifungal activity of *Burkholderia contaminans* strain MS14. *FEMS Microbiol. Lett.* **2009**, *297*, 54–60. [CrossRef] [PubMed]
113. Elshafie, H.S.; Camele, I.; Racioppi, R.; Scrano, L.; Iacobellis, N.S.; Bufo, S.A. In vitro antifungal activity of *Burkholderia gladioli* pv. agaricicola against some phytopathogenic fungi. *Int. J. Mol. Sci.* **2012**, *13*, 16291–16302. [CrossRef] [PubMed]
114. Yeh, Y.-H.; Chang, Y.-H.; Huang, P.-Y.; Huang, J.-B.; Zimmerli, L. Enhanced Arabidopsis pattern-triggered immunity by overexpression of cysteine-rich receptor-like kinases. *Front. Plant Sci.* **2015**, *6*, 322. [CrossRef] [PubMed]
115. Ederli, L.; Madeo, L.; Calderini, O.; Gehring, C.; Moretti, C.; Buonaurio, R.; Paolocci, F.; Pasqualini, S. The *Arabidopsis thaliana* cysteine-rich receptor-like kinase CRK20 modulates host responses to *Pseudomonas syringae* pv. tomato DC3000 infection. *J. Plant Physiol.* **2011**, *168*, 1784–1794. [CrossRef] [PubMed]
116. Kliebenstein, D.J. Secondary metabolites and plant/environment interactions: A view through *Arabidopsis thaliana* tinged glasses. *Plant Cell Environ.* **2004**, *27*, 675–684. [CrossRef]
117. Sankhuan, D.; Ji, M.; Takanashi, S.; Imamura, Y.; Sato, S.; Supaibulwatana, K.; Otani, M.; Nakano, M. Induction of dwarf and early flowering phenotypes in *Tricyrtis* sp. by ectopic expression of LEAFY from *Arabidopsis thaliana*. *Plant Biotechnol.* **2022**, *39*, 205–208. [CrossRef] [PubMed]
118. Takagi, H.; Hempton, A.K.; Imaizumi, T. Photoperiodic flowering in Arabidopsis: Multilayered regulatory mechanisms of CONSTANS and the florigen FLOWERING LOCUS T. *Plant Commun.* **2023**, *4*, 100552. [CrossRef] [PubMed]
119. Jung, J.-H.; Lee, H.-J.; Ryu, J.Y.; Park, C.-M. SPL3/4/5 integrate developmental aging and photoperiodic signals into the FT-FD module in Arabidopsis flowering. *Mol. Plant* **2016**, *9*, 1647–1659. [CrossRef]
120. Ogawara, T.; Higashi, K.; Kamada, H.; Ezura, H. Ethylene advances the transition from vegetative growth to flowering in *Arabidopsis thaliana*. *J. Plant Physiol.* **2003**, *160*, 1335–1340. [CrossRef]
121. Adams, D.O.; Yang, S.F. Ethylene biosynthesis: Identification of 1-aminocyclopropane-1-carboxylic acid as an intermediate in the conversion of methionine to ethylene. *Proc. Natl. Acad. Sci. USA* **1979**, *76*, 170–174. [CrossRef]
122. Nadeem, S.M.; Zahir, Z.A.; Naveed, M.; Ashraf, M. Microbial ACC-deaminase: Prospects and applications for inducing salt tolerance in plants. *Crit. Rew. Plant Sci.* **2010**, *29*, 360–393. [CrossRef]
123. Poupin, M.J.; Greve, M.; Carmona, V.; Pinedo, I. A complex molecular interplay of auxin and ethylene signaling pathways is involved in Arabidopsis growth promotion by *Burkholderia phytofirmans* PsJN. *Front. Plant Sci.* **2016**, *7*, 492. [CrossRef]
124. Gowtham, H.G.; Singh, S.B.; Shilpa, N.; Aiyaz, M.; Nataraj, K.; Udayashankar, A.C.; Amruthesh, K.N.; Murali, M.; Poczai, P.; Gafur, A. Insight into recent progress and perspectives in improvement of antioxidant machinery upon PGPR augmentation in plants under drought stress: A review. *Antioxidants* **2022**, *11*, 1763. [CrossRef] [PubMed]
125. Hu, Q.-J.; Chen, M.-X.; Song, T.; Cheng, C.-L.; Tian, Y.; Hu, J.; Zhang, J.-H. Spermidine enhanced the antioxidant capacity of rice seeds during seed aging. *Plant Growth Regul.* **2020**, *91*, 397–406. [CrossRef]
126. Wang, C.-Y.; Chen, Y.-W.; Hou, C.-Y. Antioxidant and antibacterial activity of seven predominant terpenoids. *Int. J. Food Prop.* **2019**, *22*, 230–238. [CrossRef]
127. Wani, A.B.; Chadar, H.; Wani, A.H.; Singh, S.; Upadhyay, N. Salicylic acid to decrease plant stress. *Environ. Chem. Lett.* **2017**, *15*, 15101–15123. [CrossRef]
128. Im Kim, J.; Baek, D.; Park, H.C.; Chun, H.J.; Oh, D.-H.; Lee, M.K.; Cha, J.-Y.; Kim, W.-Y.; Kim, M.C.; Chung, W.S. Overexpression of Arabidopsis YUCCA6 in potato results in high-auxin developmental phenotypes and enhanced resistance to water deficit. *Mol. Plant* **2013**, *6*, 337–349. [CrossRef] [PubMed]
129. Mir, A.R.; Alam, P.; Hayat, S. Auxin regulates growth, photosynthetic efficiency and mitigates copper induced toxicity via modulation of nutrient status, sugar metabolism and antioxidant potential in *Brassica juncea*. *Plant Physiol. Biochem.* **2022**, *185*, 244–259. [CrossRef]
130. Weisburg, W.G.; Barns, S.M.; Pelletier, D.A.; Lane, D.J. 16S ribosomal DNA amplification for phylogenetic study. *J. Bacteriol.* **1991**, *173*, 697–703. [CrossRef]

131. Altschul, S.F.; Gish, W.; Miller, W.; Myers, E.W.; Lipman, D.J. Basic local alignment search tool. *J. Mol. Biol.* **1990**, *215*, 403–410. [CrossRef]
132. De Coster, W.; D'hert, S.; Schultz, D.T.; Cruts, M.; Van Broeckhoven, C. NanoPack: Visualizing and processing long-read sequencing data. *Bioinformatics* **2018**, *34*, 2666–2669. [CrossRef]
133. Bankevich, A.; Nurk, S.; Antipov, D.; Gurevich, A.A.; Dvorkin, M.; Kulikov, A.S.; Lesin, V.M.; Nikolenko, S.I.; Pham, S.; Prjibelski, A.D. SPAdes: A new genome assembly algorithm and its applications to single-cell sequencing. *J. Comput. Biol.* **2012**, *19*, 455–477. [CrossRef] [PubMed]
134. Majoros, W.H.; Pertea, M.; Salzberg, S.L. TigrScan and GlimmerHMM: Two open source ab initio eukaryotic gene-finders. *Bioinformatics* **2004**, *20*, 2878–2879. [CrossRef]
135. Lagesen, K.; Hallin, P.; Rødland, E.A.; Stærfeldt, H.-H.; Rognes, T.; Ussery, D.W. RNAmmer: Consistent and rapid annotation of ribosomal RNA genes. *Nucleic Acids Res.* **2007**, *35*, 3100–3108. [CrossRef]
136. Lowe, T.M.; Eddy, S.R. tRNAscan-SE: A program for improved detection of transfer RNA genes in genomic sequence. *Nucleic Acids Res.* **1997**, *25*, 955–964. [CrossRef] [PubMed]
137. Chen, T.-W.; Gan, R.R.; Wu, T.H.; Huang, P.-J.; Lee, C.-Y.; Chen, Y.-Y.M.; Chen, C.-C.; Tang, P. FastAnnotator—An efficient transcript annotation web tool. In *BMC Genom: 2012*; Springer: Berlin/Heidelberg, Germany, 2012; pp. 1–8.
138. Nautiyal, C.S. An efficient microbiological growth medium for screening phosphate solubilizing microorganisms. *FEMS Microbiol. Lett.* **1999**, *170*, 265–270. [CrossRef]
139. Ahmad, F.; Ahmad, I.; Khan, M. Screening of free-living rhizospheric bacteria for their multiple plant growth promoting activities. *Microbiol. Res.* **2008**, *163*, 173–181. [CrossRef] [PubMed]
140. Tsai, S.-H.; Hsiao, Y.-C.; Chang, P.E.; Kuo, C.-E.; Lai, M.-C.; Chuang, H.-W. Exploring the biologically active metabolites produced by *Bacillus cereus* for plant growth promotion, heat stress tolerance, and resistance to bacterial soft rot in Arabidopsis. *Metabolites* **2023**, *13*, 676. [CrossRef]
141. Chang, P.E.; Wu, Y.-H.; Tai, C.-Y.; Lin, I.-H.; Wang, W.-D.; Tseng, T.-S.; Chuang, H.-W. Examining the Transcriptomic and Biochemical Signatures of *Bacillus subtilis* Strains: Impacts on Plant Growth and Abiotic Stress Tolerance. *Int. J. Mol. Sci.* **2023**, *24*, 13720. [CrossRef]
142. Kurniawan, A.; Chuang, H.W. Rhizobacterial Bacillus mycoides functions in stimulating the antioxidant defence system and multiple phytohormone signalling pathways to regulate plant growth and stress tolerance. *J. Appl. Microbiol.* **2022**, *132*, 1260–1274. [CrossRef] [PubMed]
143. Parcy, F.; Valon, C.; Raynal, M.; Gaubier-Comella, P.; Delseny, M.; Giraudat, J. Regulation of gene expression programs during Arabidopsis seed development: Roles of the ABI3 locus and of endogenous abscisic acid. *Plant Cell* **1994**, *6*, 1567–1582. [PubMed]

International Journal of
Molecular Sciences

MDPI

Review

Enhancing the Resilience of Agroecosystems Through Improved Rhizosphere Processes: A Strategic Review

Waleed Asghar [1,*], Kelly D. Craven [1], Jacob R. Swenson [1], Ryota Kataoka [2], Ahmad Mahmood [3] and Júlia Gomes Farias [4]

1 Department of Biochemistry and Molecular Biology, Oklahoma State University, Stillwater, OK 74078, USA; kelly.craven@okstate.edu (K.D.C.); jswenso@okstate.edu (J.R.S.)
2 Department of Environmental Sciences, Faculty of Life & Environmental Sciences, University of Yamanashi, Yamanashi 400-0016, Japan; rkataoka@yamanashi.ac.jp
3 Departments of Climate Change and Soil and Environmental Sciences, Muhammad Nawaz Shareef-University of Agriculture, Multan 60000, Pakistan; ahmad.mahmood@mnsuam.edu.pk
4 USDA-ARS, US Arid Land Agricultural Research Center, 21881 North Cardon Lane Maricopa, Maricopa, AZ 85138, USA; julia.stiles@usda.gov
* Correspondence: wasghar@okstate.edu

Abstract: As farming practices evolve and climate conditions shift, achieving sustainable food production for a growing global population requires innovative strategies to optimize environmentally friendly practices and minimize ecological impacts. Agroecosystems, which integrate agricultural practices with the surrounding environment, play a vital role in maintaining ecological balance and ensuring food security. Rhizosphere management has emerged as a pivotal approach to enhancing crop yields, reducing reliance on synthetic fertilizers, and supporting sustainable agriculture. The rhizosphere, a dynamic zone surrounding plant roots, hosts intense microbial activity fueled by root exudates. These exudates, along with practices such as green manure application and intercropping, significantly influence the soil's microbial community structure. Beneficial plant-associated microbes, including *Trichoderma* spp., *Penicillium* spp., *Aspergillus* spp., and *Bacillus* spp., play a crucial role in improving nutrient cycling and promoting plant health, yet their interactions within the rhizosphere remain inadequately understood. This review explores how integrating beneficial microbes, green manures, and intercropping enhances rhizosphere processes to rebuild microbial communities, sequester carbon, and reduce greenhouse gas emissions. These practices not only contribute to maintaining soil health but also foster positive plant–microbe–rhizosphere interactions that benefit entire ecosystems. By implementing such strategies alongside sound policy measures, sustainable cropping systems can be developed to address predicted climate challenges. Strengthening agroecosystem resilience through improved rhizosphere processes is essential for ensuring food security and environmental sustainability in the future. In conclusion, using these rhizosphere-driven processes, we could develop more sustainable and resilient agricultural systems that ensure food security and environmental preservation amidst changing climate situations.

Keywords: sustainable agroecosystems; rhizosphere process; soil microbes; green manure; intercropping

1. Introduction

The swift increase in the global population has markedly heightened the need for energy, food, and other essential goods, leading to the expansion of agricultural land use. Agroecosystems are intricate, evolving environments in which agricultural activities engage with natural ecosystems, affecting soil health, biodiversity, and ecosystem services. Recent studies highlight the significance of agroecosystems in enhancing resistance to climate change, boosting soil health, and fostering biodiversity [1,2]. For example, soil is a complex ecosystem and is one of our most important resources due to its many vital functions, such as providing food, fuel, and fiber; recycling nutrients that turn into beneficial nutrients for crop and plant necessities; organic matter decomposition, including plants and animal residues; regulation of water supply and quality; and provision of soil microorganisms, which are important for soil health and play vital roles in agroecosystems [3]. However, poor farming techniques, urbanization, and inadequate soil management practices have had negative impacts on agricultural soil, such as reduced soil fertility, the loss of beneficial soil microbiota, and the migration of soil organisms [4]. The top layer of soil surrounding plant roots, known as the rhizosphere, is home to a diverse range of microbes that play vital roles in promoting plant growth, organic matter degradation, and maintaining soil sustainability. However, in the lack of proper soil management, coupled with inappropriate agricultural practices, the quality of rhizosphere soil deteriorates, and its microbial communities weaken [5,6].

The rhizosphere serves as a microbial hotspot owing to the exudates emitted by plant roots, which supply vital nutrients for microbial activities [7]. Dynamics, for example, root zones, plant growth phases, plant species, and environmental conditions, change the diversity and composition of soil microbial communities (e.g., bacterial and fungi) in the rhizosphere [8]. Multiple species can compose microbial groups, demonstrating comparable mechanisms of action yet differing in their tolerance to plant cultivars and various environmental stresses. The microorganisms found in the rhizosphere include fungi, bacteria, and nematodes [9]. Large numbers of this community have an impartial effect on soil and plants, but some microbes are directly or indirectly connected with the plant–soil systems and provide nutrients from the rhizosphere to the plant. Additionally, rhizosphere microbiota contains microorganisms with both positive and negative effects on plant growth and soil health. Harmful microbes include pathogenic fungi, nematodes, and bacteria, while beneficial microbes, such as PGPM, nitrogen-fixing bacteria, and mycorrhizal fungi, contribute to plant nutrition and soil health [9,10]. However, beneficial microbes provide essential plant nutrients and contribute to lowering greenhouse gas emissions, improving nutrient availability, suppressing soil-borne plant diseases, and balancing nutrient cycling [11].

Conversely, many authors have pointed out that the use of various agricultural techniques, overuse of mineral fertilizers, agrochemicals, and farming practices could reduce soil fertility and disturb the rhizosphere microbiota, resulting in decreased soil health and crop production [12,13]. In general, overuse of agrochemicals (i.e., fungicides, herbicides, nematicides, and insecticides) is being used non-judiciously, which has a harmful effect on soil-positive microbiota in the soil as well as the overall health and quality of the soil [14,15]. This impact is mostly related to changes in microbial parameters such as enzymatic activity and biomass, which are the most important features of soil health and are important tools in monitoring soil quality. Eighty to ninety percent of soil processes are controlled by soil microorganisms and are strongly linked with plant and soil health, in addition to the role of microorganisms in the suppression of soil-borne diseases, nutrient cycling, soil resilience, and maintenance of soil sustainability [16,17]. Nevertheless, several authors reported that green manure is an alternate way to sustain agriculture and maintain soil health. The integration of the entire plant, containing more nutrients than crop straw, into the soil enhances soil fertility by adding more organic matter. Green manure improves crop yield and growth, reduces pests and insects, suppresses plant diseases, and increases soil enzyme activities, which regulate nutrient fluxes and stimulate plant growth [18–20]. Therefore, a sustainable solution must be sought by utilizing all available resources to modulate the rhizospheric process for sustainably enhancing agroecosystem productivity.

It is important for long-term sustainability in agricultural systems to keep agroecosystems stable and improve rhizosphere processes that make soil more resilient. This includes keeping and expanding the diversity of microorganisms in the rhizosphere. The rhizosphere is crucial for facilitating nutrient cycling, enhancing soil structure, and promoting plant health. The primary goal of this review is to investigate several methods for improving agricultural ecosystem resilience through rhizosphere activity optimization. We can implement these practices by utilizing beneficial microbes, green manures, and intercropping. These strategies jointly enhance synergistic communications between soil microorganisms and plant roots, creating an environment favorable for soil health and plant growth. The findings of this review improve comprehension of sustainable agriculture techniques, highlighting the importance of natural and organic resources in boosting crop production while preserving ecological balance.

2. The Rhizosphere: Its Mechanism and Roles

The rhizosphere refers to the area surrounding the root zone [18,19]. This zone is considered essential for biological activities because many exudates produced by plant roots provide nutrients for soil microorganisms. Thus, this region is vital for disease resistance, nutrient cycling, and plant growth [20]. The rhizosphere zone has a central role in the food chain of soil microorganisms. The microbial food chain mainly includes dead plant matter and living plant roots, which are the primary source of nutrients [21], and carbon from living plant tissues and root exudates. These are crucial for plant health and growth as they support disease suppression, protection, and nutrient uptake [22]. The activity of beneficial microorganisms in the rhizosphere not only provides essential nutrients to plants but also influences the quality and quantity of root exudates [22,23]. Furthermore, various metabolites exuded by plant roots into the soil primarily shape and attract the microbial community composition in the rhizosphere, facilitating interactions with both biotic and abiotic factors. Plants frequently adjust the diversity of these microbes according to the benefits they provide for growth and health [24]. In the rhizosphere, plants engage in a wide range of interactions with soil-dwelling organisms, including exploitation, competition, neutrality, mutualism, and commensalism. These diverse interactions are essential for plants to adapt to varying environmental conditions, which is crucial for their survival and agricultural productivity. In the rhizosphere, plants host a diverse collection of microbes, and they can inhabit the inner tissues (endosphere), and surface tissues (phylloplane and rhizoplane). Collectively, these microorganisms are referred to as the plant microbiome [25,26]. The plant and soil microbiome is critical for stimulating plant development, improving soil quality and fertility, and supporting agricultural sustainability by strengthening resilience to environmental stressors, nutrient cycling, and disease-fighting [27,28]. The interplay between roots and their surrounding soil constitutes one of the most prolific and diverse ecosystems in nature. The rhizosphere, the tiny and thin soil layer around the roots, has a vast population of microbes fueled by carbon-rich compounds exuded during photosynthesis and provides essential nourishment for the recruited microorganisms [29]. Overall, the rhizosphere is a kind of microenvironment wherein plant root development and nutrient uptake, soil properties, and microbe activities interact in a corresponding manner (Figure 1).

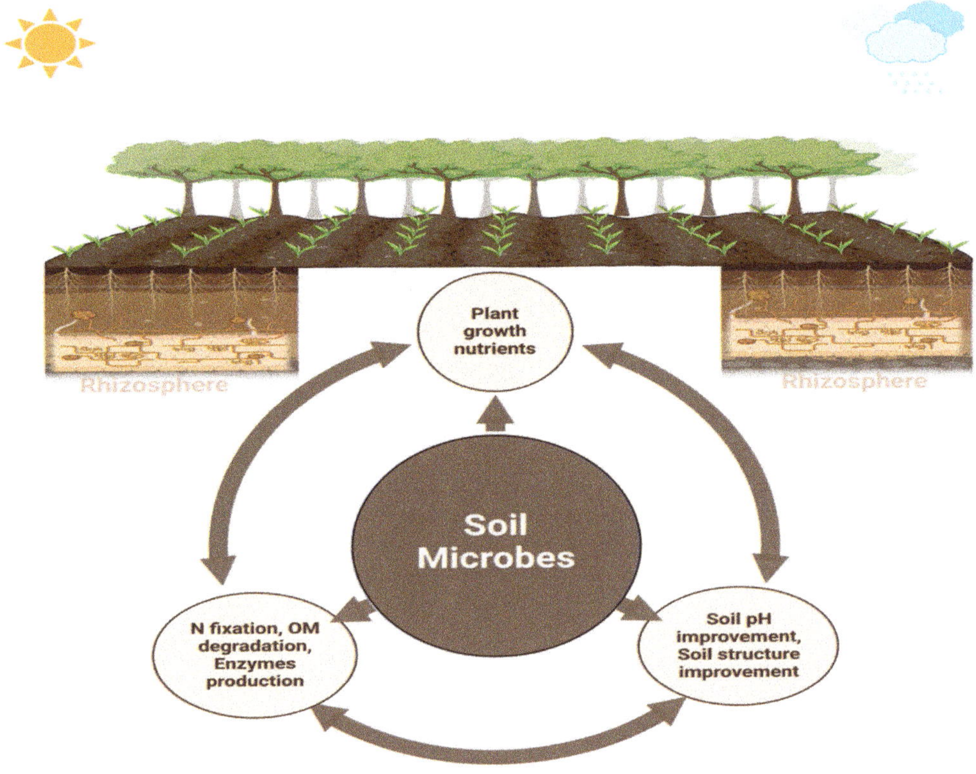

Figure 1. The rhizosphere is an active region where plant roots engage with soil microorganisms. Root exudates affect microbial activity, augmenting nutrient cycling, promoting plant growth, increasing enzyme production, and benefiting soil health. Beneficial microbes enhance nutrient absorption, improve the structure of the soil, and safeguard plants against diseases; hence, they play a vital role in the stability of agroecosystems.

2.1. *Rhizosphere Processes: Nutrient Retention in Agroecosystems*

The rhizosphere hosts diverse microbial groups that play critical roles in nutrient cycling, promoting plant growth, and protecting plants from pathogens, as well as from biotic and abiotic stresses [21]. Changes in rhizosphere processes directly contribute to nutrient retention in agroecosystems. Through plant–soil interactions, plants and their residues provide organic carbon (C) to the soil. The rate of soil carbon input depends primarily on the rate of plant growth, which is driven by photosynthesis. Root-derived carbon input, in particular, is a key regulatory factor in the plant–soil interaction within the rhizosphere, as it has a more direct impact on soil processes than carbon input from shoots [22]. This root-derived carbon plays a significant role in soil carbon sequestration. The amount of carbon that can be stored as soil organic matter (SOM) depends on the balance between carbon input and the rate at which plant materials decompose. Decomposition is the process by which microorganisms break down organic molecules into their inorganic components, such as nutrients and CO_2. Carbon dioxide, produced during microbial respiration, is released from the soil back into the atmosphere. However, some decomposition products are protected from further breakdown, playing a critical role in forming stable SOM, which is essential for long-term carbon sequestration in agroecosystems [22]. As a result, the rhizosphere process is particularly crucial for nutrient cycling, stabilizing SOM, making nutrients available for plant growth, and ultimately handling soil microorganisms. On the other hand, maintaining and increasing soil and its process through utilizing carbonized materials such as "Biochar" have long-term stability [23]. Biochars can reduce the emission of N_2O and CH_4 from the soil [24]. Biochar comprises a stable C compound as a derivative of gasification and pyrolysis and is available either as a pelletized or powder form. Char is ubiquitous in natural soils, containing as much as 35% of total organic C (TOC) in fire-impacted soils. Laird [25] reported that the application of biochar into the soil was found to enhance soil fertility and significantly increase soil organic carbon (SOC) and nutrients, which leads to the improvement in soil health. SOC consists of a complex mix of partially decomposed substances, including polysaccharides, lignin, aliphatic biopolymers, tannins, lipids, proteins, and amino sugars derived from plant litter, as well as microbial and faunal biomass [26]. However, SOC has been classified into slow, active, and inert pools, depending upon the decomposition rate in soil [27]. Furthermore, biochar largely supports the SOC pool for stabilization, while green manure incorporation adds to the more bio-accessible fraction of the soil C reserve. This is because biochar decomposes typically slower than uncharred biomass, e.g., green manure incorporation and fresh crop residues [28]. However, green manure or fresh crop residues in agricultural fields, on the other hand, can be used as a SOC storage/sequestration strategy by supporting cover crops; conservation tillage provides benefits including nutrient cycling, control of surface runoff of water, wind erosion, and crop production [29]. Yazdanpanah [30] reported that applying green manure or plat residues improved the stability and total porosity of soil aggregates and increased the soil C pool, but this depended on the type and application rate of soil amendments. However, Kenney et al. [31] reported that the removal of crop residues (e.g., corn stover and *Zea mays* L.) decreased the soil C pool. Therefore, the following strategy should be considered to achieve higher soil C sequestration via soil amendments produced from green manure/crop residues or biochar. The indirect effects of using these amendments should be considered in the context of production/control of soil GHG, environmental health, and economic cost benefits of their use. As per our knowledge, the combined effect of biochar and crop residues has rarely been investigated except in one recent study by Nguyen et al. [32], and it should be considered in the near future. Overall, biochar seems to have a greater positive impact on total soil SC, while the incorporation of crop residues/green manures can increase the microbial biomass C of soil, enhance the rhizosphere process, and reduce carbon dioxide emission.

2.2. Rhizosphere Engineering: Turning to Sustainably Increases the Agroecosystem and Its Productivity

The rhizosphere is a hotspot of microbes because many exudates are produced by plant roots, which are the primary source of nutrients for essential microbial processes and sustain the agroecosystems and their functions. The rhizosphere contains diverse microbial groups that perform various functions and affect nutrient cycling and plant growth promotion, and this portion is affected via different strategies such as crop rotation, green manuring, and sustainable use of PGPM and intercropping. However, through the addition of PGPM, Asghar and Kataoka [33] found that plant growth-promoting fungi changed the fungal community composition, enhanced the phosphatase and glycosidase enzyme activities in the soil, and stimulated the lettuce and brassica plant growth. Further, Kataoka et al. [34] reported that incorporating leguminous green manure (Hairy Vetch, *Vicia villosa* L.) enhances the soil fungal biomass and diversity and provides a healthier environment by producing soil phosphatase enzymes. Furthermore, rhizosphere engineering and its processes are explained with different strategies. In short, these interventions improve the rhizospheric processes and sustainably increase the agroecosystem and its productivity. On the other side of the agroecosystems and maintaining the rhizosphere process, biochar has emerged as a crucial innovation in scientific research, offering multiple benefits for sustainable agriculture, rhizosphere management, and environmental health [35]. Biochar is a type of charcoal produced through the combustion of organic feedstock with limited oxygen. It is considered a valuable soil conditioner and an effective means of carbon sequestration, which helps mitigate global warming and climate change [36]. Biochar decomposes slowly, making it highly durable when applied to soil, where it can remain for the long term [37]. However, generally, biochar is derived from waste materials such as animal manure, forest residues, and agricultural byproducts [38]. Through scientific processes, these waste materials are transformed into valuable products that can directly or indirectly improve soil health and promote plant growth. Biochar enhances crop yield, improves soil quality, and maintains soil health by preserving biochemical properties [39]. In terms of soil biological properties, Uzoma et al. [40] found that biochar positively influences soil microbes and their activity, leading to increased crop yields. However, the effect of biochar depends on the specific soil type and specific crops. Another study by Lu et al. [41] highlighted that biochar's porous structure creates a favorable niche for soil microbiota, boosting microbial populations. Conversely, biochar derived from rice straw has been shown to reduce certain microbial communities, such as *Ascomycota* fungi and *Actinobacteria*, though the abundance and diversity of soil microbes may vary with different types of biochar [42]. The application of biochar not only alters soil microbial communities but also affects nutrient cycling, which can have a direct impact on rhizospheric processes and agroecosystem productivity.

3. Role of Soil Microbes in the Rhizosphere Processes

The rhizosphere is home to various soil microbes, including fungi, bacteria, protozoa, and nematodes, which play essential roles in nutrient cycling, plant growth, and maintaining agroecosystems. These microbes facilitate the breakdown of organic matter and nitrogen fixation, making nutrients available to plants [4]. In agriculture, two groups of soil microorganisms are particularly important—rhizosphere fungi and bacteria—due to their critical functions. Fungi and bacteria thrive in soil, adapting to a wide range of environments, including harsh and high-salt conditions [43,44]. Fungi in the rhizosphere include pathogenic, saprotrophic, and mutualistic species, with mycorrhizal and plant-growth-promoting fungi being especially important for increasing nutrient availability, promoting plant growth and stress tolerance, and improving soil structure and disease resistance [45]. In contrast, the main rhizosphere bacterial phyla are *Firmicutes*, *Actinobacteria*, *Acidobacteria*, and *Proteobacteria* [46]. Plant growth-promoting rhizobacteria (PGPR) are crucial for enhancing plant growth, improving disease resistance, responding to carbon inputs and nutrient uptake, and assisting in phytoremediation [47–49]. In recent decades, a substantial body of literature has emerged demonstrating the activities of microbial species in plant growth promotion, biocontrol of phytopathogens, and nutrient cycling, highlighting their potential for development as alternative or supplementary agrochemicals in sustainable agriculture.

3.1. The Role of Microbes in Agroecosystems

Rhizosphere microorganisms, such as fungi and bacteria, not only act as pathogens in agricultural plants but also facilitate plant growth, increase nutrient availability, and control plant diseases through their role as biocontrol agents [50]. Recent research has highlighted the utilization of rhizosphere microorganisms to enhance plant development and mitigate diseases, thereby reducing reliance on pesticides and herbicides [51]. These microorganisms establish symbiotic associations with plant roots, exemplified by mycorrhizal fungi, which augment water and nutrient absorption, especially phosphorus. Furthermore, specific bacterial and fungal species might inhibit detrimental infections, thereby reducing dependence on chemical pesticides. Furthermore, these microbes help in the degradation of pollutants, thereby improving soil quality and resilience [49,52]. In comparison, the excessive application of chemical fertilizers and pesticides has detrimental environmental consequences, and growing reliance on these substances is concerning. Commercially available as biofertilizers and biocontrol agents, a wide range of microorganisms are utilized in agricultural production and the management of agroecosystems through rhizosphere processes that chemical fertilizers often disrupt. The diversity and activity of soil microbial communities are essential for preserving soil structure, enhancing plant growth, and sustaining ecosystem stability, rendering these microorganisms crucial for sustainable agricultural operations [53]. Therefore, enhancing the resilience of agroecosystems through rhizosphere microbes is crucial for promoting sustainable agriculture.

3.2. Plant Growth Promotion via Microbes

The application of chemical fertilizers to meet the food demand for a growing human population has led to many unpredicted environmental concerns related to soil–plant health and agroecosystems [54]. To protect soil–plant health and the environment, there is a need to develop and adopt new biological techniques that are not harmful to agroecosystems, soil, and plant health. Many researchers and scientists have already reported that microbes related to biofertilizers are functional and could be an alternative to chemical fertilizers, provide maximum environmental benefits, and improve crop yield and soil microbial biomass. In this context, microbe-related biofertilizers are capable and could be used in an eco-friendly manner to enhance crop production and maintain agroecosystems [55]. Microbe-related biofertilizers are defined as biological inoculants of beneficial fungal and bacterial strain mobilizers that can increase crop productivity and sustain soil biodiversity [56]. Mazid and Khan [56] reported that these inoculants, applicable directly or indirectly to soil and plant roots, are cost-effective and supply important nutrients for plants, enhancing soil nutritional composition, structure, and fertility.

Moreover, biofertilizer-related microbes are commonly known as plant growth-promoting microbes (PGPM). There are various fungal and bacterial strains that have already been reported and used commercially to improve plant growth, soil conditions, and the nutritional demand of plants. Such bacterial and fungal strains are *Azotobacter*, *Pseudomonas*, *Azospirillum*, *Rhizobium*, *Streptomyces*, *Bacillus* spp., *Trichoderma*, *Penicillium*, *Phoma*, *Fusarium*, and *Aspergillus* spp., and these strains have been gaining global attention due to their potential to control soil diseases, promote plant growth, and maintain agroecosystems [57–59]. The significant role and potential activities of PGPM are shown in Table 1.

Table 1. Potential role and activities of PGPM in the rhizosphere for sustainable agriculture.

PGPM	Function and Role	References
Penicillium spp.	Promotion of plant growth, increase systemic resistance, and confrontation to *Pseudomonas syringae* pv. Tomato	[60,61]
Trichoderma spp.	Plant growth promotion, production of indole-3-acetic acid IAA, production of siderophores, resistance against soil-borne disease	[33,62]
Aspergillus spp.	Induces systemic resistance against necrotrophic fungus, plant growth promotion and phytohormones production	[63]
Purpureocillium spp.	Plant growth promotion and siderophores and enzyme production	[64]
Talaromyces spp.	Promotion of plant growth, phosphate solubilization, siderophores production	[65]

Table 1. *Cont.*

PGPM	Function and Role	References
Bacillus spp.	Promotion of plant growth, Indole-3-acetic acid (IAA) and phosphate solubilization, siderophore production	[52]
Azotobacter spp.	Ammonia production, Indole-3-acetic acid, growth enhancement of canola and lettuce plant, Phosphorus solubilization	[23,66]
Pseudomonas spp.	Growth improvement of canola and tomato plants, hormones production such as IAA, production of siderophore	[67,68]
Azospirillum spp.	Indole-3-acetic acid (IAA) production, siderophore and exo-polysaccharides, and promote canola plant growth	[67,68]
Streptomyces spp.	*Streptomyces* inhibit *Curvularia oryzae* and volatile organic compounds (VOCs) releasing, promote plant growth as well as activities of defense-related enzymes	[69]

3.3. Singling Between Plants and Microbes in the Rhizosphere

The rhizosphere contains four main components: soil, microorganisms, roots, and their interactions. These components collectively influence a range of physical, chemical, and biological processes that affect nutrient use, plant development, and overall plant health. The primary goal of rhizosphere signaling is the interaction between plants and the symbiotic microbes in their surrounding area, which promotes the growth of various rhizosphere microbial communities that have a beneficial impact on plant productivity [70]. A robust reciprocal influence marks the interaction between the plant and the rhizosphere microbiome (Figure 1), where they interact through the exchange of signaling molecules produced and recognized by both the plant and its related microbiosystem [71]. Disease-free and asymptomatic plants have complex relationships with their rhizosphere microbes, which improve plant performance and maintain the agroforestry system [72,73]. Plants change the pH level, the structure of the soil, the amount of oxygen that is available, and the energy source that comes from carbon-rich exudates. These changes have an effect on the microbiome around the plants' roots. Many of the chemically varying primary and secondary metabolites found in plant root exudates have bioactive effects on microorganisms, influencing their composition and function [74]. The exudation of carbon from plant roots, which accounts for almost one-third of the carbon produced by photosynthesis, significantly influences the results of chemical interactions at both the individual and community levels [75]. Root exudates serve as a principal means of communication between plants and their living environment, enabling several reactions, including nutrient uptake, competition for resources, signaling across species, the attraction of microorganisms, and several other interactions [76]. Furthermore, sugars, amino acids, organic acids, phenolic compounds, and secondary metabolites like coumarins, glucosinolates, benzoxazinoids, camalexin, and triterpenes are the main organic parts that come from exudate [77,78]. Plant species cultivate a unique microbial community in their rhizosphere by creating a varied carbon-rich environment. With more evidence, primary metabolites, including glucose, organic acids, and amino acids, supply carbon and energy for microbial proliferation, promoting beneficial microorganisms and inhibiting pathogens. Secondary metabolites, like flavonoids, phenolics, and terpenoids, often have bioactive properties, like the ability to kill bacteria or transfer information, that affect only the microbes in the soil and plants, and these changes in microbial community structure can significantly impact soil and agroecosystem [79–81]. This community provides many adaptive benefits to the plant host by influencing the composition of microorganisms and adjusting their advantageous characteristics [78]. Additionally, chemicals made from root exudates play a key role in creating a stress-resistant microbiome that helps plants deal with abiotic stresses like not getting enough food, being sick, and drought. The identification of stress-derived metabolomics and microbiota is a viable approach to addressing both abiotic and biotic limitations [82–84]. However, we have not thoroughly investigated the positive impacts of root-enriched microbial species supported by specialized metabolites produced from root exudate [85]. This paragraph of the manuscript on a review of current research on the plant rhizosphere microbiota and the compounds produced by root exudates for nutrient acquisition as depicted in the conceptual (Figure 2).

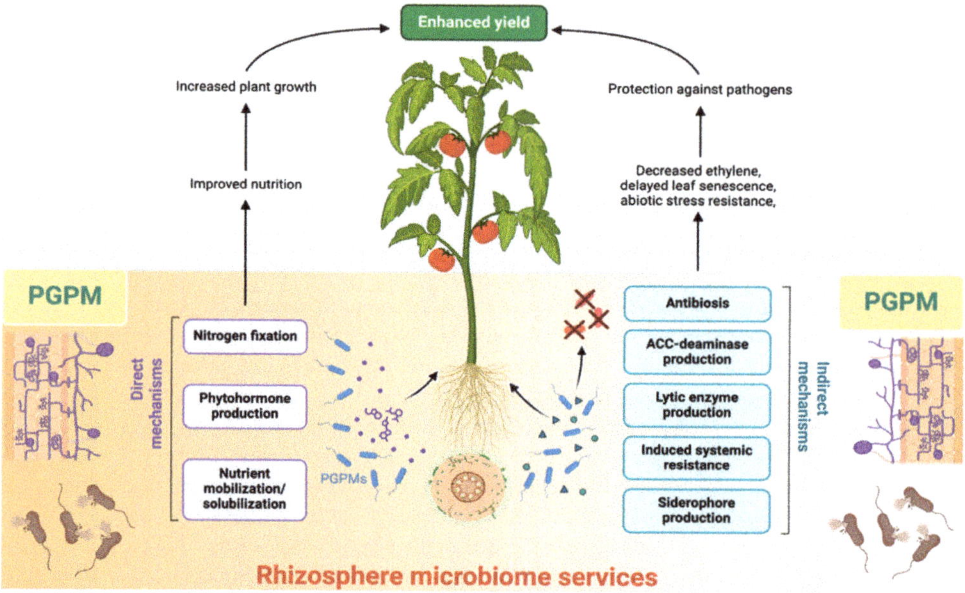

Figure 2. Soils in agroecosystems serve as a primary nutrient reservoir for crops, supplying necessary nutrients and water while also functioning as the most significant microbial reproductive bank for the rhizosphere microbiome. The plant offers many microhabitats, including both the endosphere and rhizosphere, for the development of the rhizosphere and plant microbiome. Beneficial plant–microbe interactions facilitate the rhizosphere microbiome's services. Microbiome services improve plant growth, stress tolerance, and disease resistance, as well as host resilience to abiotic challenges such as cold, salinity, and drought, resulting in biomass production increasing up to fourfold and enhancing rhizosphere processes through their direct and indirect mechanisms. We created the conceptual diagram to provide an overview of rhizosphere microbiome services.

4. The Role of Green Manures in Agroecosystems

Green manures represent a promising strategy for sustaining agroecosystems and improving the soil health and rhizosphere processes. Green manure can reduce chemical fertilizers in agroecosystems, ultimately improving soil health and quality [86]. Green manure provides various benefits to the agroecosystem and soil, such as soil covering, reduced soil temperature, organic matter improvement, increased water infiltration, and reduced weed infestation [87]. Moreover, green manure has environmental and economic benefits such as food production, maintaining biodiversity, soil carbon sequestration, and soil retention [88]. However, to enhance the resilience of agroecosystems and maintain soil health, planting green manure is an essential and valuable practice and works to reduce soil-borne diseases through biofumigation. Green manure provides better soil health and biodiversity and promotes the beneficial fungal genera that break the soil-borne pathogen's life cycle that is connected with the crop genotype [89]. Many types of green manures, including leguminous and non-leguminous, have multiple functions and benefits for soil and plant health, such as incorporating leguminous green manure to enhance wheat yield and production due to the mineralization of inorganic nitrogen and stimulating biological fixation [90]. Furthermore, Kataoka et al. [34] also reported that incorporating leguminous green manure (Hairy Vetch, *Vicia villosa* L.) enhances the soil fungal biomass and diversity and provides a healthier environment by producing soil phosphatase enzymes. Therefore, it could be assumed that planting and incorporating different green manures could maintain the agroecosystems and rhizosphere processes by enhancing biological processes and increasing organic matter. From another point of view, green manure may alter the soil N and C availability. For example, Hairy Vetch (a leguminous plant) fixes N_2, whereas barley (a non-leguminous plant) has high biomass productivity and thus adds more C, which influences ecosystem functioning and microbial groups in soil [91]. In comparison, Chavarría et al. [91] reported that green manure incorporation can improve agroecosystem services by improving nutrient retention capacity, accelerating soil organic matter, and reducing greenhouse gases in agricultural soils. However, applying alternative management, such as green manuring practices, stimulates the improvement of soil-diverse microbial communities, accelerating the specific enzyme activities related to P and C [92]. Thus, the relationship between microbial community composition and their activity stimulates plant growth and soil quality.

4.1. Green Manure and Their Beneficial Effects on the Rhizosphere Processes

The application of green manure is an additional practice for enhancing the resilience of agroecosystems and rhizosphere processes and increasing soil's productive capacity through nutrient availability. This application is geared towards maintaining agroecosystems and long-term environmental sustainability by enhancing processes such as nutrient cycling and preserving and maintaining soil biodiversity and the physical and chemical stability of the soil. Incorporating green manure can positively affect soil microbial communities by directly adding nutrients after incorporation and indirectly by changing plant and soil properties [93,94]. Directly, the decomposition of green manure offers an immediate supply of carbon, nitrogen, and other vital elements; hence, it enhances microbial activity and fosters the proliferation of beneficial bacteria [95]. The mineralization of organic matter from green manure produces labile chemicals that act as substrates for soil microorganisms, thereby boosting microbial biomass and diversity [96]. The incorporation of green manure indirectly modifies the soil's physical and chemical characteristics, including pH, soil structure, and water retention, fostering a conducive environment for microbial colonization and activity [97]. At the same time, Sun et al. [45] reported that the plant could affect soil microbial diversity and community by releasing root exudates. The decomposing residues of brassica green manure release glucosinolates, which assist in controlling parasitic nematodes [98]. Furthermore, the exclusive use of grass species may improve the soil nutrient profile compared to monoculture. The advantages of green manure designate it as a sustainable practice capable of providing various agro-ecological services; however, effective management of green manure is essential to substantially enhance soil carbon stocks and mitigate climate change, particularly in semi-arid Mediterranean regions, by decreasing greenhouse gas emissions [99,100]. Furthermore, many studies have proven that planting and incorporating green manure have multiple beneficial effects on soil health and the rhizosphere process [86,101]. The different types of green manure and their beneficial effects on the rhizosphere are shown in Table 2.

Table 2. Green manure and its beneficial effects.

Green Manure	Their Beneficial Effects	References
Oats, Hairy Vetch	Nutrient availability, soil organic matter, enhancing of soil enzymes and fungal biomass and soil inorganic N and P, reducing soil erosion	[34,86,102]
Winter Rye, Brassica	Biofumigation, reduction in soil-borne diseases, enhancement of soil enzymes and nitrogen fixation, as well as improvement in soil biodiversity and structure	[86,87,102]
Mustard, Barley	Soil and ground covering, nutrient cycling, and increasing microbial biomass, as well as weed suppression, Increasing fungi and bacteria ratio in soil	[20,87]
Red clover, Peas, Oats	Enhancement of biological activity in rhizosphere, nitrogen fixation and reduction of soil erosion, improve biological health, increase fungal abundance and soil enzyme activity	[103,104]
Cereal rye, Annual Ryegrass	Enhancement of SOM, control weed growth, control N_2O emissions, NO^{3-} reduction, C sequestration, and sustain cash crop yields	[89,104–106]
Sunn hemp, Alfalfa, Cowpea	Reduce erosion of soil, nitrogen fixation, increase SOM, improve soil biodiversity, and weeds suppress	[105,107]
Oilseed Radish, Winter Wheat, Rapeseed	Reduce soil erosion, SOM improvement, loosen topsoil, enhance soil biodiversity, and suppress pest	[102,105,107]

4.2. Green Manure and Their Effects on Soil Health

Soil health is a vital component of sustainable agriculture, signifying the soil's ability to operate as a living ecosystem that promotes plant productivity, preserving environmental quality, and supporting biodiversity [108]. Healthy soils exhibit an appropriate combination of physical, chemical, and biological attributes, encompassing adequate organic matter, nutrient accessibility, optimal structure, and vibrant microbial communities. Beneficial microorganisms, including nitrogen-fixing bacteria, mycorrhizal fungi, and decomposers, are vital to nutrient cycling, organic waste breakdown, and disease suppression [109]. Soil health is a primary requirement for agriculture and plant growth, but most agricultural soils are poor in organic nutrients and organic matter due to the high applications of inorganic or chemical-based fertilizers to achieve high crop yields. Different agricultural techniques should be adopted; otherwise, it can become very hazardous and lead to a lack of organic matter. Improving soil health through chemical, physical, and biological properties is attributed to the integrated application of green manure and organic application due to better nutrient uptake and preservation of soil health [110]. A variety of green manures, including grains, root crops, legumes, and oil crops, can serve as vegetative covers, each offering different benefits for enhancing soil health. Therefore, growing and incorporating green manure in agricultural lands helps maintain and preserve soil health and soil nutrient availability and ultimately increases soil organic matter content. The incorporation of green manure provides soil nutrients and organic matter to the soil, and organic matter plays an important role in soil biodiversity, soil microbiota, and the arrangement of soil aggregates to improve the soil structure [111,112]. Therefore, incorporating green manure leads to an increased organic matter content, which has an important role in soil health and quality. Furthermore, green manure suppresses soil-borne diseases through biofumigation. Soil-borne diseases also affect soil health, reduce crop productivity, and negatively affect soil biodiversity. Many soil-borne pathogens have negative effects on plant growth and crop production, such as *Phytophthora* spp., *Fusarium* spp., *Verticillium* spp., and *Rhizoctonia* spp. Soil-borne diseases reduce agricultural yield by as much as 50–70%, including those of wheat, vegetables, and cotton [113,114]. However, the practices of green manure in agricultural lands produce significant benefits such as increased soil nutrient availability, enhanced soil organic carbon, reduced soil compaction, improved particle aggregation and soil structure, as well as strengthened microbe activity, diversity, abundance, and suppression of soil-borne diseases.

To protect crops from soil-borne diseases, many fumigants and fungicides need to be used regularly. Later, it was noted that the use of fumigants and fungicides or any other chemical-based product causes ecological problems such as human health hazards and reduced beneficial microbes in the soil biodiversity, directly affecting soil health [115]. Consequently, organically, alternative strategies should be adopted to reduce soil-borne diseases, preserve soil health, and improve crop production. Alternatively, incorporation of green manure and cover crops are useful strategies for protecting soil health from soil-borne diseases and improving soil and plant health through biofumigation, providing nutrients and increasing beneficial microbial activity. Hao et al. [116] reported that broccoli (*B. oleracea*) grown as green manure reduces the *S. minor sclerotia* in the rhizosphere. Whalen [117] also reported that Fusarium wilt is another soil-borne disease that reduces watermelon production in the southeastern United States. The soil-borne disease Fusarium wilts is due to *Fusarium oxysporum*, which attacks the crop of watermelon and reduces the annual number of fruits, directly affecting the annual yield of watermelon production. For the management of Fusarium wilt disease, Hairy Vetch (*Vicia villosa* L.) green manure was incorporated. After incorporating Hairy Vetch, the occurrence of Fusarium wilt disease in the rhizosphere of the watermelon field was reduced, leading to an increased annual watermelon yield by 45% [118]. Papp et al. [119] demonstrated that the cultivation of green manures influences soil microorganism environments, which are essential for sustaining soil functions and ecological system sustainability because they participate in nutrient cycling and organic matter turnover, providing multiple services. Researchers have observed that the incorporation of green manure modifies the composition and dynamics of soil fungus and bacterial communities and encourages beneficial microorganisms. Ntalli and Caboni [120] observed that certain cruciferous green manures, including canola (*Brassica napus* L.) and rape (*Brassica rapa* L.), upon termination, release isothiocyanates (ITCs) via the hydrolysis of glucosinolates (GSLs) during the decomposition of their tissues, serving as effective biofumigants against various soil-borne pathogens and pests. The utilization of cruciferous species as green manure may facilitate the natural management of possible soil-borne diseases. Furthermore, Waisen et al. [121] demonstrated that brown mustard (*Brassica juncea* L.) and radish (*Raphanus raphanistrum* subsp. sativus) serve as excellent biofumigant crops that protect against plant–parasitic nematodes while keeping the soil healthy and maintaining the strength of the nematode population structure strong. Aydınlı and Mennan [122] also found that planting arugula (*Eruca sativa*) and radish (*R. sativus*) as winter crops before planting plants that are easily damaged by the root-knot nematode Meloidogyne arenaria may lower the number of eggs laid, the gall index, and the damage that occurs, while also increasing crop yields. Leguminous green manure can substantially enrich the soil with nitrogen throughout its growth and can acidify the rhizosphere by enhancing the absorption of insoluble phosphorus into the soil [123,124]. Leguminous green manure can fix atmospheric nitrogen via symbiosis with rhizobia in root nodules [125]. Furthermore, the fixed nitrogen can be allocated to intercropped non-leguminous plants within mixed cropping systems, or it may succeed crops in rotational practices. Biological nitrogen-fixing (BNF) systems may decrease the reliance on commercial nitrogen fertilizers [126,127]. Further, Scavo et al. [128] reported that the existence of *Trifolium subterraneum* for three straight years resulted in a significant rise in nitric nitrogen, ammoniacal nitrogen, and nitrogen cycle bacteria. Campiglia et al. [129] observed similar wheat yields when they used underground clover as living mulch in intercropping systems. In addition, Guardia et al. [130] showed that the mitigating effect of the legume green manure (vetch) mostly comes from less synthetic nitrogen being added to the next cash crop, less indirect N_2O emissions from NO_3^--leaching, and more carbon being stored because photosynthesis is higher.

More specifically, green manures are essential for improving soil health via a mix of physical, chemical, and biological processes that promote sustainable agroecosystems. The integration of green manure biomass augments soil organic matter, thereby enhancing soil structure, increasing water retention, and facilitating aeration, which fosters optimal circumstances for root growth [95,96]. This organic supplement serves as a nutrient reservoir, progressively releasing vital macronutrients such as nitrogen, phosphate, and potassium throughout decomposition and thus diminishing dependence on synthetic fertilizers. Green manures increase the diversity and activity of microbes in the soil. They do this by creating a dynamic food web that supports good microbes and stops soil-borne diseases through competitive exclusion and antibiotic synthesis. Furthermore, green manures facilitate carbon sequestration by integrating plant-derived carbon into stable soil organic components, thus reducing greenhouse gas emissions and enhancing soil resistance to climate change [131,132]. Adding leguminous green manures also helps fix nitrogen, and the breakdown of bioactive chemicals like glucosinolates in cruciferous green manures acts as natural biofumigants, reducing the number of pests and diseases that can affect the crop. Integrating green manures into cropping systems enhances soil health and bolsters agroecosystem sustainability, establishing it as a fundamental practice for climate-resilient agriculture [18]. Therefore, green manure and cover crops should be adopted to protect soil health, reduce soil-borne diseases as well as refine water use efficiency and many more beneficial impacts, as depicted in the conceptual (Figure 3).

Figure 3. Various green manures serve to improve soil health. The diagram illustrates the roles of leguminous and non-leguminous green manures in agroecosystems. Leguminous green manures,

including clover, Hairy Vetch, and peas, improve soil health chiefly by facilitating biological nitrogen fixation; hence, they diminish the reliance on synthetic fertilizers and augment nitrogen availability for succeeding crops. Non-leguminous green manures, including mustard, rye, and radish, enhance organic matter content, improve soil structure, and increase carbon sequestration. Both varieties of green manures enhance soil biodiversity by promoting microbial activity and nurturing beneficial microbial populations, such as nitrogen-fixing bacteria and mycorrhizal fungi. Furthermore, they recycle nutrients, inhibit soil-borne diseases, and enhance nutrient availability, promoting a nutrient-dense and biologically active soil ecosystem. The cumulative impacts lead to increased soil fertility, greater resilience against erosion and climate stress, and extended productivity of agroecosystems.

5. Role of Cropping Systems in Agroecosystems

Understanding the value of agroecosystems requires knowledge of soil, its different properties, and the processes that maintain its health. Unfortunately, the soil is currently being degraded rapidly due to extensive application of chemical-based fertilizers, lack of agricultural techniques, and irregular cropping are rapidly degrading the soil, but what is most concerning is that soil is a non-renewable resource at a human temporal scale [89]. Therefore, scientists and researchers have been trying to adopt alternative strategies to protect soil health, ultimately enhancing agroecosystems and rhizosphere processes. Cropping systems, such as intercropping and crop rotation practices, could serve this purpose in agricultural practices, thereby positively influencing soil health and agroecosystems [133]. Cropping systems are adopted to maximize the crop yields of agroecosystems as well as to maintain agroecosystems through soil health and quality preservation. In contrast, the minimization of chemical-based products directly or indirectly affects agroecosystems and rhizosphere processes [134]. With the primary goal of soil health maintenance to ensure the long-term stability of agroecosystems and the high production of crops, we need to adopt agronomic practices such as intercropping or crop rotations. Intercropping practices can improve agroecosystems by reducing the use of chemical-based fertilizers and soil pollution, preserving soil microbial diversity.

5.1. Intercropping and Their Effects on Rhizosphere Processes and Soil Health

Intercropping, which involves the cultivation of two or more crops in a field concurrently, is a useful strategy in agronomic practice. Different crops, such as cover and cash crops, grow at the same time. The practice has gained global attention due to its beneficial effects on the enhancement of agroecosystems and soil health. In brief, intercropping practices can maintain agroecosystems and soil health by the reduced use of chemical-based fertilizers [135], enhancing soil-plant functions [136], suppressing the occurrence of soil-borne diseases [133], increasing soil nutrient and organic matter contents [137], and promoting the beneficial soil microorganisms, which support the rhizosphere processes [138]. Cong et al. [139] reported that intercropping systems, including that of faba beans, corn, and wheat, had about 11%, 4%, and 23% higher below-ground plant biomass as well as increased organic carbon (C) and nitrogen (N) contents than those in plant species rotations found in Gansu, China. Furthermore, de Medeiros et al. [140] reported that intercropping with beans and pigeon peas considerably minimizes black root rot (*Scytalidium lignicola*) in cassava by up to 50% when matched with cassava in monoculture systems, as well as enhanced the rhizosphere process, that is, increasing the soil enzymes activities, microbial biomass, and soil nutrients related to organic C and N. Increasing the use of intercropping with different plant species have been typically interconnected with the enhancement of rhizosphere processes, providing environmental benefits [141]. For example, leguminous species are advantageous for soil by providing nitrogen-fixing microorganisms and enhancing soil enzyme activities. On the other hand, non-leguminous species usually provide soil cover and nitrogen availability; such intercropping self-regulates soil nitrogen levels to improve soil nutrient use and reduce the C footprint [142]. Furthermore, Graf et al. [143] also reported that compared to sole cropping, intercropping practices on *Dactylis glomerata* and *Medicago sativa* have led to increased shoot biomass and nitrous oxide (N_2O) production rate, which suggests that understanding the intercropping strategies is helpful for the maintenance of agroecosystems and soil health. Intercropping is regaining agronomical and ecological importance because of its ability to self-regulate soil nutrients and preserve soil health. An overview of intercropping in agroecosystems and its beneficial interactions based on Brooker et al. [144] and their outcomes are shown in Figure 4.

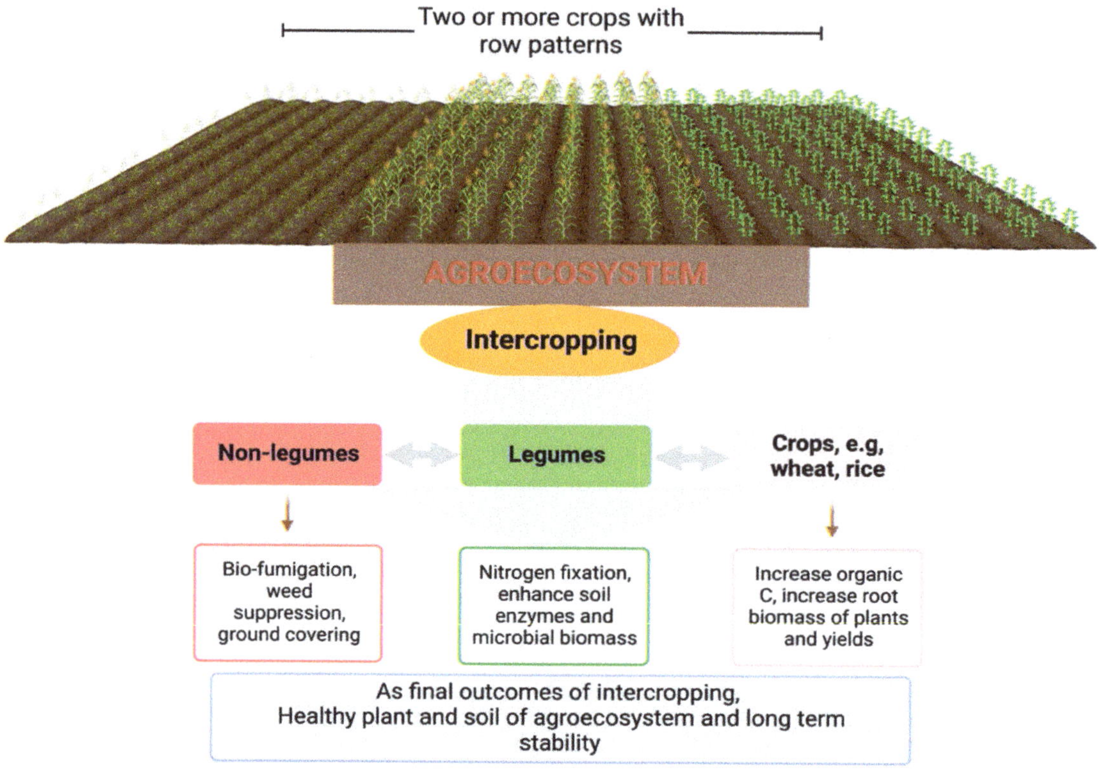

Figure 4. An overview of intercropping systems highlights their beneficial impacts on cash crops and long-term agroecosystem sustainability. Intercropping improves nutrient use efficiency, enhances soil health, promotes biodiversity, reduces pest and disease incidence, and increases crop resilience, contributing to stable yields and sustainable agricultural systems.

Furthermore, legume and cereal intercropping systems have been extensively employed to address limitations on natural resources, promote sustainable agricultural development, and guarantee food security due to their economic and ecological advantages [145]. Faba/wheat intercropping decreased nitrogen fertilizer application by 5–16% while simultaneously enhancing wheat production by 15–25% [146]. However, Luo et al. [147] reported that a 40 percent reduction in synthetic nitrogen input, coupled with soybean intercropping, might sustain sugarcane productivity and promote sustainable soil management. Many research studies have demonstrated that below-ground facilitative interactions among species can improve nutrient availability and utilization. Ref. [148] revealed that the root interactions between maize and faba beans, along with the root exudates from maize, significantly improved the symbiotic N_2 fixation and nodulation in faba beans grown alongside maize. Intercropping can make crop roots release more organic acids and control the mineralization of organic phosphorus, the dissolution of inorganic phosphorus, or both. This makes more phosphorus available in the soil. Recent research shows that growing cereals and legumes together improves metabolic function and changes the structure of the soil's microbial community through interactions between species [149]. According to Zhang et al. [150], elevated carbon uptake by maize (*Zea mays* L.) promoted nutrient cycling by altering the abundance of functional groups in soil microbes and improving the stability and complexity of microbial networks in soybean/maize intercropping systems. Teshita et al. [151] proposed that, particularly in low-nitrogen soils, intercropping maize with alfalfa (*Medicago sativa* L.) could enhance the structural complexity of the soil food web. Moreover, intercropping improves agricultural sustainability by utilizing crop diversity to exploit synergistic interactions across species [152]. This method improves crop yield and nitrogen utilization efficiency through similar and beneficial interactions. The rhizosphere of leguminous plants cultivates distinct microbial communities, including nitrogen-fixing bacterium (nifh) populations, via root exudates in legume-based intercropping systems, enhancing ecological benefits such as nitrogen fixation and phosphorus mobilization [153,154]. Within these approaches, maize/soybean intercropping is notable for its effective transfer of fixed nitrogen from soybean to maize, hence improving nitrogen efficiency and yields [155]. Multiple factors, such as cropping configurations and fertilization methods, affect how well maize and soybean intercropping systems work. Improving microbes is crucial for nutrient absorption and yield maximization [156,157]. Understanding how rhizosphere microbial communities react, especially in terms of their ability to fix nitrogen, would help improve intercropping patterns and fertilizer management, which would lead to more sustainable farming and food security.

5.2. Crop Rotation and Their Effects on Rhizosphere Processes and Soil

Crop rotation is a conventional and effective method for maintaining agroecosystems, managing biodiversity by improving soil health, suppressing disease and pest occurrences, and hence enhancing crop yields [158]. The efficiency and value of crop rotation are dependent upon various elements, including the types of crops utilized in the rotation, the sequence and frequency of specific crops, the duration of rotation, the agronomic history of the agricultural land, and the properties of the soil [159,160]. Recent studies reported that farmers had used crop rotation through traditional methods to manage crop pests and diseases and manage soil productivity [89,161]. Usually, these crop rotations involve growing three or four different types of crops in a sequence. However, with increasing food demand and agricultural production in recent years, many farmers grow just one or two crop species, with pesticides and mineral fertilizers contributing to compensate for the lack of crop rotations [162]. The crop rotations are not always helpful and also decline the soil quality, and crop production has been reported with the number of crops grown linked with short rotation or continuous cropping, including soybean, sugarcane, maize, wheat, and oilseed rape [163]. Interestingly, in crop rotation, oilseed rape is very important for the rotation and even highly profitable when grown as a break crop for cereals. Crop rotations with oilseed rape and maize affect the dynamics and structure of soil plant-associated microbial communities and are known to be beneficial for soil health via suppression of soil-borne diseases and plant growth promotion [164]. Microbial species living nearby or associated with plants are directly influenced by the root's architecture and the chemical characteristics of root exudates when we rotate with maize or oilseed rape crops [165]. The direct influence of plant roots around soil is known as the rhizosphere. The rhizosphere is the habitat of plant-beneficial microorganisms and plant pathogens [109]. With the research evidence of Benitez, Osborne, and Lehman [164], crop rotations promote beneficial plant growth-promoting microbes, affect the soil microbial community in the rhizosphere, and influence maize seedling growth characteristics.

Furthermore, rotating crops with grain legumes can significantly improve the yield and protein content of subsequent wheat crops, owing to the increased nitrogen availability in the soil from legume biological fixation [90]. Leguminous crops (such as peas and chickpeas) and various chickpea genotypes (cultivars) in rotation can alter soil functional microbial populations and affect the productivity of pulse crops and subsequent wheat crops. Different crops can produce root exudates and distinct residues that enhance soil microbial diversity and activity, as well as microbial biomass, nitrogen, and carbon cycling [160,166,167]. Modifications in crop rotation duration and frequency of identical crops throughout time might influence the prevalence of root rot diseases and improve crop production and soil health. Short series rotations exhibit heightened sensitivity to host-specific diseases, resulting in poorer yields compared to lengthier series rotations [162,168]. In addition, Lupwayi et al. [169] found that the wheat phase of a five-year rotation exhibited greater microbial biomass in both rhizosphere and bulk soil compared to the three-year rotation of the wheat phase, likely due to increased carbon inputs from crop residues. In Canada, particularly in the western part, two phases of pea in a four-year rotation increased soil nitrogen levels, whereas three legume phases markedly altered the function and composition of the rhizosphere bacterial community in comparison to continuous wheat cultivation and growth [170]. Increasing the rate of one crop in rotation can be detrimental to soil health. For example, Bainard et al. [171] observed that a higher pulse phase in rotation caused host-specific fungal pathogens to build up in the soil, which could make rotation less beneficial for crop yield and soil health. Meanwhile, an additional crucial factor in crop rotation design is the ability of soil-borne diseases to utilize alternate crops as hosts or to remain dormant in the soil for extended periods, as well as the response of these crops to disease. Utilizing non-host plants in crop rotations to manage diseases transmitted by the soil is critical for mitigating yield losses, particularly given that many pathogens can persist in the soil for extended periods as spores or other latent forms in the absence of their preferred host plants [162,172]. For example, Nayyar et al. [173] reported that Fusarium root rot in peas cultivated in rotation on the Canadian prairie was associated with a restricted soil microbial community and reduced populations of arbuscular mycorrhizal fungi and beneficial bacteria. In certain instances, continuous cropping with enhanced crop diversification elevated the population of antagonistic soil microorganisms, hence lowering the populations of soil pathogens and reducing the "take-all" effect in wheat. Typically, incorporating three or more crops within a cropping system can improve soil health and optimize the productivity of crops [89,162]. However, the benefits of crop rotation are practical; understanding the specific effects of crop systems and how they affect subsequent crops will encourage and adopt a worthy crop rotation, which will promote soil health and increase crop productivity. These findings make it easier to create customized interventions that use microbial populations to maintain soil health, decrease the emissions of greenhouse gases (GHGs) and crop production, and the efficient use of resources in intercropping systems. According to references, we propose intercropping in agroecosystems, as depicted in the conceptual (Figure 5).

Figure 5. We present a conceptual depiction of the integration of crop rotation systems, tracing their journey from challenges to outcomes. Crop rotation systems that incorporate cash and legume crops can sustain agricultural yields, increase farmers' income, and minimize greenhouse gas emissions through legumes' biological nitrogen fixation, which partially replaces synthetic nitrogen inputs. Incorporating legumes into crop rotations can improve soil health by promoting microbial and enzyme activity, optimizing nutrient cycling, and enhancing carbon sequestration.

6. Indicators for Assessing Soil Health in Agroecosystems

Soil health typically refers to the capacity of soil to operate as a crucial living system that supports biological productivity, preserves both soil and plant health, enhances environmental quality, and maintains an agroecosystem [174]. Soil is a highly intricate and complex, multifunctional system including gaseous, liquid, and solid elements that interact through various chemical, physical, and biological processes. Healthy soil sustains agroecosystems and facilitates the provision of essential services to ecosystems [175]. In order to evaluate and sustain agroecosystems, it is essential to consider the biological, physical, and chemical functions and traits of the soil, particularly the biological ones that serve as sensitive indicators of soil health and quality [176]. Microbiological and biochemical markers show that the variety and activity of soil microbes are important for the long-term health of agroecosystems because they keep soil health functions going, like cycling carbon and nutrients [177]. Microbial indicators exhibit more sensitivity than physical traits to environmental alterations such as soil usage and management; hence, they enable early detection and forecasting of changes and disturbances in environmental sustainability [178]. Related studies recognize soil microbial biomass, which includes fungi, bacteria, and algae, as a primary biological indicator of soil health and as a crucial source of nutrient cycling and delivery relative to plant demand [179]. Additionally, through photosynthesis, plants fix and transfer carbon as carbohydrates into the food web, making it one of the most critical biological processes on Earth [177]. Both agricultural and forestry soils widely use soil respiration and microbial biomass as bio-indicators of soil health.

We expect biochemical indicators to effectively integrate the combined impact on soil's chemical, physical, and biological processes and characteristics, making them suitable for diverse management and environmental conditions. This paragraph aims to elucidate and augment our focus on soil quality and related research, particularly with biochemical soil quality indicators such as enzymatic and microbial activity [180]. Due to their role in nutrient cycling through microbial processes, soil enzyme activities frequently serve as sensitive indicators of soil biochemical quality and as measures of ecological quality. They can effectively respond to both anthropogenic and natural alterations in soil and are easily quantifiable [181]. However, soil enzymes are important in the functioning of soil ecosystems, including nutrient and carbon cycling and maintenance [182]. For example, soil enzymes have an important role in C (β-glucosidase and galactosidase), N (urease), and P (acid phosphatases and alkaline) cycling, and these enzymes play crucial roles in the breakdown of OM and release nutrients, which are important for soil health and plant growth promotion. Dick and Tabatabai [183] reported that soil enzymes serve as reliable indicators of soil quality because of their close association with soil biological processes, their ease of measurement, and their swift response to alterations in soil management. In summary, soil enzymes are valuable tools for assessing both long and short-term changes in soil quality and management practices.

Furthermore, indicators of soil health for sustaining agroecosystems need to be associated with soil processes and responsive to alterations in management and environmental conditions. Soil biological traits, such as microbial biomass and its activities, serve as sensitive and rapid indicators that capture responses and data from diverse environmental conditions [184,185]. We use soil biological metrics, such as microbial density, activity, and biomass, as indicators of soil health [186]. Furthermore, a specific drawback of employing soil microbial characteristics as soil health indicators is the technological limitations encountered in the research of soil microbial populations. Molecular techniques that are more advanced, especially next-generation sequencing (NGS) methods (such as shotgun sequencing for structural and amplicon sequencing and functional microbial diversity investigations), have made it possible to study the part of soil microbial communities that cannot be grown in a lab. This is because most soil microorganisms do not grow on media or in a lab setting. However, while analyzing the data, we must acknowledge the limitations of these novel methodologies. In conclusion, we can use soil to focus on agroecosystem maintenance and enhance rhizosphere processes; soil health indicators are used to inform management techniques, optimize fertilizer application, and promote sustainable crop productivity.

7. Integrated Agricultural Practices for Improving Rhizospheric Processes

Integrated agricultural practices are essential for improving rhizospheric processes, which are important for sustainable crop yield and soil health. Integrated agricultural practices established soil properties (biological, chemical, and physical properties) that influence nutrient cycling and soil microbial community composition and structure and enhance the rhizospheric process [187]. However, to reduce adverse effects, integrative biofertilizers, crop rotation, green manuring, intercropping, and mulching practices should be promoted and tested worldwide to improve the rhizosphere processes and agroecosystems resilience. The disturbance of rhizosphere processes and depletion of organic matter are restraining agricultural productivity around the globe via improper agricultural practices. In short, integrated agricultural practices (IAP) are based on the integration of different crops, livestock wastes, and fertilizers into production systems that, through valuable management practices, maintain a high level of soil fertility, productivity, and quality, reduce external inputs of agrochemical and fertilizers, and enhance the functions of soil inside biological cycles and its processes. Mixed-crop livestock and other available resources, such as seed primed—the basis of integrated agricultural practices—allow the most efficient and effective use of natural resources and biological cycles to improve below-ground activities [188].

Sarkar et al. [189] reported that microbial-assisted nutrient management is a sustainable, eco-friendly, and cost-effective option under integrated agricultural systems. The usage of less amount of agrochemicals provides support organically or naturally to protect the environment from nutrient denitrification or runoff, land degradation, and soil pollution. Duarah et al. [190] stated that reducing the amount of synthetic NPK fertilizer application and seed priming (with bacteria) can significantly improve nitrogen-use efficiency. However, Entesari et al. [191] witnessed that bio-priming seeds of soybean with *Trichoderma* sp., prior to planting, enhanced the nutrient status of the crop and improved the crop yield. Phosphorus (P), an essential nutrient, is often fixed in the soil and forms complexes with metals, limiting its availability to plants. However, certain fungal and bacterial communities have a trait called P solubilization, which allows them to convert unavailable or fixed phosphorus into forms that plants can absorb. Kim et al. [192] reported that incorporating green manure can help re-mobilize phosphorus and make non-exchangeable potassium available, circulating nutrients more efficiently. This integrated approach significantly enhances phosphorus availability, a key element for plant growth and the activity of soil microorganisms. Integrated Agricultural Practices (IAP) provide numerous advantages by carefully selecting crops and utilizing available resources to develop creative, sustainable systems. These practices effectively achieve several critical objectives: minimizing disease and insect problems, reducing inputs and energy requirements, and lowering the reliance on agrochemicals and pesticides. Additionally, IAP mitigates risks associated with climate variability and economic fluctuations, making agricultural systems more resilient. Improved rhizosphere processes foster healthy soils, which are vital for sustaining ecosystem services that benefit plants, animals, and humans alike. By optimizing rhizospheric functions, agroecosystems can better support nutrient cycling, enhance soil biodiversity, and maintain productivity, ensuring the sustainable production of food and fiber. Ultimately, this integrated approach is essential for strengthening agroecosystem resilience, safeguarding long-term agricultural sustainability, and addressing the challenges of future global food security.

8. Concluding Remarks and Future Directions

Based on the evidence, recognizing the importance of rhizosphere processes and their interactions with microorganisms is essential for promoting sustainable agroecosystems. The rhizosphere fosters various microbial populations that augment nutrient accessibility, regulate pathogens, and promote soil structure. Some important things that must be performed to rebuild microbial communities, improve soil health, and make nutrient cycling easier are making green manure, intercropping, and using microbial inoculants. These solutions diminish dependence on synthetic fertilizers, augment carbon sequestration, and improve plant resilience to abiotic stress, fostering steady agricultural yields and economic advantages. Optimizing rhizosphere functions enables agroecosystems to enhance food and fiber production, preserve soil fertility, and ensure long-term sustainability, fostering resistance to environmental challenges and safeguarding global food security. However, the findings from this review provide valuable insights to agronomists, soil scientists, and crop managers on optimizing rhizosphere processes to improve agroecosystem resilience. In closing, the rhizosphere will be central to future ecological innovations in agriculture, driving both resilience and sustainability. However, further experimental research is required to fully unlock the prospect of the rhizosphere. Future studies should focus on integrating advanced technologies, such as microbial genomics and precision agriculture, to optimize rhizosphere management for more resilient and sustainable agroecosystems. The development of smart agricultural strategies, coupled with policy support, will be vital for addressing the growing challenges posed by climate change and soil degradation.

Author Contributions: W.A. and R.K. designed and wrote the manuscript. K.D.C. and R.K. reviewed and supervision. A.M., J.R.S., and J.G.F. edited and wrote the manuscript. All authors have read and agreed to the published version of the manuscript.

Funding: This work has not received any funding or grant.

Institutional Review Board Statement: Not applicable.

Informed Consent Statement: Not applicable.

Data Availability Statement: No data was used for the research described in the article.

Acknowledgments: Figures for the review manuscript were created by BioRender (https://biorender.com).

Conflicts of Interest: The authors declare no conflicts of interest. The authors declare that this article is original research and has not been submitted previously to any journal. We hereby transfer the copyright ownership to the International Journal of Molecular Sciences for due process of review and editing. Additionally, the authors listed below declare no conflicts of interest or competing interests.

References

1. Cardinale, B.J.; Duffy, J.E.; Gonzalez, A.; Hooper, D.U.; Perrings, C.; Venail, P.; Narwani, A.; Mace, G.M.; Tilman, D.; Wardle, D.A. Biodiversity loss and its impact on humanity. *Nature* **2012**, *486*, 59–67. [CrossRef]
2. Tahat, M.M.; Alananbeh, K.M.; Othman, Y.A.; Leskovar, D.I. Soil health and sustainable agriculture. *Sustainability* **2020**, *12*, 4859. [CrossRef]
3. Hatfield, J.L.; Sauer, T.J.; Cruse, R.M. Chapter one—soil: The forgotten piece of the water, food, energy nexus. *Advances in agronomy* **2017**, *143*, 1–46.
4. Omotayo, O.P.; Babalola, O.O. Resident rhizosphere microbiome's ecological dynamics and conservation: Towards achieving the envisioned Sustainable Development Goals, a review. *Int. Soil Water Conserv. Res.* **2020**, *9*, 127–142. [CrossRef]
5. Liang, H.; Wang, X.; Yan, J.; Luo, L. Characterizing the intra-vineyard variation of soil bacterial and fungal communities. *Front. Microbiol.* **2019**, *10*, 1239.
6. Schmidt, J.E.; Vannette, R.L.; Igwe, A.; Blundell, R.; Casteel, C.L.; Gaudin, A.C. Effects of agricultural management on rhizosphere microbial structure and function in processing tomato plants. *AEM* **2019**, *85*, e01064-19. [CrossRef]
7. Asghar, W.; Kataoka, R. Co-application of Green Manure and *Trichoderma* spp. Induced Plant Growth Promotion by Nutrient Improvement and Increased Fungal Biomass in Soil. *Agric. Res.* **2024**, *13*, 779–788. [CrossRef]
8. Asghar, W.; Craven, K.D.; Kataoka, R.; Mahmood, A.; Asghar, N.; Raza, T.; Iftikhar, F. The application of *Trichoderma* spp., an old but new useful fungus, in sustainable soil health intensification: A comprehensive strategy for addressing challenges. *Plant Stress* **2024**, *12*, 100455. [CrossRef]
9. Raaijmakers, J.M.; Paulitz, T.C.; Steinberg, C.; Alabouvette, C.; Moënne-Loccoz, Y. The rhizosphere: A playground and battlefield for soilborne pathogens and beneficial microorganisms. *Plant Soil* **2009**, *321*, 341–361. [CrossRef]
10. Sturz, A.; Carter, M.; Johnston, H. A review of plant disease, pathogen interactions and microbial antagonism under conservation tillage in temperate humid agriculture. *Soil Tillage Res.* **1997**, *41*, 169–189. [CrossRef]
11. Wu, J.; Zhu, J.; Zhang, D.; Cheng, H.; Hao, B.; Cao, A.; Yan, D.; Wang, Q.; Li, Y. Beneficial effect on the soil microenvironment of *Trichoderma* applied after fumigation for cucumber production. *PLoS ONE* **2022**, *17*, e0266347. [CrossRef] [PubMed]
12. Jiang, S.-Q.; Yu, Y.-N.; Gao, R.-W.; Wang, H.; Zhang, J.; Li, R.; Long, X.-H.; Shen, Q.-R.; Chen, W.; Cai, F. High-throughput absolute quantification sequencing reveals the effect of different fertilizer applications on bacterial community in a tomato cultivated coastal saline soil. *Sci. Total Environ.* **2019**, *687*, 601–609. [CrossRef] [PubMed]
13. Bajsa, N.; Morel, M.A.; Braña, V.; Castro-Sowinski, S. The effect of agricultural practices on resident soil microbial communities: Focus on biocontrol and biofertilization. *Mol. Microb. Ecol. Rhizosphere* **2013**, *2*, 687–700.
14. Meena, H.; Meena, R.S.; Rajput, B.S.; Kumar, S. Response of bio-regulators to morphology and yield of clusterbean [*Cyamopsis tetragonoloba* (L.) Taub.] under different sowing environments. *J. Appl. Nat. Sci.* **2016**, *8*, 715–718. [CrossRef]
15. Dubey, R.K.; Dubey, P.K.; Chaurasia, R.; Singh, H.B.; Abhilash, P.C. Sustainable agronomic practices for enhancing the soil quality and yield of *Cicer arietinum* L. under diverse agroecosystems. *J. Environ. Manag.* **2020**, *262*, 110284. [CrossRef]
16. Bruinsma, M.; Kowalchuk, G.; Van Veen, J. Effects of genetically modified plants on microbial communities and processes in soil. *Biol. Fertil. Soils* **2003**, *37*, 329–337. [CrossRef]
17. Sharma, S.K.; Ramesh, A.; Sharma, M.P.; Joshi, O.P.; Govaerts, B.; Steenwerth, K.L.; Karlen, D.L. Microbial community structure and diversity as indicators for evaluating soil quality. In *Biodiversity, Biofuels, Agroforestry and Conservation Agriculture*; Springer: Dordrecht, The Netherlands, 2010; pp. 317–358.

18. Asghar, W.; Kataoka, R. Green manure incorporation accelerates enzyme activity, plant growth, and changes in the fungal community of soil. *Arch. Microbiol.* **2022**, *204*, 7. [CrossRef] [PubMed]
19. Yang, L.; Zhou, X.; Liao, Y.; Lu, Y.; Nie, J.; Cao, W. Co-incorporation of rice straw and green manure benefits rice yield and nutrient uptake. *Crop Sci.* **2019**, *59*, 749–759. [CrossRef]
20. Khan, M.I.; Gwon, H.S.; Alam, M.A.; Song, H.J.; Das, S.; Kim, P.J. Short term effects of different green manure amendments on the composition of main microbial groups and microbial activity of a submerged rice cropping system. *Appl. Soil Ecol.* **2020**, *147*, 103400. [CrossRef]
21. Hakim, S.; Naqqash, T.; Nawaz, M.S.; Laraib, I.; Siddique, M.J.; Zia, R.; Mirza, M.S.; Imran, A. Rhizosphere Engineering With Plant Growth-Promoting Microorganisms for Agriculture and Ecological Sustainability. *Front. Sustain. Food Syst.* **2021**, *5*, 16. [CrossRef]
22. de Graaff, M.-A. Interactions Between Plants and Soil Nutrient Cycling Under Elevated CO_2. Ph.D. Thesis, Wageningen University and Research, Wageningen, The Netherlands, 2007.
23. Joseph, S.; Downie, A.; Munroe, P.; Crosky, A.; Lehmann, J. Biochar for carbon sequestration, reduction of greenhouse gas emissions and enhancement of soil fertility; a review of the materials science. In Proceedings of the Australian combustion symposium, Darwin, Australia, 26–29 November 2023; pp. 130–133.
24. Yanai, Y.; Toyota, K.; Okazaki, M. Effects of charcoal addition on N_2O emissions from soil resulting from rewetting air-dried soil in short-term laboratory experiments. *Soil Sci. Plant Nutr.* **2007**, *53*, 181–188. [CrossRef]
25. Laird, D.A. The charcoal vision: A win–win–win scenario for simultaneously producing bioenergy, permanently sequestering carbon, while improving soil and water quality. *Agron. J.* **2008**, *100*, 178–181. [CrossRef]
26. Stockmann, U.; Adams, M.A.; Crawford, J.W.; Field, D.J.; Henakaarchchi, N.; Jenkins, M.; Minasny, B.; McBratney, A.B.; De Courcelles, V.d.R.; Singh, K.; et al. The knowns, known unknowns and unknowns of sequestration of soil organic carbon. *Agric. Ecosyst. Environ.* **2013**, *164*, 80–99. [CrossRef]
27. Yuan, P.; Wang, J.; Pan, Y.; Shen, B.; Wu, C. Review of biochar for the management of contaminated soil: Preparation, application and prospect. *Sci. Total Environ.* **2019**, *659*, 473–490. [CrossRef] [PubMed]
28. Paustian, K.; Lehmann, J.; Ogle, S.; Reay, D.; Robertson, G.P.; Smith, P. Climate-smart soils. *Nature* **2016**, *532*, 49–57. [CrossRef] [PubMed]
29. Majumder, S.; Neogi, S.; Dutta, T.; Powel, M.A.; Banik, P. The impact of biochar on soil carbon sequestration: Meta-analytical approach to evaluating environmental and economic advantages. *J. Environ. Manag.* **2019**, *250*, 109466. [CrossRef] [PubMed]
30. Yazdanpanah, N. CO_2 emission and structural characteristics of two calcareous soils amended with municipal solid waste and plant residue. *Solid Earth* **2016**, *7*, 105–114. [CrossRef]
31. Kenney, I.; Blanco-Canqui, H.; Presley, D.R.; Rice, C.W.; Janssen, K.; Olson, B. Soil and crop response to stover removal from rainfed and irrigated corn. *Gcb Bioenergy* **2015**, *7*, 219–230. [CrossRef]
32. Nguyen, D.H.; Scheer, C.; Rowlings, D.W.; Grace, P.R. Rice husk biochar and crop residue amendment in subtropical cropping soils: Effect on biomass production, nitrogen use efficiency and greenhouse gas emissions. *Biol. Fertil. Soils* **2016**, *52*, 261–270. [CrossRef]
33. Asghar, W.; Kataoka, R. Effect of co-application of *Trichoderma* spp. with organic composts on plant growth enhancement, soil enzymes and fungal community in soil. *Arch. Microbiol.* **2021**, *203*, 4281–4291. [CrossRef] [PubMed]
34. Kataoka, R.; Nagasaka, K.; Tanaka, Y.; Yamamura, H.; Shinohara, S.; Haramoto, E.; Hayakawa, M.; Sakamoto, Y. Hairy vetch (*Vicia villosa*), as a green manure, increases fungal biomass, fungal community composition, and phosphatase activity in soil. *Appl. Soil Ecol.* **2017**, *117*, 16–20. [CrossRef]
35. Ayaz, M.; Feizienė, D.; Tilvikienė, V.; Akhtar, K.; Stulpinaitė, U.; Iqbal, R. Biochar Role in the Sustainability of Agriculture and Environment. *Sustainability* **2021**, *13*, 1330. [CrossRef]
36. Lehmann, J.; Joseph, S. *Biochar for Environmental Management: An Introduction*; Routledge: Oxfordshire, UK, 2015.
37. Lu, L.; Yu, W.; Wang, Y.; Zhang, K.; Zhu, X.; Zhang, Y.; Wu, Y.; Ullah, H.; Xiao, X.; Chen, B. Application of biochar-based materials in environmental remediation: From multi-level structures to specific devices. *Biochar* **2020**, *2*, 1–31. [CrossRef]
38. Brewer, C.E.; Chuang, V.J.; Masiello, C.A.; Gonnermann, H.; Gao, X.; Dugan, B.; Driver, L.E.; Panzacchi, P.; Zygourakis, K.; Davies, C.A. New approaches to measuring biochar density and porosity. *Biomass Bioenergy* **2014**, *66*, 176–185. [CrossRef]
39. Arif, M.; Ali, S.; Ilyas, M.; Riaz, M.; Akhtar, K.; Ali, K.; Adnan, M.; Fahad, S.; Khan, I.; Shah, S. Enhancing phosphorus availability, soil organic carbon, maize productivity and farm profitability through biochar and organic–inorganic fertilizers in an irrigated maize agroecosystem under semi-arid climate. *Soil Use Manag.* **2021**, *37*, 104–119. [CrossRef]
40. Uzoma, K.; Inoue, M.; Andry, H.; Fujimaki, H.; Zahoor, A.; Nishihara, E. Effect of cow manure biochar on maize productivity under sandy soil condition. *Soil Use Manag.* **2011**, *27*, 205–212. [CrossRef]
41. Lu, H.; Yan, M.; Wong, M.H.; Mo, W.Y.; Wang, Y.; Chen, X.W.; Wang, J.-J. Effects of biochar on soil microbial community and functional genes of a landfill cover three years after ecological restoration. *Sci. Total Environ.* **2020**, *717*, 137133. [CrossRef] [PubMed]

42. Farrell, M.; Kuhn, T.K.; Macdonald, L.M.; Maddern, T.M.; Murphy, D.V.; Hall, P.A.; Singh, B.P.; Baumann, K.; Krull, E.S.; Baldock, J.A. Microbial utilisation of biochar-derived carbon. *Sci. Total Environ.* **2013**, *465*, 288–297. [CrossRef]
43. Marschner, P.; Crowley, D.; Rengel, Z. Rhizosphere interactions between microorganisms and plants govern iron and phosphorus acquisition along the root axis–model and research methods. *Soil Biol. Biochem.* **2011**, *43*, 883–894. [CrossRef]
44. Hinsinger, P.; Bengough, A.G.; Vetterlein, D.; Young, I.M. Rhizosphere: Biophysics, biogeochemistry and ecological relevance. *Plant Soil* **2009**, *321*, 117–152. [CrossRef]
45. Sun, R.; Zhang, X.-X.; Guo, X.; Wang, D.; Chu, H. Bacterial diversity in soils subjected to long-term chemical fertilization can be more stably maintained with the addition of livestock manure than wheat straw. *Soil Biol. Biochem.* **2015**, *88*, 9–18. [CrossRef]
46. Mendes, R.; Kruijt, M.; De Bruijn, I.; Dekkers, E.; van der Voort, M.; Schneider, J.H.; Piceno, Y.M.; DeSantis, T.Z.; Andersen, G.L.; Bakker, P.A.; et al. Deciphering the rhizosphere microbiome for disease-suppressive bacteria. *Science* **2011**, *332*, 1097–1100. [CrossRef] [PubMed]
47. Lagos, L.; Maruyama, F.; Nannipieri, P.; Mora, M.; Ogram, A.; Jorquera, M. Current overview on the study of bacteria in the rhizosphere by modern molecular techniques: A mini-review. *J. Soil Sci. Plant Nutr.* **2015**, *15*, 504–523.
48. Peiffer, J.A.; Spor, A.; Koren, O.; Jin, Z.; Tringe, S.G.; Dangl, J.L.; Buckler, E.S.; Ley, R.E. Diversity and heritability of the maize rhizosphere microbiome under field conditions. *Proc. Natl. Acad. Sci. USA* **2013**, *110*, 6548–6553. [CrossRef]
49. Ito, K.; Mahmood, A.; Kataoka, R.; Takagi, K. Dichlorodiphenyltrichloroethane (DDT) degradation by *Streptomyces* sp. isolated from DDT contaminated soil. *Bioremediation J.* **2020**, *25*, 148–158. [CrossRef]
50. Asghar, W.; Kataoka, R. Fungal volatiles from green manure-incorporated soils promote the growth of lettuce (*Lactuca sativa*) and mediate antifungal activity against *Fusarium oxysporum* In Vitro. *Plant Soil* **2023**, 1–12. [CrossRef]
51. Selvasekaran, P.; Chidambaram, R. Agriculturally Important Fungi for Crop Protection. In *Agriculturally Important Fungi for Sustainable Agriculture*; Springer: Cham, Switzerland, 2020; pp. 1–53.
52. Mahmood, A.; Kataoka, R. Metabolite profiling reveals a complex response of plants to application of plant growth-promoting endophytic bacteria. *Microbiol. Res.* **2020**, *234*, 126421. [CrossRef]
53. Mahmood, A.; Iguchi, R.; Kataoka, R. Multifunctional food waste fertilizer having the capability of Fusarium-growth inhibition and phosphate solubility: A new horizon of food waste recycle using microorganisms. *Waste Manag.* **2019**, *94*, 77–84. [CrossRef] [PubMed]
54. Da Costa, P.B.; Beneduzi, A.; de Souza, R.; Schoenfeld, R.; Vargas, L.K.; Passaglia, L.M. The effects of different fertilization conditions on bacterial plant growth promoting traits: Guidelines for directed bacterial prospection and testing. *Plant Soil* **2013**, *368*, 267–280. [CrossRef]
55. Kour, D.; Rana, K.L.; Yadav, A.N.; Yadav, N.; Kumar, M.; Kumar, V.; Vyas, P.; Dhaliwal, H.S.; Saxena, A.K. Microbial biofertilizers: Bioresources and eco-friendly technologies for agricultural and environmental sustainability. *Biocatal. Agric. Biotechnol.* **2020**, *23*, 101487. [CrossRef]
56. Mazid, M.; Khan, T.A. Future of bio-fertilizers in Indian agriculture: An overview. *Int. J. Agric. Food Res.* **2015**, *3*, 10–23. [CrossRef]
57. Murali, M.; Amruthesh, K.; Sudisha, J.; Niranjana, S.; Shetty, H. Screening for plant growth promoting fungi and their ability for growth promotion and induction of resistance in pearl millet against downy mildew disease. *J. Phytol.* **2012**, *4*, 30–36.
58. Murali, M.; Amruthesh, K.N. Plant Growth-promoting Fungus Penicillium oxalicum enhances plant growth and induces resistance in pearl millet against downy mildew disease. *J. Phytopathol.* **2015**, *163*, 743–754. [CrossRef]
59. Sayyed, R.; Chincholkar, S.; Reddy, M.; Gangurde, N.; Patel, P. Siderophore producing PGPR for crop nutrition and phytopathogen suppression. In *Bacteria in Agrobiology: Disease Management*; Springer: Berlin/Heidelberg, Germany, 2013; pp. 449–471.
60. Hossain, M.M.; Sultana, F.; Kubota, M.; Koyama, H.; Hyakumachi, M. The plant growth-promoting fungus Penicillium simplicissimum GP17-2 induces resistance in Arabidopsis thaliana by activation of multiple defense signals. *Plant Cell Physiol.* **2007**, *48*, 1724–1736. [CrossRef]
61. Hossain, M.M.; Sultana, F.; Miyazawa, M.; Hyakumachi, M. The plant growth-promoting fungus Penicillium spp. GP15-1 enhances growth and confers protection against damping-off and anthracnose in the cucumber. *J. Oleo Sci.* **2014**, *63*, 391–400. [CrossRef] [PubMed]
62. Abdel-lateif, K.S. *Trichoderma* as biological control weapon against soil borne plant pathogens. *Afr. J. Biotechnol.* **2017**, *16*, 2299–2306.
63. Salas-Marina, M.A.; Silva-Flores, M.A.; Cervantes-Badillo, M.G.; Rosales-Saavedra, M.T.; Islas-Osuna, M.A.; Casas-Flores, S. The plant growth-promoting fungus Aspergillus ustus promotes growth and induces resistance against different lifestyle pathogens in Arabidopsis thaliana. *J. Microbiol. Biotechnol.* **2011**, *21*, 686–696. [CrossRef] [PubMed]
64. Cavello, I.A.; Crespo, J.M.; García, S.S.; Zapiola, J.M.; Luna, M.F.; Cavalitto, S.F. Plant growth promotion activity of keratinolytic fungi growing on a recalcitrant waste known as "Hair Waste". *Biotechnol. Res. Int.* **2015**, *2015*, 952921. [CrossRef] [PubMed]

65. Yamagiwa, Y.; Inagaki, Y.; Ichinose, Y.; Toyoda, K.; Hyakumachi, M.; Shiraishi, T. Talaromyces wortmannii FS2 emits β-caryphyllene, which promotes plant growth and induces resistance. *J. Gen. Plant Pathol.* **2011**, *77*, 336–341. [CrossRef]
66. Kumar, V.; Behl, R.K.; Narula, N. Establishment of phosphate-solubilizing strains of Azotobacter chroococcum in the rhizosphere and their effect on wheat cultivars under green house conditions. *Microbiol. Res.* **2001**, *156*, 87–93. [CrossRef] [PubMed]
67. Ahemad, M.; Khan, M.S. Alleviation of fungicide-induced phytotoxicity in greengram [*Vigna radiata* (L.) Wilczek] using fungicide-tolerant and plant growth promoting Pseudomonas strain. *Saudi J. Biol. Sci.* **2012**, *19*, 451–459. [CrossRef]
68. Singh, J.S. Plant growth promoting rhizobacteria: Potential microbes for sustainable agriculture. *Resonance* **2013**, *18*, 275–281. [CrossRef]
69. Sunpapao, A.; Chairin, T.; Ito, S.-i. The biocontrol by *Streptomyces* and *Trichoderma* of leaf spot disease caused by Curvularia oryzae in oil palm seedlings. *Biocontrol* **2018**, *123*, 36–42. [CrossRef]
70. Jamil, F.; Mukhtar, H.; Fouillaud, M.; Dufossé, L. Rhizosphere signaling: Insights into plant–rhizomicrobiome interactions for sustainable agronomy. *Microorganisms* **2022**, *10*, 899. [CrossRef] [PubMed]
71. Solomon, W.; Janda, T.; Molnár, Z. Unveiling the significance of rhizosphere: Implications for plant growth, stress response, and sustainable agriculture. *Plant Physiology and Biochemistry* **2023**, 108290. [CrossRef]
72. Hassani, M.A.; Durán, P.; Hacquard, S. Microbial interactions within the plant holobiont. *Microbiome* **2018**, *6*, 1–17. [CrossRef] [PubMed]
73. Asghar, W.; Kondo, S.; Iguchi, R.; Mahmood, A.; Kataoka, R. Agricultural utilization of unused resources: Liquid food waste material as a new source of plant growth-promoting microbes. *Agronomy* **2020**, *10*, 954. [CrossRef]
74. Jacoby, R.P.; Koprivova, A.; Kopriva, S. Pinpointing secondary metabolites that shape the composition and function of the plant microbiome. *J. Exp. Bot.* **2021**, *72*, 57–69. [CrossRef] [PubMed]
75. O'Banion, B.S.; O'Neal, L.; Alexandre, G.; Lebeis, S.L. Bridging the gap between single-strain and community-level plant-microbe chemical interactions. *Mol. Plant-Microbe Interact.* **2020**, *33*, 124–134. [CrossRef]
76. Rizaludin, M.S.; Stopnisek, N.; Raaijmakers, J.M.; Garbeva, P. The chemistry of stress: Understanding the 'cry for help' of plant roots. *Metabolites* **2021**, *11*, 357. [CrossRef] [PubMed]
77. Jacoby, R.P.; Chen, L.; Schwier, M.; Koprivova, A.; Kopriva, S. Recent advances in the role of plant metabolites in shaping the root microbiome. *F1000Research* **2020**, *9*, F1000. [CrossRef] [PubMed]
78. Trivedi, P.; Leach, J.E.; Tringe, S.G.; Sa, T.; Singh, B.K. Plant–microbiome interactions: From community assembly to plant health. *Nat. Rev. Microbiol.* **2020**, *18*, 607–621. [CrossRef] [PubMed]
79. Bais, H.P.; Weir, T.L.; Perry, L.G.; Gilroy, S.; Vivanco, J.M. The role of root exudates in rhizosphere interactions with plants and other organisms. *Annu. Rev. Plant Biol.* **2006**, *57*, 233–266. [CrossRef] [PubMed]
80. Salam, M.T.B.; Mahmood, A.; Asghar, W.; Ito, K.; Kataoka, R. Phytomicrobiomes: A Potential Approach for Sustainable Pesticide Biodegradation. *Applied Sciences* **2024**, *14*, 2740. [CrossRef]
81. Zhalnina, K.; Louie, K.B.; Hao, Z.; Mansoori, N.; Da Rocha, U.N.; Shi, S.; Cho, H.; Karaoz, U.; Loqué, D.; Bowen, B.P. Dynamic root exudate chemistry and microbial substrate preferences drive patterns in rhizosphere microbial community assembly. *Nat. Microbiol.* **2018**, *3*, 470–480. [CrossRef] [PubMed]
82. Mönchgesang, S.; Strehmel, N.; Schmidt, S.; Westphal, L.; Taruttis, F.; Müller, E.; Herklotz, S.; Neumann, S.; Scheel, D. Natural variation of root exudates in Arabidopsis thaliana-linking metabolomic and genomic data. *Sci. Rep.* **2016**, *6*, 29033. [CrossRef]
83. Mohanram, S.; Kumar, P. Rhizosphere microbiome: Revisiting the synergy of plant-microbe interactions. *Ann. Microbiol.* **2019**, *69*, 307–320. [CrossRef]
84. Ab Rahman, S.F.S.; Singh, E.; Pieterse, C.M.; Schenk, P.M. Emerging microbial biocontrol strategies for plant pathogens. *Plant Sci.* **2018**, *267*, 102–111. [CrossRef]
85. Hong, Y.; Zhou, Q.; Hao, Y.; Huang, A.C. Crafting the plant root metabolome for improved microbe-assisted stress resilience. *New Phytol.* **2022**, *234*, 1945–1950. [CrossRef] [PubMed]
86. Toom, M.; Tamm, S.; Talgre, L.; Tamm, I.; Tamm, Ü.; Narits, L.; Hiiesalu, I.; Mäe, A.; Lauringson, E. The effect of cover crops on the yield of spring barley in Estonia. *Agriculture* **2019**, *9*, 172. [CrossRef]
87. Florentín, M.A.; Peñalva, M.; Calegari, A.; Derpsch, R.; McDonald, M. Green manure/cover crops and crop rotation in conservation agriculture on small farms. *Integr. Crop Manag.* **2010**, *12*, 97.
88. Ntakirutimana, L.; Li, F.; Huang, X.; Wang, S.; Yin, C. Green manure planting incentive measures of local authorities and farmers' perceptions of the utilization of rotation fallow for sustainable agriculture in Guangxi, China. *Sustainability* **2019**, *11*, 2723. [CrossRef]
89. Yang, T.; Siddique, K.H.; Liu, K. Cropping systems in agriculture and their impact on soil health—A review. *Glob. Ecol. Conserv.* **2020**, *23*, e01118. [CrossRef]

90. Gan, Y.; Miller, P.; McConkey, B.; Zentner, R.; Stevenson, F.; McDonald, C. Influence of diverse cropping sequences on durum wheat yield and protein in the semiarid northern Great Plains. *Agron. J.* **2003**, *95*, 245–252. [CrossRef]
91. Chavarria, D.N.; Verdenelli, R.A.; Serri, D.L.; Restovich, S.B.; Andriulo, A.E.; Meriles, J.M.; Vargas-Gil, S. Effect of cover crops on microbial community structure and related enzyme activities and macronutrient availability. *Eur. J. Soil Biol.* **2016**, *76*, 74–82. [CrossRef]
92. Balota, E.L.; Calegari, A.; Nakatani, A.S.; Coyne, M.S. Benefits of winter cover crops and no-tillage for microbial parameters in a Brazilian Oxisol: A long-term study. *Agric Ecosyst. Environ.* **2014**, *197*, 31–40. [CrossRef]
93. Zhang, X.; Zhang, R.; Gao, J.; Wang, X.; Fan, F.; Ma, X.; Yin, H.; Zhang, C.; Feng, K.; Deng, Y. Thirty-one years of rice-rice-green manure rotations shape the rhizosphere microbial community and enrich beneficial bacteria. *Soil Biol. Biochem.* **2017**, *104*, 208–217. [CrossRef]
94. Khan, M.I.; Hwang, H.Y.; Kim, G.W.; Kim, P.J.; Das, S. Microbial responses to temperature sensitivity of soil respiration in a dry fallow cover cropping and submerged rice mono-cropping system. *Appl. Soil Ecol.* **2018**, *128*, 98–108. [CrossRef]
95. Creamer, N.G.; Baldwin, K.R. An evaluation of summer cover crops for use in vegetable production systems in North Carolina. *HortScience* **2000**, *35*, 600–603. [CrossRef]
96. Singh, B.K.; Trivedi, P.; Egidi, E.; Macdonald, C.A.; Delgado-Baquerizo, M. Crop microbiome and sustainable agriculture. *Nat. Rev. Microbiol.* **2020**, *18*, 601–602. [CrossRef]
97. Blanco-Canqui, H.; Ruis, S.J. No-tillage and soil physical environment. *Geoderma* **2018**, *326*, 164–200. [CrossRef]
98. Quintarelli, V.; Radicetti, E.; Allevato, E.; Stazi, S.R.; Haider, G.; Abideen, Z.; Bibi, S.; Jamal, A.; Mancinelli, R. Cover crops for sustainable cropping systems: A review. *Agriculture* **2022**, *12*, 2076. [CrossRef]
99. Lazcano, C.; Gonzalez-Maldonado, N.; Yao, E.H.; Wong, C.T.; Merrilees, J.J.; Falcone, M.; Peterson, J.D.; Casassa, L.F.; Decock, C. Sheep grazing as a strategy to manage cover crops in Mediterranean vineyards: Short-term effects on soil C, N and greenhouse gas (N_2O, CH_4, CO_2) emissions. *Agric. Ecosyst. Environ.* **2022**, *327*, 107825. [CrossRef]
100. Fiorini, A.; Maris, S.C.; Abalos, D.; Amaducci, S.; Tabaglio, V. Combining no-till with rye (*Secale cereale* L.) cover crop mitigates nitrous oxide emissions without decreasing yield. *Soil Tillage Res.* **2020**, *196*, 104442. [CrossRef]
101. Doltra, J.; Olesen, J.E. The role of catch crops in the ecological intensification of spring cereals in organic farming under Nordic climate. *Eur. J. Agron.* **2013**, *44*, 98–108. [CrossRef]
102. Sharma, P.; Singh, A.; Kahlon, C.S.; Brar, A.S.; Grover, K.K.; Dia, M.; Steiner, R.L. The role of cover crops towards sustainable soil health and agriculture—A review paper. *Am. J. Plant Sci.* **2018**, *9*, 1935–1951. [CrossRef]
103. Clark, A. *Managing Cover Crops Profitably*; Diane Publishing: Collingdale, PA, USA, 2008.
104. Thapa, V.R.; Ghimire, R.; Acosta-Martínez, V.; Marsalis, M.A.; Schipanski, M.E. Cover crop biomass and species composition affect soil microbial community structure and enzyme activities in semiarid cropping systems. *Appl. Soil Ecol.* **2021**, *157*, 103735. [CrossRef]
105. Adetunji, A.T.; Ncube, B.; Mulidzi, R.; Lewu, F.B. Management impact and benefit of cover crops on soil quality: A review. *Soil Tillage Res.* **2020**, *204*, 104717. [CrossRef]
106. Boselli, R.; Fiorini, A.; Santelli, S.; Ardenti, F.; Capra, F.; Maris, S.C.; Tabaglio, V. Cover crops during transition to no-till maintain yield and enhance soil fertility in intensive agro-ecosystems. *Field Crops Res.* **2020**, *255*, 107871. [CrossRef]
107. Coombs, C.; Lauzon, J.D.; Deen, B.; Van Eerd, L.L. Legume cover crop management on nitrogen dynamics and yield in grain corn systems. *Field Crops Res* **2017**, *201*, 75–85. [CrossRef]
108. Doran, J.W.; Zeiss, M.R. Soil health and sustainability: Managing the biotic component of soil quality. *Appl Soil Ecol* **2000**, *15*, 3–11. [CrossRef]
109. Philippot, L.; Raaijmakers, J.M.; Lemanceau, P.; Van Der Putten, W.H. Going back to the roots: The microbial ecology of the rhizosphere. *Nat. Rev. Microbiol.* **2013**, *11*, 789–799. [CrossRef]
110. Patra, D.; Anwar, M.; Chand, S. Integrated nutrient management and waste recycling for restoring soil fertility and productivity in Japanese mint and mustard sequence in Uttar Pradesh, India. *Agric. Ecosyst. Environ.* **2000**, *80*, 267–275. [CrossRef]
111. Annabi, M.; Houot, S.; Francou, C.; Poitrenaud, M.; Bissonnais, Y.L. Soil aggregate stability improvement with urban composts of different maturities. *Soil Sci. Soc. Am. J.* **2007**, *71*, 413–423. [CrossRef]
112. Belachew, T.; Abera, Y. Effect of green manuring in combination with nitrogen on soil fertility and yield of bread wheat (*Triticum aestivum*) under double cropping system of Sinanadinsho, Southeast Ethiopia. *J. Biodivers. Environ. Sci.* **2011**, *1*, 1–11.
113. Baysal-Gurel, F.; Gardener, B.; Miller, S.A. Soil Borne Disease Management in Organic Vegetable Production. *Org. Agri* **2012**.
114. Panth, M.; Hassler, S.C.; Baysal-Gurel, F. Methods for management of soilborne diseases in crop production. *Agriculture* **2020**, *10*, 16. [CrossRef]
115. Mokhtar, M.; El-Mougy, N. Bio-compost application for controlling soil-borne plant pathogens—A review. *Int. J. Eng. Innov. Technol.* **2014**, *4*, 2277–3754.

116. Hao, J.; Subbarao, K.V.; Koike, S.T. Effects of broccoli rotation on lettuce drop caused by Sclerotinia minor and on the population density of sclerotia in soil. *Plant Dis.* **2003**, *87*, 159–166. [CrossRef] [PubMed]
117. Whalen, J. *Crop Profile for Watermelon in Maryland*; Online. Natl. Info. System for the Regional IPM Centers; North Carolina State University: Raleigh, NC, USA, 1999.
118. Keinath, A.P.; Hassell, R.L.; Everts, K.L.; Zhou, X.-G. Cover crops of hybrid common vetch reduce Fusarium wilt of seedless watermelon in the eastern United States. *Plant Health Prog.* **2010**, *11*, 8. [CrossRef]
119. Papp, R.; Marinari, S.; Moscatelli, M.C.; van der Heijden, M.; Wittwer, R.; Campiglia, E.; Radicetti, E.; Mancinelli, R.; Fradgley, N.; Pearce, B.; et al. Short-term changes in soil biochemical properties as affected by subsidiary crop cultivation in four European pedo-climatic zones. *Soil Tillage Res.* **2018**, *180*, 126–136. [CrossRef]
120. Ntalli, N.; Caboni, P. A review of isothiocyanates biofumigation activity on plant parasitic nematodes. *Phytochem. Rev.* **2017**, *16*, 827–834. [CrossRef]
121. Waisen, P.; Cheng, Z.; Sipes, B.S.; Wang, K.-H. Biofumigation effects of brassicaceous cover crops on soil health in cucurbit agroecosystems in Hawaii, USA. *Pedosphere* **2022**, *32*, 521–531. [CrossRef]
122. Aydınlı, G.; Mennan, S. Biofumigation studies by using Raphanus sativus and Eruca sativa as a winter cycle crops to control root-knot nematodes. *Braz. Arch. Biol. Technol.* **2018**, *61*, e18180249. [CrossRef]
123. Gabriel, J.; Quemada, M. Replacing bare fallow with cover crops in a maize cropping system: Yield, N uptake and fertiliser fate. *Eur. J. Agron.* **2011**, *34*, 133–143. [CrossRef]
124. Mukherjee, R.; Sen, S. Agricultural sustainability through nitrogen fixation: Approaches and techniques. *Harvest* **2021**, *6*, 48–55.
125. Wang, Y.; Lambers, H. Root-released organic anions in response to low phosphorus availability: Recent progress, challenges and future perspectives. *Plant Soil* **2020**, *447*, 135–156. [CrossRef]
126. Meena, R.S.; Vijayakumar, V.; Yadav, G.S.; Mitran, T. Response and interaction of Bradyrhizobium japonicum and arbuscular mycorrhizal fungi in the soybean rhizosphere. *Plant Growth Regul.* **2018**, *84*, 207–223. [CrossRef]
127. Lai, H.; Gao, F.; Su, H.; Zheng, P.; Li, Y.; Yao, H. Nitrogen distribution and soil microbial community characteristics in a legume–cereal intercropping system: A review. *Agronomy* **2022**, *12*, 1900. [CrossRef]
128. Scavo, A.; Restuccia, A.; Lombardo, S.; Fontanazza, S.; Abbate, C.; Pandino, G.; Anastasi, U.; Onofri, A.; Mauromicale, G. Improving soil health, weed management and nitrogen dynamics by Trifolium subterraneum cover cropping. *Agron. Sustain. Dev.* **2020**, *40*, 18. [CrossRef]
129. Campiglia, E.; Mancinelli, R.; Radicetti, E.; Baresel, J.P. Evaluating spatial arrangement for durum wheat (*Triticum durum* Desf.) and subclover (*Trifolium subterraneum* L.) intercropping systems. *Field Crops Res.* **2014**, *169*, 49–57. [CrossRef]
130. Guardia, G.; Aguilera, E.; Vallejo, A.; Sanz-Cobena, A.; Alonso-Ayuso, M.; Quemada, M. Effective climate change mitigation through cover cropping and integrated fertilization: A global warming potential assessment from a 10-year field experiment. *J. Clean. Prod.* **2019**, *241*, 118307. [CrossRef]
131. Berendsen, R.L.; Pieterse, C.M.; Bakker, P.A. The rhizosphere microbiome and plant health. *Trends Plant Sc* **2012**, *17*, 478–486. [CrossRef] [PubMed]
132. Snapp, S.S.; Swinton, S.M.; Labarta, R.; Mutch, D.; Black, J.R.; Leep, R.; Nyiraneza, J.; O'neil, K. Evaluating cover crops for benefits, costs and performance within cropping system niches. *Agron. J.* **2005**, *97*, 322–332. [CrossRef]
133. Vukicevich, E.; Lowery, T.; Bowen, P.; Úrbez-Torres, J.R.; Hart, M. Cover crops to increase soil microbial diversity and mitigate decline in perennial agriculture. A review. *Agron. Sustain. Dev.* **2016**, *36*, 48. [CrossRef]
134. Fargione, J.E.; Bassett, S.; Boucher, T.; Bridgham, S.D.; Conant, R.T.; Cook-Patton, S.C.; Ellis, P.W.; Falcucci, A.; Fourqurean, J.W.; Gopalakrishna, T. Natural climate solutions for the United States. *Sci. Adv.* **2018**, *4*, eaat1869. [CrossRef] [PubMed]
135. Lemaire, G.; Franzluebbers, A.; de Faccio Carvalho, P.C.; Dedieu, B. Integrated crop–livestock systems: Strategies to achieve synergy between agricultural production and environmental quality. *Agric. Ecosyst. Environ.* **2014**, *190*, 4–8. [CrossRef]
136. Bukovsky-Reyes, S.; Isaac, M.E.; Blesh, J. Effects of intercropping and soil properties on root functional traits of cover crops. *Agric. Ecosyst. Environ.* **2019**, *285*, 106614. [CrossRef]
137. Hinsinger, P.; Betencourt, E.; Bernard, L.; Brauman, A.; Plassard, C.; Shen, J.; Tang, X.; Zhang, F. P for two, sharing a scarce resource: Soil phosphorus acquisition in the rhizosphere of intercropped species. *Plant Physiol.* **2011**, *156*, 1078–1086. [CrossRef] [PubMed]
138. Sun, F.; Pan, K.; Olatunji, O.A.; Li, Z.; Chen, W.; Zhang, A.; Song, D.; Sun, X.; Huang, D.; Tan, X. Specific legumes allay drought effects on soil microbial food web activities of the focal species in agroecosystem. *Plant Soil* **2019**, *437*, 455–471. [CrossRef]
139. Cong, W.F.; Hoffland, E.; Li, L.; Six, J.; Sun, J.H.; Bao, X.G.; Zhang, F.S.; Van Der Werf, W. Intercropping enhances soil carbon and nitrogen. *Glob. Chang. Biol.* **2015**, *21*, 1715–1726. [CrossRef] [PubMed]
140. de Medeiros, E.V.; Notaro, K.d.A.; de Barros, J.A.; Duda, G.P.; Moraes, M.d.C.H.d.S.; Ambrósio, M.M.d.Q.; Negreiros, A.M.P.; Sales Júnior, R. Soils from intercropped fields have a higher capacity to suppress black root rot in cassava, caused by Scytalidium lignicola. *J. Phytopathol.* **2019**, *167*, 209–217. [CrossRef]

141. Matson, P.A.; Parton, W.J.; Power, A.G.; Swift, M.J. Agricultural intensification and ecosystem properties. *Science* **1997**, *277*, 504–509. [CrossRef] [PubMed]
142. de Araújo Santos, G.A.; Moitinho, M.R.; de Oliveira Silva, B.; Xavier, C.V.; Teixeira, D.D.B.; Corá, J.E.; Júnior, N.L.S. Effects of long-term no-tillage systems with different succession cropping strategies on the variation of soil CO_2 emission. *Sci. Total Environ.* **2019**, *686*, 413–424. [CrossRef]
143. Graf, D.R.; Saghaï, A.; Zhao, M.; Carlsson, G.; Jones, C.M.; Hallin, S. Lucerne (Medicago sativa) alters N_2O-reducing communities associated with cocksfoot (Dactylis glomerata) roots and promotes N_2O production in intercropping in a greenhouse experiment. *Soil Biol. Biochem.* **2019**, *137*, 107547. [CrossRef]
144. Brooker, R.W.; Bennett, A.E.; Cong, W.F.; Daniell, T.J.; George, T.S.; Hallett, P.D.; Hawes, C.; Iannetta, P.P.; Jones, H.G.; Karley, A.J. Improving intercropping: A synthesis of research in agronomy, plant physiology and ecology. *New Phytol.* **2015**, *206*, 107–117. [CrossRef] [PubMed]
145. Chai, Q.; Nemecek, T.; Liang, C.; Zhao, C.; Yu, A.; Coulter, J.A.; Wang, Y.; Hu, F.; Wang, L.; Siddique, K.H. Integrated farming with intercropping increases food production while reducing environmental footprint. *PNAS* **2021**, *118*, e2106382118. [CrossRef]
146. Xiao, J.; Yin, X.; Ren, J.; Zhang, M.; Tang, L.; Zheng, Y. Complementation drives higher growth rate and yield of wheat and saves nitrogen fertilizer in wheat and faba bean intercropping. *Field Crops Res.* **2018**, *221*, 119–129. [CrossRef]
147. Luo, S.; Yu, L.; Liu, Y.; Zhang, Y.; Yang, W.; Li, Z.; Wang, J. Effects of reduced nitrogen input on productivity and N_2O emissions in a sugarcane/soybean intercropping system. *Eur J Agron* **2016**, *81*, 78–85. [CrossRef]
148. Li, B.; Li, Y.-Y.; Wu, H.-M.; Zhang, F.-F.; Li, C.-J.; Li, X.-X.; Lambers, H.; Li, L. Root exudates drive interspecific facilitation by enhancing nodulation and N_2 fixation. *PNAS* **2016**, *113*, 6496–6501. [CrossRef] [PubMed]
149. Gong, X.; Dang, K.; Lv, S.; Zhao, G.; Wang, H.; Feng, B. Interspecific competition and nitrogen application alter soil ecoenzymatic stoichiometry, microbial nutrient status, and improve grain yield in broomcorn millet/mung bean intercropping systems. *Field Crops Res.* **2021**, *270*, 108227. [CrossRef]
150. Zhang, G.; Yang, H.; Zhang, W.; Bezemer, T.M.; Liang, W.; Li, Q.; Li, L. Interspecific interactions between crops influence soil functional groups and networks in a maize/soybean intercropping system. *Agric. Ecosyst. Environ.* **2023**, *355*, 108595. [CrossRef]
151. Teshita, A.; Feng, Y.; Qian, R.; Wang, X.; Khan, W.; Gao, Y. Alfalfa and maize intercropping enhances soil nematode structure and food web complexity in low-nitrogen soils. *Appl. Soil Ecol.* **2023**, *186*, 104809. [CrossRef]
152. Ma, H.; Zhou, J.; Ge, J.; Nie, J.; Zhao, J.; Xue, Z.; Hu, Y.; Yang, Y.; Peixoto, L.; Zang, H. Intercropping improves soil ecosystem multifunctionality through enhanced available nutrients but depends on regional factors. *Plant Soil* **2022**, *480*, 71–84. [CrossRef]
153. Benitez-Alfonso, Y.; Soanes, B.K.; Zimba, S.; Sinanaj, B.; German, L.; Sharma, V.; Bohra, A.; Kolesnikova, A.; Dunn, J.A.; Martin, A.C. Enhancing climate change resilience in agricultural crops. *Curr. Biol.* **2023**, *33*, R1246–R1261. [CrossRef] [PubMed]
154. Mugi-Ngenga, E.; Bastiaans, L.; Zingore, S.; Anten, N.; Giller, K. The role of nitrogen fixation and crop N dynamics on performance and legacy effects of maize-grain legumes intercrops on smallholder farms in Tanzania. *Eur. J. Agron.* **2022**, *141*, 126617. [CrossRef]
155. Du, Q.; Zhou, L.; Chen, P.; Liu, X.; Song, C.; Yang, F.; Wang, X.; Liu, W.; Sun, X.; Du, J. Relay-intercropping soybean with maize maintains soil fertility and increases nitrogen recovery efficiency by reducing nitrogen input. *Crop J.* **2020**, *8*, 140–152. [CrossRef]
156. Li, C.; Hoffland, E.; Kuyper, T.W.; Yu, Y.; Zhang, C.; Li, H.; Zhang, F.; van der Werf, W. Syndromes of production in intercropping impact yield gains. *Nat. Plants* **2020**, *6*, 653–660. [CrossRef] [PubMed]
157. Wang, N.; Wang, T.; Chen, Y.; Wang, M.; Lu, Q.; Wang, K.; Dou, Z.; Chi, Z.; Qiu, W.; Dai, J. Microbiome convergence enables siderophore-secreting-rhizobacteria to improve iron nutrition and yield of peanut intercropped with maize. *Nat. Commun.* **2024**, *15*, 839. [CrossRef] [PubMed]
158. Barbieri, P.; Pellerin, S.; Seufert, V.; Nesme, T. Changes in crop rotations would impact food production in an organically farmed world. *Nat. Sustain.* **2019**, *2*, 378–385. [CrossRef]
159. Tiemann, L.; Grandy, A.S.; Atkinson, E.; Marin-Spiotta, E.; McDaniel, M. Crop rotational diversity enhances belowground communities and functions in an agroecosystem. *Ecol. lett.* **2015**, *18*, 761–771. [CrossRef] [PubMed]
160. Li, X.; Jousset, A.; de Boer, W.; Carrión, V.J.; Zhang, T.; Wang, X.; Kuramae, E.E. Legacy of land use history determines reprogramming of plant physiology by soil microbiome. *ISME J.* **2019**, *13*, 738–751. [CrossRef] [PubMed]
161. Hilton, S.; Bennett, A.J.; Keane, G.; Bending, G.D.; Chandler, D.; Stobart, R.; Mills, P. Impact of shortened crop rotation of oilseed rape on soil and rhizosphere microbial diversity in relation to yield decline. *PLoS ONE* **2013**, *8*, e59859. [CrossRef] [PubMed]
162. Bennett, A.J.; Bending, G.D.; Chandler, D.; Hilton, S.; Mills, P. Meeting the demand for crop production: The challenge of yield decline in crops grown in short rotations. *Biol. Rev.* **2012**, *87*, 52–71. [CrossRef] [PubMed]
163. Crookston, R.; Kurle, J.; Copeland, P.; Ford, J.; Lueschen, W. Rotational cropping sequence affects yield of corn and soybean. *Agron. J.* **1991**, *83*, 108–113. [CrossRef]
164. Benitez, M.-S.; Osborne, S.L.; Lehman, R.M. Previous crop and rotation history effects on maize seedling health and associated rhizosphere microbiome. *Sci. Rep.* **2017**, *7*, 1–13. [CrossRef]

165. Berg, G.; Smalla, K. Plant species and soil type cooperatively shape the structure and function of microbial communities in the rhizosphere. *FEMS Microbiol. Ecol.* **2009**, *68*, 1–13. [CrossRef] [PubMed]
166. Yang, C.; Hamel, C.; Gan, Y.; Vujanovic, V. Pyrosequencing reveals how pulses influence rhizobacterial communities with feedback on wheat growth in the semiarid Prairie. *Plant Soil* **2013**, *367*, 493–505. [CrossRef]
167. Gurr, G.M.; Lu, Z.; Zheng, X.; Xu, H.; Zhu, P.; Chen, G.; Yao, X.; Cheng, J.; Zhu, Z.; Catindig, J.L. Multi-country evidence that crop diversification promotes ecological intensification of agriculture. *Nat. Plants* **2016**, *2*, 1–4. [CrossRef]
168. Vilich, V. Crop rotation with pure stands and mixtures of barley and wheat to control stem and root rot diseases. *Crop Prot.* **1993**, *12*, 373–379. [CrossRef]
169. Lupwayi, N.Z.; Larney, F.J.; Blackshaw, R.E.; Pearson, D.C.; Eastman, A.H. Soil microbial biomass and its relationship with yields of irrigated wheat under long-term conservation management. *Soil Sci.* **2018**, *183*, 179–187. [CrossRef]
170. Hamel, C.; Gan, Y.; Sokolski, S.; Bainard, L.D. High frequency cropping of pulses modifies soil nitrogen level and the rhizosphere bacterial microbiome in 4-year rotation systems of the semiarid prairie. *Appl. Soil Ecol.* **2018**, *126*, 47–56. [CrossRef]
171. Bainard, L.D.; Navarro-Borrell, A.; Hamel, C.; Braun, K.; Hanson, K.; Gan, Y. Increasing the frequency of pulses in crop rotations reduces soil fungal diversity and increases the proportion of fungal pathotrophs in a semiarid agroecosystem. *Agric Ecosyst Environ* **2017**, *240*, 206–214. [CrossRef]
172. Merz, U.; Falloon, R.E. Powdery scab of potato—increased knowledge of pathogen biology and disease epidemiology for effective disease management. *Potato Res.* **2009**, *52*, 17–37. [CrossRef]
173. Nayyar, A.; Hamel, C.; Lafond, G.; Gossen, B.D.; Hanson, K.; Germida, J. Soil microbial quality associated with yield reduction in continuous-pea. *Appl. Soil Ecol.* **2009**, *43*, 115–121. [CrossRef]
174. Urra, J.; Alkorta, I.; Garbisu, C. Potential benefits and risks for soil health derived from the use of organic amendments in agriculture. *Agronomy* **2019**, *9*, 542. [CrossRef]
175. Qaswar, M.; Ahmed, W.; Jing, H.; Hongzhu, F.; Xiaojun, S.; Xianjun, J.; Kailou, L.; Yongmei, X.; Zhongqun, H.; Asghar, W. Soil carbon (C), nitrogen (N) and phosphorus (P) stoichiometry drives phosphorus lability in paddy soil under long-term fertilization: A fractionation and path analysis study. *PLoS ONE* **2019**, *14*, e0218195. [CrossRef]
176. Bünemann, E.K.; Bongiorno, G.; Bai, Z.; Creamer, R.E.; De Deyn, G.; de Goede, R.; Fleskens, L.; Geissen, V.; Kuyper, T.W.; Mäder, P. Soil quality—A critical review. *Soil Biol. Biochem.* **2018**, *120*, 105–125. [CrossRef]
177. Cardoso, E.J.B.N.; Vasconcellos, R.L.F.; Bini, D.; Miyauchi, M.Y.H.; Santos, C.A.d.; Alves, P.R.L.; Paula, A.M.d.; Nakatani, A.S.; Pereira, J.d.M.; Nogueira, M.A. Soil health: Looking for suitable indicators. What should be considered to assess the effects of use and management on soil health? *Sci. Agric.* **2013**, *70*, 274–289. [CrossRef]
178. Marchiori Junior, M.; Melo, W. Carbon, microbial biomass carbon and enzyme activity of a soil under natural forest, grassland and cotton culture. *Rev. Bras. De Cienc. Do Solo* **1999**, *23*, 257–263.
179. Sicardi, M.; Garía-Préchac, F.; Frioni, L. Soil microbial indicators sensitive to land use conversion from pastures to commercial Eucalyptus grandis (*Hill ex Maiden*) plantations in Uruguay. *Appl. Soil Ecol.* **2004**, *27*, 125–133. [CrossRef]
180. Karaca, A.; Cetin, S.C.; Turgay, O.C.; Kizilkaya, R. Soil enzymes as indication of soil quality. In *Soil Enzymology*; Springer: Berlin/Heidelberg, Germany, 2010; pp. 119–148.
181. Nannipieri, P.; Kandeler, E.; Ruggiero, P. Enzyme activities and microbiological and biochemical processes in soil. *Enzym. Environ.* **2002**, 1–33.
182. Bandick, A.K.; Dick, R.P. Field management effects on soil enzyme activities. *Soil Biol. Biochem.* **1999**, *31*, 1471–1479. [CrossRef]
183. Dick, W.; Tabatabai, M.A. *Significance and Potential Uses of Soil Enzymes*; Soil microbial ecology: Applications in agricultural and environmental management; Marcel Dekker Inc.: New York, NY, USA, 1992; pp. 95–127.
184. Barrutia, O.; Garbisu, C.; Epelde, L.; Sampedro, M.; Goicolea, M.; Becerril, J. Plant tolerance to diesel minimizes its impact on soil microbial characteristics during rhizoremediation of diesel-contaminated soils. *Sci. Total Environ.* **2011**, *409*, 4087–4093. [CrossRef] [PubMed]
185. Galende, M.A.; Becerril, J.M.; Barrutia, O.; Artetxe, U.; Garbisu, C.; Hernández, A. Field assessment of the effectiveness of organic amendments for aided phytostabilization of a Pb–Zn contaminated mine soil. *J. Geochem. Explor.* **2014**, *145*, 181–189. [CrossRef]
186. Pardo, T.; Clemente, R.; Epelde, L.; Garbisu, C.; Bernal, M. Evaluation of the phytostabilisation efficiency in a trace elements contaminated soil using soil health indicators. *J. Hazard. Mater.* **2014**, *268*, 68–76. [CrossRef] [PubMed]
187. Dubey, R.K.; Dubey, P.K.; Chaurasia, R.; Rao, C.S.; Abhilash, P.C. Impact of Integrated Agronomic Practices on Soil Fertility and Respiration on the Indo-Gangetic Plain of North India. *Agronomy* **2021**, *11*, 402. [CrossRef]
188. Viaux, P.; Rieu, C. *Integrated Farming Systems and Sustainable Agriculture in France*; Monographs-British Crop Protection Council: Hampshire, UK, 1995; p. 297.
189. Sarkar, D.; Rakshit, A.; Al-Turki, A.I.; Sayyed, R.; Datta, R. Connecting bio-priming approach with integrated nutrient management for improved nutrient use efficiency in crop species. *Agriculture* **2021**, *11*, 372. [CrossRef]
190. Duarah, I.; Deka, M.; Saikia, N.; Boruah, H.D. Phosphate solubilizers enhance NPK fertilizer use efficiency in rice and legume cultivation. *3 Biotech* **2011**, *1*, 227–238. [CrossRef] [PubMed]

191. Entesari, M.; Sharifzadeh, F.; Ahmadzadeh, M.; Farhangfar, M. Seed biopriming with *Trichoderma* species and Pseudomonas fluorescent on growth parameters, enzymes activity and nutritional status of soybean. *Int. J. Agron. Plant Prod.* **2013**, *4*, 610–619.
192. Kim, P.J.; Chung, D.Y.; Malo, D. Characteristics of phosphorus accumulation in soils under organic and conventional farming in plastic film houses in Korea. *Soil Sci. Plant Nutr.* **2001**, *47*, 281–289. [CrossRef]

MDPI AG
Grosspeteranlage 5
4052 Basel
Switzerland
Tel.: +41 61 683 77 34

International Journal of Molecular Sciences Editorial Office
E-mail: ijms@mdpi.com
www.mdpi.com/journal/ijms

www.ingramcontent.com/pod-product-compliance
Lightning Source LLC
Chambersburg PA
CBHW061608210926
51850CB00009B/2251